New Methods and Applications for UAVs

New Methods and Applications for UAVs

Guest Editors

Andrzej Łukaszewicz
Carlos Tavares Calafate
Wojciech Giernacki

MDPI

Basel • Beijing • Wuhan • Barcelona • Belgrade • Novi Sad • Cluj • Manchester

Guest Editors
Andrzej Łukaszewicz
Faculty of Mechanical Engineering
Bialystok University of Technology
Bialystok
Poland

Carlos Tavares Calafate
Computer Engineering Department (DISCA)
Universitat Politècnica de València (UPV)
Valencia
Spain

Wojciech Giernacki
Faculty of Control, Robotics and Electrical Engineering
Poznan University of Technology
Poznan
Poland

Editorial Office
MDPI AG
Grosspeteranlage 5
4052 Basel, Switzerland

This is a reprint of the Special Issue, published open access by the journal *Sensors* (ISSN 1424-8220), freely accessible at: https://www.mdpi.com/journal/sensors/special_issues/Applications_UAVs.

For citation purposes, cite each article independently as indicated on the article page online and as indicated below:

Lastname, A.A.; Lastname, B.B. Article Title. *Journal Name* **Year**, *Volume Number*, Page Range.

ISBN 978-3-7258-5023-5 (Hbk)
ISBN 978-3-7258-5024-2 (PDF)
https://doi.org/10.3390/books978-3-7258-5024-2

Contents

About the Editors

Andrzej Łukaszewicz

Andrzej Łukaszewicz, Ph.D., holds the position of Assistant Professor at the Institute of Mechanical Engineering in the Faculty of Mechanical Engineering at Bialystok University of Technology, Poland (October 1997–present). He holds a Ph.D. degree in solid mechanics from Warsaw University of Technology in Warsaw, Poland. He is interested in UAV design, CAx, CAD, topology optimization, lattice structures, the mechanics of materials, fracture mechanics and materials testing. His main research area is mechanical engineering. He has co-authored more than 190 publications in journals, conference proceedings and book chapters, and has reviewed more than 100 papers in 28 international journals. He is also a member of the Topical Advisory Panel of six international journals.

Carlos Tavares Calafate

Carlos Tavares Calafate is a Full Professor in the Department of Computer Engineering at the Technical University of Valencia (UPV), Spain. He graduated with honors in Electrical and Computer Engineering from the University of Oporto (Portugal) in 2001. He received his Cum Laude Ph.D. in Informatics from the Technical University of Valencia in 2006, where he has worked since 2002. His research interests include ad hoc and vehicular networks, UAVs, smart cities and the IoT, QoS, network protocols, video streaming and network security. To date, he has published more than 500 articles, several of which have appeared in journals such as *IEEE Transactions on Vehicular Technology, IEEE Transactions on Mobile Computing, IEEE/ACM Transactions on Networking, Elsevier Ad Hoc Networks,* and *IEEE Communications Magazine*. He is an Associate Editor for several international journals by publishers including Elsevier, Hindawi, MDPI, IET, and SAGE.

Wojciech Giernacki

Wojciech Giernacki, Ph.D., D.Sc. (Eng.), is an Associate Professor at the Institute of Robotics and Machine Intelligence, Poznan University of Technology. He obtained a Ph.D. in control engineering and robotics from Poznan University of Technology in 2011, and a D.Sc. in 2019. He founded and is head of the PUT AeroLab and Unmanned Aerial Vehicles Research Group, as well as the Division of Control and Optimization at the Institute of Robotics and Machine Intelligence. His scientific interests are focused around issues related to UAVs, especially their robust and adaptive control and optimization techniques, as well as sensor data fusion.

Preface

This collection is dedicated to presenting the current state of the art in UAV research. It explores the application of advanced methods, algorithms and technologies, based on tools such as artificial intelligence, neural networks and sensors.

This Special Issue Reprint provides a platform for scientists and engineers to present the latest advances, challenges and opportunities in the development of UAV sensor systems.

Its main goal is to emphasize the potential of integrating numerical simulations and design methods to achieve improved autonomous unmanned flying platforms that are suitable for a range of applications, and to foster knowledge exchange and innovative research on effective UAV design, control and exploitation.

The Editors would like to thank the authors of the nine papers featured in the Special Issue "New Methods and Applications for UAVs" for their contributions to UAV advancement.

Andrzej Łukaszewicz, Carlos Tavares Calafate, and Wojciech Giernacki
Guest Editors

sensors

MDPI

Review

Review on Type of Sensors and Detection Method of Anti-Collision System of Unmanned Aerial Vehicle

Navaneetha Krishna Chandran [1], Mohammed Thariq Hameed Sultan [1,2,3,*], Andrzej Łukaszewicz [4,*], Farah Syazwani Shahar [2], Andriy Holovatyy [5] and Wojciech Giernacki [6]

1 Laboratory of Biocomposite Technology, Institute of Tropical Forestry and Forest Products (INTROP), University Putra Malaysia, Serdang 43400, Selangor Darul Ehsan, Malaysia; navaneethakrishna93@gmail.com
2 Department of Aerospace Engineering, Faculty of Engineering, University Putra Malaysia, Serdang 43400, Selangor Darul Ehsan, Malaysia; farahsyazwani@upm.edu.my
3 Aerospace Malaysia Innovation Centre (944751-A), Prime Minister's Department, MIGHT Partnership Hub, Jalan Impact, Cyberjaya 63000, Selangor Darul Ehsan, Malaysia
4 Institute of Mechanical Engineering, Faculty of Mechanical Engineering, Bialystok University of Technology, 15-351 Bialystok, Poland
5 Department of Computer-Aided Design Systems, Lviv Polytechnic National University, 79013 Lviv, Ukraine; andrii.i.holovatyi@lpnu.ua
6 Institute of Robotics and Machine Intelligence, Faculty of Control, Robotics and Electrical Engineering, Poznan University of Technology, 60-965 Poznan, Poland; wojciech.giernacki@put.poznan.pl
* Correspondence: thariq@upm.edu.my (M.T.H.S.); a.lukaszewicz@pb.edu.pl (A.Ł.)

Abstract: Unmanned aerial vehicle (UAV) usage is increasing drastically worldwide as UAVs are used in various industries for many applications, such as inspection, logistics, agriculture, and many more. This is because performing a task using UAV makes the job more efficient and reduces the workload needed. However, for a UAV to be operated manually or autonomously, the UAV must be equipped with proper safety features. An anti-collision system is one of the most crucial and fundamental safety features that UAVs must be equipped with. The anti-collision system allows the UAV to maintain a safe distance from any obstacles. The anti-collision technologies are of crucial relevance to assure the survival and safety of UAVs. Anti-collision of UAVs can be varied in the aspect of sensor usage and the system's working principle. This article provides a comprehensive overview of anti-collision technologies for UAVs. It also presents drone safety laws and regulations that prevent a collision at the policy level. The process of anti-collision technologies is studied from three aspects: Obstacle detection, collision prediction, and collision avoidance. A detailed overview and comparison of the methods of each element and an analysis of their advantages and disadvantages have been provided. In addition, the future trends of UAV anti-collision technologies from the viewpoint of fast obstacle detection and wireless networking are presented.

Keywords: anti-collision methods; detection system; sensors; unmanned aerial vehicle

1. Introduction

General Visual Inspection (GVI) is a typical approach for quality control, data collection, and analysis. It involves using basic human senses such as vision, hearing, touch, smell, and non-specialized inspection equipment. Unmanned aerial systems (UAS), also known as UAVs, are being developed for automated visual inspection and monitoring in various industrial applications [1]. These systems consist of UAVs outfitted with the appropriate payload and sensors for the job at hand [2].

Sensor and measurement reliance is crucial for UAV operations and functionality, as they serve as indispensable resources to ensure the safety and security of UAVs. Since UAVs operate autonomously without a pilot's input, a series of sensors and systems are required for the UAVs to position themselves. Usually, UAVs use a global positioning system (GPS) to position themselves. However, GPS input will not always be accurate, especially when the UAV has to be equipped with sensors such as rangefinders, which are very useful when the UAV flies at low altitudes. The investigation of the quadcopter control problem came to a standstill until relatively recently, since the control of four separate motor-based propulsion systems was nearly impossible without modern electronic equipment. These technologies have only become increasingly sophisticated, versatile, quick, and affordable in the past several decades.

Due to the intricacy of the issue, controlling a quadcopter is a topic that is both intriguing and important. The fact that the system has just four inputs (the angular velocity of the propellers) despite having six degrees of freedom (three rotational axes and three transnational axes) gives the system the quality of being under-actuated [3]. Even though some of them have more than six inputs, they all have the same number of axes to manipulate, meaning they are all under-actuated. This is because all those inputs can only directly control the three rotation axes, not the translation axis [4].

Additionally, the dynamics on which this form of UAV operates give freedom in movement and robustness towards propulsion problems. This sort of UAV is ideal for reconnaissance missions. As an illustration, control algorithms may be programmed so that a UAV can keep its stability even if fifty percent of the propellers that control one axis of rotation stop working correctly. On the other hand, since it is an airborne vehicle, the frictions of the chassis are almost non-existent, and the control algorithm is responsible for handling the damping.

A UAV's level of autonomy is defined by its ability to perform a set of activities without direct human intervention [5]. Different kinds of onboard sensors allow unmanned vehicles to make autonomous decisions in real time [6–8]. Demand for unmanned vehicles is rising fast because of the minimal danger to human life, enhanced durability for more extended missions, and accessibility in challenging terrains. Still, one of the most difficult problems to address is planning their course in unpredictable situations [9–11]. The necessity for an onboard system to prevent accidents with objects and other vehicles is apparent, given their autonomy and the distances they may travel from base stations or their operators [12,13].

Whether a vehicle is autonomous or not, it must include a collision avoidance system. Several potential causes of collisions include operator/driver error, machinery failure, and adverse environmental factors. According to statistics provided by planecrashinfo.com, over 58% of fatal aviation crashes occurred due to human mistakes between January 1960 and December 2015 [14]. To reduce the need for human input, the autopilot may be upgraded with features like object recognition, collision avoidance, and route planning. Methods of intelligent autonomous collision avoidance have the potential to contribute to making aircrafts even safer and saving lives.

The exponential growth in UAVs using in public spaces has made a necessity for sophisticated and highly dependable collision avoidance systems evident and incontestable from the public safety perspective. UAVs can access risky or inaccessible locations without risking human lives. Therefore UAVs should be built to operate independently and avoid crashing into anything while in flight [15]. Precision agriculture is an application of UAVs that has been increasing rapidly worldwide. Precision agriculture is expanding quickly in commercial goods and research and development applications. In order to correctly account for the geographical and temporal fluctuations of crop and soil components, this revolutionary trend is redefining the crop management system and placing a higher focus on data collecting and analysis, whether in real-time or offline.

Figure 1 shows the basic architecture of an anti-collision system that will be implemented in a vehicle. Anti-collision systems consist of two major parts: the input and output [15]. These parts can also be recognized as perspective and action. Any system designed to prevent accidents from happening must begin with perception, or more specifically, obstacle detection [16]. At this stage, sensors gather information about the surrounding area and locate any hazards. However, the active part comes after the perspective, where once the threat has been detected, the situation will be analyzed by the computation of the control system of the UAVs. As a result, the actuators will implement proper countermeasures to avoid the hazard [17].

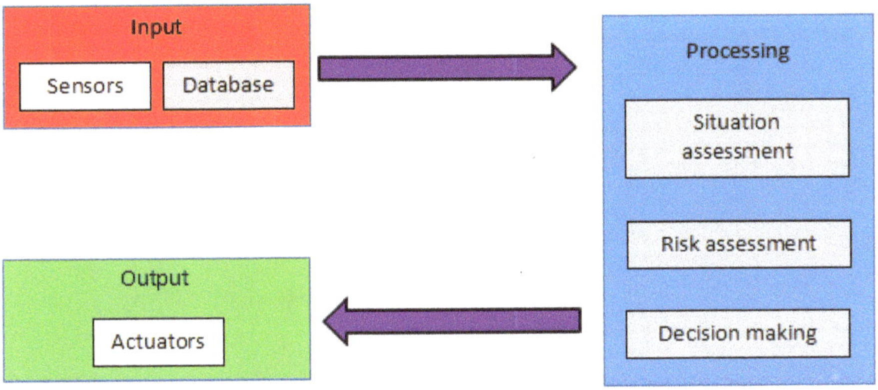

Figure 1. Anti-collision system general architecture.

Sensors come in a wide variety, but they may be broken down into two broad categories: active and passive. The backscatter is measured by an active sensor with its own source that sends out a beam of light or a wave. On the other hand, passive sensors can only estimate the energy emitted by an item, such as sunlight reflected off the object. Anti-collision systems use a total of four different approaches in detecting the hazards, which are geometric (using the UAV's and obstacles' positions and velocities to reformat nodes, typically via trajectory simulation), force-field (manipulating attractive and repulsive forces to avoid collisions), optimized (using the known parameters of obstacles to find the most efficient route), and sense-and-avoid (making avoidance decisions at runtime based on sensing the environment) [18,19].

The complexity of collision avoidance systems may vary from as simple as alerting the vehicle's pilot to be involved to wholly or partly taking control of the system on its own to prevent the accident [20]. For an unmanned vehicle to travel without direct human intervention, it must be equipped with several specialized systems that identify obstacles, prevent collisions, plan routes, determine their exact location, and implement the necessary controls [21]. Multiple UAVs provide substantial benefits over single UAVs. They are in high demand for a wide range of applications, including military and commercial usage, search and rescue, traffic monitoring, threat detection (particularly near borders), and atmospheric research [22–24]. UAVs may struggle to complete missions in a demanding dynamic environment due to cargo restrictions, power constraints, poor vision due to weather, and difficulties in remote monitoring. To ensure unmanned vehicles' success and safe navigation, the robotics community is working tirelessly to overcome these difficulties and deliver the technical level fit for challenging settings [25–28].

One of the most challenging problems for autonomous vehicles is detecting and avoiding collisions with objects, which becomes much more critical in dynamic situations with several UAVs and moving obstacles [29]. Sensing is the initial process in which the system takes data from its immediate environment. When an impediment enters the system's field of view, the detection stage performs a risk assessment. To prevent a possible collision, the collision avoidance module calculates how much of a detour has to be made from the original route. Once the system has completed its calculations, it will execute the appropriate move to escape the danger safely.

2. Obstacle Detection Sensors

The drone needs a "perspective model" of its environment to avoid crashing into obstacles [30,31]. To do this, the UAV must have a perception unit consisting of one or more sensors [32]. Sensors, like imaging sensors of varying resolutions, are crucial components of remote sensing systems. Sensors may be used in a wide variety of contexts. LiDAR, visible cameras, thermal or infrared cameras, and solid-state or mechanical devices are all examples of sensors that may be used for monitoring [27,33]. The sensors that have been used for the anti-collision system are majorly categorized into two, which are active sensors and passive sensors. In Figure 2, the categorization of the anti-collision system sensors is shown.

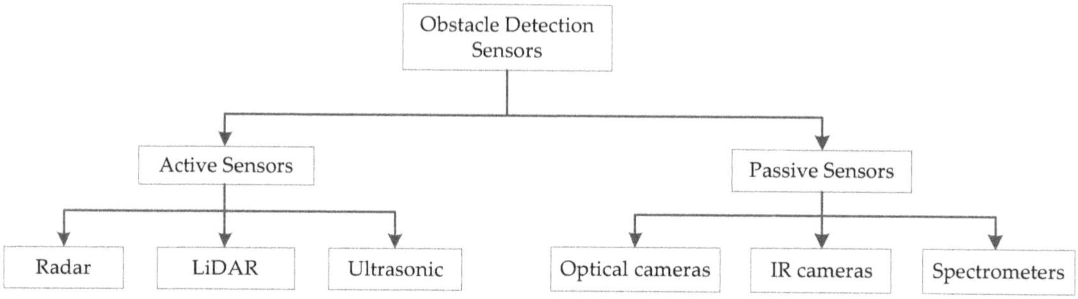

Figure 2. Categorization of anti-collision system sensors.

2.1. Active Sensors

Sensing using active sensors involves emitting radiation and then detecting the reflected radiation. All the necessary components, including the source and the detector, are built within an active sensor. A sensor works by having a transmitter send out some signal (light, electricity, sound) that then gets reflected off of whatever it is being used to detect [34,35]. Most of these sensors operate in the spectrum's microwave range, allowing them to penetrate the atmosphere under most circumstances. The metrics of interest of the obstacles, such as distance and angles, may be adequately returned by such sensors since they have a short reaction time, need less processing power, can scan more significant regions quickly, and are less impacted by weather and lighting conditions. In [36], the authors use MMW radar. In their setup, things are detected and followed by watching radar echoes and figuring out how far away they are from the vehicle. Different distances and weather conditions are also used to conclude the performance. Despite the allure, radar-based solutions are either too costly or too heavy to be practical on more miniature robots, such as battery-powered UAVs [37,38].

2.1.1. Radar

A radar sensor transmits a radio wave that will be reflected back to the sensor after hitting an object. The distance between the object and the radar is determined by timing how long it takes the signal to return. Despite their high cost, airborne radar systems are often used for their precision to provide data. Both continuous-wave and pulsed-wave radars exist, with the former emitting a steady stream of linearly modulated (or frequency-modulated) signals and the latter emitting intense but brief bursts of signals; however, both types have blind spots [39]. As a bonus, radars could also track the objects' speeds and other motion data. For instance, the radar may determine an object's velocity by measuring how much the frequency of its echo or bounced-off signal changes as it approaches the radar [40].

Using a compact radar, the authors of [40] could get range data in real time, regardless of the weather. The system incorporates a compact radar sensor and an OCAS (obstacle collision avoidance system) computer. OCAS utilizes radar data such as obstacle velocity, azimuth angles, and range to determine avoidance criteria and provide orders to the flight controller to execute the appropriate maneuver to prevent collisions. The findings indicated that with the set safety margins, the likelihood of successfully avoiding a crash is more than 85%, even if there is an inaccuracy in the radar data.

The benefits of integrating radar sensors into UAVs for obstacle identification and for detecting and calculating additional aspects of the observed obstruction, such as the velocity of the obstacle and the angular information utilizing multichannel radars, are thoroughly explored by the authors in [41]. Experiments reveal that with forward-looking radars, with the radar's simultaneous multi-target range capabilities, it is possible to identify targets across an extensive angular range of 60 degrees in azimuth. For their suggested autonomous collision avoidance system, the authors of [41] used Ultra-Wideband (UWB) collocated MIMO radar. Radar cognition's capacity to modify the waveform of ultra-wideband multiple-input multiple-output radar transmissions for better detection and, by extension, to steer the UAV by giving an estimate of the collision locations is a significant advantage.

2.1.2. LiDAR

One may compare the operation of a light detection and ranging (LiDAR) sensor to that of a radar. One half of a LiDAR sensor fires laser pulses at the surface(s), while the other half scans their reflection and calculates distance based on how long each pulse takes to return. Rapid and precise data collection is achieved using LiDAR. LiDAR sensors have shrunk in size and shed weight over the years, making it possible to put them on mini and small UAVs [42,43]. LiDAR-based systems are more cost-effective than radar systems, particularly those using 1D and 2D LiDAR sensors.

The designed system was successfully field tested by the authors of [44] using a variety of laser scanners installed on a vehicle, which are laser radars ranging in three dimensions. Regarding 3D mapping and 3D obstacle detection, 3D LiDARs are as standard as it gets in the sensor world [45,46]. Since LiDAR is constantly being moved and ranged, the gathered data is prone to motion distortion, which makes using these devices challenging. To get around this, as proposed by the authors of [45], additional sensors may be used with LiDAR. Only 3D LiDARs allow for precise assessment of an object's posture.

2.1.3. Ultrasonic

To determine an item's distance, ultrasonic sensors transmit sound waves and then analyze the echoes they receive [47]. The sound waves produced are outside the range humans can hear (25 to 50 kilohertz) [48]. Compared to other types of range sensors, ultrasonic sensors are both more affordable and widely accessible. The object's transparency does not affect ultrasonic sensors, unlike LiDARs. Unlike ultrasonic sensors, which are color-blind, LiDARs have trouble identifying transparent materials like glass. However, the sonic sensor will not provide accurate readings if the item reflects the sound wave in the opposite direction than the receiver or if the substance has the properties of absorbing sound.

Like radars and LiDARs, this method relies on emitting a wave, waiting for the reflected wave to return, and then calculating the distance based on the time difference between the two. Compared to other types of range sensors, ultrasonic sensors are both more accessible and more affordable. Since each sensor in Table 1 has its advantages and disadvantages compared to the others, it is clear that more than one sensor can be employed to provide complete protection against the collision avoidance issue. Multiple sensors may be utilized to cover a greater area and eliminate blind spots, or different kinds of sensors can be fused to create a super sensor whose weaknesses cancel out those of its components.

Table 1. Comparison between the active sensors of the anti-collision system.

Sensor	Sensor Size	Power Required	Accuracy	Range	Weather Condition	Light Sensitivity	Cost
Radar	Large	High	High	Long	Not Affected	No	High
LiDar	Small	Low	Medium	Medium	Affected	No	Medium
Ultrasonic	Small	Low	Low	Short	Slightly Affected	No	Low

According to Table 1, the LiDAR and ultrasonic sensors, which can be used in the UAV's anti-collision system, are smaller than radar. This makes the ultrasonic and LiDAR the ideal method of obstacle sensing for small UAVs, as they are less in weight, reducing the UAV's payload. In addition, the power consumption by ultrasonic and LiDAR is also low compared to radar. However, the accuracy and range of the radar are highest compared to ultrasonic and LiDAR, which makes the radar suitable for use in large UAVs that fly at high altitudes. On the other hand, the radar is not affected by weather conditions, but the LiDAR is affected, while ultrasonic is slightly affected by the weather condition. Last but not least, the cost of an ultrasonic sensor is the lowest compared to radar and LiDAR, which makes it more affordable.

2.2. *Passive Sensors*

The energy the seen items or landscape gives off is measured using passive sensors. Optical cameras, infrared (IR) cameras, and spectrometers are the most common types of passive sensors now used in sensing applications [49]. Wide varieties of cameras, each optimized for a specific wavelength, exist. The authors of [50] offer a system for acoustic signal tracking and real-time vehicle identification. The result is obtained by isolating the resilient spatial characteristics from the noisy input and then processing them using sequential state estimation. They provide empirical acoustic data to back up the suggested technique.

In contrast, thermal or infrared cameras operate in the infrared light range and have a larger wavelength than the visible light range. Therefore, the primary distinction between the two is that visual cameras use visible light to create a picture, while thermal cameras use infrared radiation. Ordinary cameras struggle when light levels are low, while IR cameras thrive [51]. It takes more computational resources since an additional algorithm is required to extract points of interest in addition to the algorithm already needed to calculate the range and other characteristics of the barriers [52]. Vision cameras are susceptible to environmental factors, including sunlight, fog, and rain, in addition to the field-of-view restrictions imposed by the sensor being employed [53,54].

2.2.1. Optical

Taking pictures of the world around us is the foundation of visual sensors and cameras, which then utilize those pictures to extract information. There are three main types of optical cameras: monocular, stereo, and event-based [55–57]. Using cameras has several advantages, including their compact size, lightweight, low power consumption, adaptability, and simple mounting. Some drawbacks of employing such sensors include their sensitivity to lighting and background color changes and their need for clear weather. When any of these conditions are present, the recorded image's quality plummets, significantly influencing the final product.

According to [58], a monocular camera may be used to identify obstacles in the path of a ground robot. Coarse obstacle identification in the bottom third of the picture is achieved by an enhanced Inverse Perspective Mapping (IPM) with a vertical plane model; however, this method is only suitable for slow-moving robots. Using stereo cameras is one method proposed by the authors of [59]. In stereo cameras, absolute depth is determined by combining internal and external camera characteristics, unlike in monocular cameras. The amount of processing power needed rises when stereo images are used. Because of the high processing cost and the need to accommodate highly complex systems with six degrees of freedom, like drones, the authors solve this problem by dividing the collected pictures into nine zones.

2.2.2. Infrared

Sensors operating in the infrared spectrum, such as those used in infrared (IR) cameras, are deployed when ambient light is scarce. They may also be used with visual cameras to compensate for the latter's lackluster performance, particularly at night. Data from a thermal camera may be analyzed by automatically determining the image's orientation by extracting fake control points due to the thermal camera's output being hazy and distorted with lesser resolution than that of an RGB camera [60].

3. Obstacle Detection Method

Both reactive and deliberative planning frameworks may be used for collision avoidance. During management by reaction, the UAV is equipped with onboard sensors to collect data about its immediate environment and behave accordingly. It facilitates instantaneous responses to changing environmental conditions. An alternative navigational strategy may be necessary if reactive control leads to a local minimum and becomes trapped there. The method of decision-making used by autonomous commercial cars will determine their level of safety and sanity. By dynamically connecting rear anti-collision elements, a driving decision network built on an actor-critic architecture has been developed to ensure safe driving. To interpret sensor data efficiently, this network considers the effects of different elements on collision prevention, such as rearward target detection, safety clearance, and vehicle roll stability. This has been accomplished by creating an improved reward function that considers these factors inside a multi-objective optimization framework. The network attempts to improve collision avoidance skills and guarantee the safety and stability of the vehicle by thoroughly examining these parameters. The force-field method, geometry, optimization-based methods, and sense-and-avoid techniques are the four main approaches to collision avoidance algorithms, as shown in Figure 3.

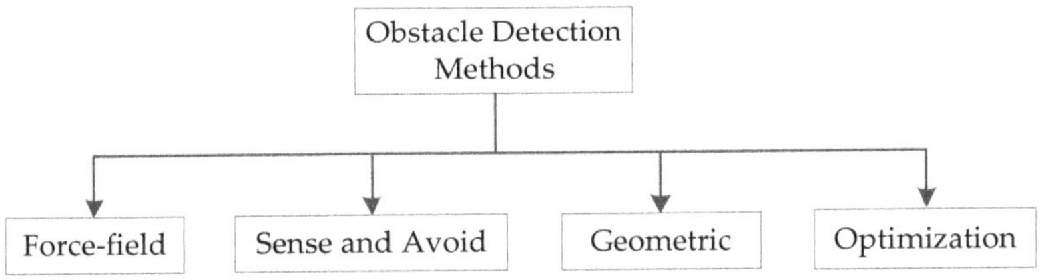

Figure 3. The main approaches to collision avoidance algorithms.

3.1. Force-Field Method

Using the idea of a repulsive or attractive force field, force-field techniques (also called potential field methods) may steer a UAV away from obstruction or draw it closer to a target [61,62]. Instead of using physical barriers, the authors of [63] propose using a potential field to surround a robot. In order to determine the shortest route between two places, the authors of [64] suggest using an artificial potential field. The points that create repulsive and attractive pressures for the robot are the obstacles and the targets, respectively.

The authors of [65] suggested a new artificial potential field technique to generate optimum collision-free paths in dynamic environments with numerous obstacles, where other UAVs are also treated as moving obstacles. This method is dubbed an improved curl-free vector field. Although simulations confirmed the method's viability, more validation in 3D settings with static and dynamic factors is required [66]. Regarding UAV navigation in 3D space, the authors of [67] describe an artificial potential field technique that has been improved to produce safe and smooth paths. By factoring in the behavior of other UAVs and their interactions, the proposed optimized artificial potential field (APF) algorithm improves the performance of standard APF algorithms. During route planning, the algorithm considers other UAVs to be moving obstacles.

A vehicle collision avoidance algorithm is provided in [68], using synthetic potential fields. The algorithm considers the relative velocities of the cars and the surrounding traffic to decide whether to slow down or speed up to pass another vehicle. This decision is based on the size and the form of the potential fields of the barriers. Too big of a time step might lead to collisions or unstable behavior, so getting it exactly right is essential. A 1D virtual force field approach is proposed for moving obstacle detection [69]. They argue that the inability to account for the barriers' mobility causes the efficiency loss seen with conventional obstacle force field approaches.

3.2. Sense and Avoid Method

In order to control the flight path of each UAV in a swarm without information about the plans of other drones, with fast response time, sense-and-avoid techniques focus on reducing the computational power required by simplifying the collision avoidance process to individual detection and avoidance of obstacles. Methods based on "Sense and Avoid" The speed with which collision avoidance can respond makes it a good tool for complex contexts. A robot or agent is outfitted with several sensing technologies, including LiDAR, sonar, and radar. Although it cannot distinguish between different objects, radar can quickly respond to anything that enters its field of view [69–71].

In [72], the authors suggest a technique for categorizing objects as static or dynamic using 2D LiDAR data. Additionally, the program can provide rough estimates of the speeds of the moving obstructions. In [73], the authors use a computer vision method to implement an animal detection and collision-avoidance system. The team has trained its system with over 2200 photos and tested it with footage of animals in traffic. In [74], the authors implement a preset neural network module in MATLAB to operate with five ultrasonic (US) sensors to triangulate and determine objects' exact location and form. They use three distinct shapes in their evaluations. To accomplish object recognition and avoidance, the inventors of [75] fused a US sensor with a binocular stereo-vision camera. Using stereo vision as the primary method, a new route is constructed via an algorithm based on the Rapidly Explored Random Tree (RRT) scheme.

3.3. Geometric Method

To ensure that the predetermined minimum distances between agents, such as UAVs, are not violated, geometric techniques depend on studying geometric features. The UAVs' separation distances and travel speeds have been used to calculate the time remaining until a collision occurs. In [76], the authors provide an analytical method for resolving the planar instance of the issue of aircraft collision. We can find closed-form analytical solutions for the best possible sets of orders to end the dispute by analyzing the trajectories' geometric properties.

In [77], conflict avoidance in a 3D environment is accomplished by using information such as the aircraft's coordinates and velocities in conjunction with a mixed geometric and collision cone technique. However, the authors depend on numerical optimization techniques for the most common scenarios and only get analytical conclusions for specific circumstances. The paper [78] investigates UAV swarms that use geometry-based collision avoidance techniques. The suggested method integrates line-of-sight vectors with relative velocity vectors to consider a formation's dynamic limitations. Each UAV may assess if the formation can be maintained while avoiding collisions by computing a collision envelope and using that information to determine the potential directions for avoiding collisions.

In [79], the authors combined geometric avoidance and the selection of start time from critical avoidance to provide a novel approach to collision avoidance based on kinematics, the risk of collisions, and navigational constraints. Instead of trying to avoid all of the barriers simultaneously, FGA may prioritize which obstacles must be avoided first, depending on how much time must pass before they can be safely passed. The authors of [80] developed a way to safely pilot UAVs from the beginning of a mission to its completion while ensuring that the vehicles stay on their intended course and avoid potential hazards. The authors offer a solution that individually tackles the system's collision avoidance control and trajectory control and then merges them via a planned movement strategy.

3.4. Optimization Method

Methods based on optimization need geospatial data for the formulation of the avoidance trajectory. Probabilistic search algorithms aim to offer the most productive locations to conduct a search, given the level of uncertainty associated with that information. Different optimization techniques, such as those inspired by ants, genetic algorithms, gradient descent-based approaches, particle swarm optimization, greedy methods, and local approximations, have been developed to handle the enormous computing demands of these algorithms.

For instance, to successfully calculate optimum collision-free search pathways for UAVs under communication-related limitations, the authors of [81] use a minimum time search method with ant colony optimization. The authors of [82] provide a prediction technique for the next UAV coordinates based on the set of probable instructions the UAV will execute in the near future. After considering the destination coordinates and the UAV's current location, the algorithm generates a cost function for the best trajectory. Using particle swarm optimization, a novel technique for autonomous vehicle route planning in the wild. This strategy uses the sensor data by giving various kinds of territory different weights, then using those weights to categorize the possible paths across the landscape.

3.5. Summary of Object Detection Method

Table 2 summarizes previous research studies on detection and anti-collision system. From Table 2, it can be concluded that the geometric detection and force field methods are suitable for long-range UAVs. However, the sense and avoid method is suitable for short-range UAVs. The compatibility of real-time detection in four detection methods allows the UAVs to analyze the surroundings and be more varied about the surrounding. The 3D compatibility in geometric, optimization, and sense-and-avoid methods allows the system to generate a 3D mapping around the surroundings, allowing the maneuvering to be more precise in the UAVs.

Table 2. Previous studies of detection and anti-collision system.

	Geometric			Sense and Avoid		Force Field		Optimization
	[78,79]	[80]	[83]	[72]	[74]	[69]	[65]	[82]
Multiple UAV Compatibility	/	/	/	/	/	/	O	/
3D Compatibility	/	/	/	/	/	O	O	/
Communication	O	/	/	/	/	O	O	/
Alternate Route Generation	/	/	/	/	O	/	/	/
Real-time Detection	/	/	/	/	/	/	/	/

/—Available. O—Not Available.

Other than these obstacle detection methods, which involve their implementation, many obstacle detection methods are being developed around the world. One obstacle detection method is neural network-based navigation. Human decisions about these types of motions may be observed in various situations, including those with randomly produced barriers and pertinent environmental data [84]. In comparison to human decision-making, the simulation results showed that the suggested method had a high estimation accuracy rate of almost 90%. In contrast to the adaptive project framework (APF) method, the neural network methodology demonstrated its usefulness by successfully navigating over obstacles without running into the local minimum problem, hence emphasizing the strength of neural network decision-making.

4. Conclusions

Analyzing this short review on the sensor type and detection method of anti-collision systems of UAVs, the selection of sensors and detection method mainly depends on the UAV type and the objective of the UAV mission. The table below presents the research gap and the stigmatization of the research review identified through the literature review.

In this context, the recommended method of detection in an anti-collision system in a UAV depends on the UAV's specification and the UAV's mission objective. Methods of obstacle detection using geometric are considered effective, where they are capable of 3 dimensions projection alternate route generation, and multiple UAV compatibility. However, they cannot communicate with ground control. The geometric object detection method basically uses input from GPS in order to position the UAV itself. This detection method is suitable in urban areas, where there will be strong GPS signals. However, strong GPS signals may not be found in rural areas, especially in plantation areas, where UAVs' applications have rapidly increased in agricultural applications. When the GPS signal strength is low, the UAV cannot position itself accurately. Hence, optimization and sense and avoid methods will be more suitable in this case than geometric object detection methods. More specifically, optimization and sense-and-avoid detection methods are suitable for UAVs that fly at low altitudes; however, the geometric is ideal for high-altitude and long-range UAVs.

On the other hand, the force field detection method is more suitable in an environment consisting of multiple UAVs, where the UAV can sense the electromagnetic emission from other UAVs. However, although the force field method is the same as the geometric method, where it is suitable for long-range UAVs, it is not suitable for urban areas because there will be a lot of electromagnetic wave interference, eventually affecting the force field detection method. This literature review gives a better understanding of the anti-collision system within a UAV. It allows the optimization of anti-collision systems according to the UAV in which the anti-collision system will be implemented.

Author Contributions: Conceptualization, N.K.C. and A.H.; writing—original draft preparation, N.K.C. and A.H.; writing—review and editing, F.S.S., M.T.H.S. and A.Ł.; visualization, N.K.C. and A.H.; supervision, M.T.H.S., A.Ł. and W.G.; project administration, M.T.H.S., F.S.S., W.G. and A.Ł.; funding acquisition, A.Ł. and M.T.H.S. All authors have read and agreed to the published version of the manuscript.

Funding: The authors would like to thank Ministry of Higher Education, Malaysia for the financial support through Higher Institution Centre of Excellence (HICoE) research grant (Vot number 6369119).

Institutional Review Board Statement: Not applicable.

Informed Consent Statement: Not applicable.

Data Availability Statement: Data sharing does not apply to this article as no new data were created or analyzed in this study.

Acknowledgments: The authors would like to thank the Department of Aerospace Engineering, Faculty of Engineering, Universiti Putra Malaysia; the Laboratory of Biocomposite Technology, Institute of Tropical Forestry and Forest Product (INTROP–HICOE), Universiti Putra, Malaysia, (HICOE); the Institute of Mechanical Engineering, Bialystok University of Technology, Poland; the Department of Computer-Aided Design Systems, Lviv Polytechnic National University, Ukraine; and also the Institute of Robotics and Machine Intelligence, Faculty of Control, Robotics and Electrical Engineering, Poznan University of Technology, Poland, for their close collaboration in this study.

Conflicts of Interest: The authors declare no conflict of interest.

References

1. Zosimovych, N. Preliminary design of a VTOL unmanned aerial system for remote sensing of landscapes. *Aeron Aero Open Access J.* **2020**, *4*, 62–67. [CrossRef]
2. Papa, U.; Ponte, S. Preliminary Design of an Unmanned Aircraft System for Aircraft General Visual Inspection. *Electronics* **2018**, *7*, 435. [CrossRef]
3. Giernacki, W.; Gośliński, J.; Goślińska, J.; Espinoza-Fraire, T.; Rao, J. Mathematical Modeling of the Coaxial Quadrotor Dynamics for Its Attitude and Altitude Control. *Energies* **2021**, *14*, 1232. [CrossRef]
4. Gheorghiță, D.; Vîntu, I.; Mirea, L.; Brăescu, C. Quadcopter control system. In Proceedings of the 2015 19th International Conference on System Theory, Control and Computing (ICSTCC), Cheile Gradistei, Romania, 14–16 October 2015; pp. 421–426.

5. Huang, H.-M. Autonomy levels for unmanned systems (ALFUS) framework: Safety and application issues. In Proceedings of the 2007 Workshop on Performance Metrics for Intelligent Systems, Washington, DC, USA, 28–30 August 2007.
6. Zhang, W.; Zelinsky, G.; Samaras, D. Real-time accurate object detection using multiple resolutions. In Proceedings of the 2007 IEEE 11th International Conference on Computer Vision, Rio De Janeiro, Brazil, 14–21 October 2007.
7. Holovatyy, A.; Teslyuk, V.; Lobur, M. VHDL-AMS model of delta-sigma modulator for A/D converter in MEMS interface circuit. Perspective Technologies and Methods In MEMS Design. In Proceedings of the MEMSTECH 2015—11th International Conference, Lviv, Ukraine, 2–6 September 2015; pp. 55–57. [CrossRef]
8. Holovatyy, A.; Lobur, M.; Teslyuk, V. VHDL-AMS model of mechanical elements of MEMS tuning fork gyroscope for the schematic level of computer-aided design. Perspective Technologies and Methods. In Proceedings of the 2008 International Conference on Perspective Technologies and Methods in MEMS Design, Lviv, Ukraine, 21–24 May 2008; pp. 138–140. [CrossRef]
9. Zhuge, C.; Cai, Y.; Tang, Z. A novel dynamic obstacle avoidance algorithm based on collision time histogram. *Chin. J. Electron.* **2017**, *6*, 522–529. [CrossRef]
10. Puchalski, R.; Giernacki, W. UAV Fault Detection Methods, State-of-the-Art. *Drones* **2022**, *6*, 330. [CrossRef]
11. Bondyra, A.; Kołodziejczak, M.; Kulikowski, R.; Giernacki, W. An Acoustic Fault Detection and Isolation System for Multirotor UAV. *Energies* **2022**, *15*, 3955. [CrossRef]
12. Chao, H.; Cao, Y.; Chen, Y. Autopilots for small fixed-wing unmanned air vehicles: A survey. In Proceedings of the 2007 International Conference on Mechatronics and Automation, Harbin, China, 5–8 August 2007.
13. Vijayavargiya, A.; Sharma, A.; Anirudh; Kumar, A.; Kumar, A.; Yadav, A.; Sharma, A.; Jangid, A.; Dubey, A. Unmanned aerial vehicle. *Imp. J. Interdiscip.* **2016**, *2*, 1747–1750.
14. Shim, D.; Chung, H.; Kim, H.J.; Sastry, S. Autonomous exploration in unknown urban environments for unmanned aerial vehicles. In Proceedings of the AIAA Guidance, Navigation, and Control Conference and Exhibit, San Francisco, CA, USA, 15–18 August 2005.
15. Mikołajczyk, T.; Mikołajewski, D.; Kłodowski, A.; Łukaszewicz, A.; Mikołajewska, E.; Paczkowski, T.; Macko, M.; Skornia, M. Energy Sources of Mobile Robot Power Systems: A Systematic Review and Comparison of Efficiency. *Appl. Sci.* **2023**, *13*, 7547. [CrossRef]
16. Zhang, A.; Zhou, D.; Yang, M.; Yang, P. Finite-time formation control for unmanned aerial vehicle swarm system with time-delay and input saturation. *IEEE Access Pract. Innov. Open Solut.* **2019**, *7*, 5853–5864. [CrossRef]
17. Yasin, J.N.; Mohamed, S.A.S.; Haghbayan, M.-H.; Heikkonen, J.; Tenhunen, H.; Plosila, J. Unmanned aerial vehicles (UAVs): Collision avoidance systems and approaches. *IIEEE Access Pract. Innov. Open Solut.* **2020**, *8*, 105139–105155. [CrossRef]
18. Mircheski, I.; Łukaszewicz, A.; Szczebiot, R. Injection process design for manufacturing of bicycle plastic bottle holder using CAx tools. *Procedia Manuf.* **2019**, *32*, 68–73. [CrossRef]
19. Kiefer, R.J.; Grimm, D.K.; Litkouhi, B.B.; Sadekar, V. Collision Avoidance System. U.S. Patent 7 245 231 2007, 6 August 2004.
20. Foka, A.; Trahanias, P. Real-time hierarchical POMDPs for autonomous robot navigation. *Robot. Auton. Syst.* **2020**, *55*, 561–571. [CrossRef]
21. Murray, R.M. Recent research in cooperative control of multivehicle systems. *J. Dyn. Syst. Meas. Control* **2007**, *129*, 571–598. [CrossRef]
22. Ladd, G.; Bland, G. Non-military applications for small UAS platforms. In Proceedings of the AIAA Infotech@Aerospace Conference, Washington, DC, USA, 6–9 April 2009.
23. He, L.; Bai, P.; Liang, X.; Zhang, J.; Wang, W. Feedback formation control of UAV swarm with multiple implicit leaders. *Aerosp. Sci. Technol.* **2018**, *72*, 327–334. [CrossRef]
24. Esfahlani, S.S. Mixed reality and remote sensing application of unmanned aerial vehicle in fire and smoke detection. *J. Ind. Inf. Integr.* **2019**, *15*, 42–49. [CrossRef]
25. Valavanis, K.P. *Unmanned Aircraft Systems: The Current State-of-the Art*; Springer: Cham, Switzerland, 2016.
26. Wargo, C.A.; Church, G.C.; Glaneueski, J.; Strout, M. Unmanned Aircraft Systems (UAS) research and future analysis. In Proceedings of the 2014 IEEE Aerospace Conference, Big Sky, MT, USA, 1–8 March 2014.
27. Horla, D.; Giernacki, W.; Báča, T.; Spurny, V.; Saska, M. AL-TUNE: A Family of Methods to Effectively Tune UAV Controllers in In-flight Conditions. *J. Intell. Robot Syst.* **2021**, *103*, 5. [CrossRef]
28. Wang, X.; Yadav, V.; Balakrishnan, S.N. Cooperative UAV formation flying with obstacle/collision avoidance. *IEEE Trans. Control. Syst. Technol. A Publ. IEEE Control. Syst. Soc.* **2007**, *15*, 672–679. [CrossRef]
29. Łukaszewicz, A.; Panas, K.; Szczebiot, R. Design process of technological line to vegetables packaging using CAx tools. In Proceedings of the 17th International Scientific Conference on Engineering for Rural Development, Jelgava, Latvia, 23–25 May 2018; pp. 871–887. [CrossRef]
30. Łukaszewicz, A.; Szafran, K.; Jóźwik, J. CAx techniques used in UAV design process. In Proceedings of the 2020 IEEE 7th International Workshop on Metrology for AeroSpace (MetroAeroSpace), Pisa, Italy, 22–24 June 2020; pp. 95–98. [CrossRef]
31. Active Sensors. (n.d.). Esa.int. Available online: https://www.esa.int/Education/7.ActiveSensors (accessed on 15 December 2022).
32. What Is Active Sensor?—Definition. Available online: https://internetofthingsagenda.techtarget.com/defisensor (accessed on 13 March 2020).
33. Blanc, C.; Aufrère, R.; Malaterre, L.; Gallice, J.; Alizon, J. Obstacle detection and tracking by millimeter wave RADAR. *IFAC Proc. Vol.* **2004**, *37*, 322–327. [CrossRef]

34. Korn, B.; Edinger, C. UAS in civil airspace: Demonstrating "sense and avoid" capabilities in flight trials. In Proceedings of the 2008 IEEE/AIAA 27th Digital Avionics Systems Conference, St. Paul, MN, USA, 26–30 October 2008.
35. Owen, M.P.; Duffy, S.M.; Edwards, M.W.M. Unmanned aircraft sense and avoid radar: Surrogate flight testing performance evaluation. In Proceedings of the 2014 IEEE Radar Conference, Cincinnati, OH, USA 19–23 May 2014.
36. Quist, E.B.; Beard, R.W. Radar odometry on fixed-wing small unmanned aircraft. *IEEE Trans. Aerosp. Electron. Syst.* **2016**, *52*, 396–410. [CrossRef]
37. Kwag, Y.K.; Chung, C.H. UAV based collision avoidance radar sensor. In Proceedings of the 2007 IEEE International Geoscience and Remote Sensing Symposium, Barcelona, Spain, 23–28 July 2007.
38. Hugler, P.; Roos, F.; Schartel, M.; Geiger, M.; Waldschmidt, C. Radar taking off: New capabilities for UAVs. *IEEE Microw. Mag.* **2018**, *19*, 43–53. [CrossRef]
39. Nijsure, Y.A.; Kaddoum, G.; Khaddaj Mallat, N.; Gagnon, G.; Gagnon, F. Cognitive chaotic UWB-MIMO detect-avoid radar for autonomous UAV navigation. *IEEE Trans. Intell. Transp. Syst. A Publ. IEEE Intell. Transp. Syst. Counc.* **2016**, *17*, 3121–3131. [CrossRef]
40. Mohamed, S.A.S.; Haghbayan, M.-H.; Westerlund, T.; Heikkonen, J.; Tenhunen, H.; Plosila, J. A survey on odometry for autonomous navigation systems. *IEEE Access Pract. Innov. Open Solut.* **2019**, *7*, 97466–97486. [CrossRef]
41. Nashashibi, F.; Bargeton, A. Laser-based vehicles tracking and classification using occlusion reasoning and confidence estimation. In Proceedings of the 2008 IEEE Intelligent Vehicles Symposium, Eindhoven, The Netherlands, 4–6 June 2008.
42. Nüchter, A.; Lingemann, K.; Hertzberg, J.; Surmann, H. 6D SLAM-3D mapping outdoor environments: 6D SLAM-3D Mapping Outdoor Environments. *J. Field Robot.* **2007**, *24*, 699–722. [CrossRef]
43. Zhang, J.; Singh, S. Visual-lidar odometry and mapping: Low-drift, robust, and fast. In Proceedings of the 2015 IEEE International Conference on Robotics and Automation (ICRA), Seattle, WA, USA, 26–30 May 2015.
44. Tahir, A.; Böling, J.; Haghbayan, M.-H.; Toivonen, H.T.; Plosila, J. Swarms of unmanned aerial vehicles—A survey. *J. Ind. Inf. Integr.* **2019**, *16*, 100106. [CrossRef]
45. Armingol, J.M.; Alfonso, J.; Aliane, N.; Clavijo, M.; Campos-Cordobés, S.; de la Escalera, A.; del Ser, J.; Fernández, J.; García, F.; Jiménez, F.; et al. *Environmental Perception for Intelligent Vehicles*; Jiménez, F., Ed.; Butterworth-Heinemann: Oxford, UK, 2018; Volume 2, pp. 23–101.
46. Wang, C.-C.R.; Lien, J.-J.J. Automatic vehicle detection using local features—A statistical approach. *IEEE Trans. Intell. Transp. Syst. A Publ. IEEE Intell. Transp. Syst. Counc.* **2008**, *9*, 83–96. [CrossRef]
47. Mizumachi, M.; Kaminuma, A.; Ono, N.; Ando, S. Robust sensing of approaching vehicles relying on acoustic cues. *Sensors* **2014**, *14*, 9546–9561. [CrossRef]
48. Kim, J.; Hong, S.; Baek, J.; Kim, E.; Lee, H. Autonomous vehicle detection system using visible and infrared camera. In Proceedings of the 2012 12th International Conference on Control, Automation and Systems, Jeju, Republic of Korea, 17–21 October 2012; pp. 630–634.
49. Kota, F.; Zsedrovits, T.; Nagy, Z. Sense-and-avoid system development on an FPGA. In Proceedings of the 2019 International Conference on Unmanned Aircraft Systems (ICUAS), Atlanta, Gam, USA, 11–14 June 2019.
50. Mcfadyen, A.; Durand-Petiteville, A.; Mejias, L. Decision strategies for automated visual collision avoidance. In Proceedings of the 2014 International Conference on Unmanned Aircraft Systems (ICUAS), Orlando, FL, USA, 27–30 May 2014.
51. Saha, S.; Natraj, A.; Waharte, S. A real-time monocular vision-based frontal obstacle detection and avoidance for low cost UAVs in GPS denied environment. In Proceedings of the 2014 IEEE International Conference on Aerospace Electronics and Remote Sensing Technology, Yogyakarta, Indonesia, 13–14 November 2014.
52. Mejias, L.; McNamara, S.; Lai, J.; Ford, J. Vision-based detection and tracking of aerial targets for UAV collision avoidance. In Proceedings of the 2010 IEEE/RSJ International Conference on Intelligent Robots and Systems, Taipei, Taiwan, 8–22 October 2010.
53. Mohamed, S.A.S.; Haghbayan, M.-H.; Heikkonen, J.; Tenhunen, H.; Plosila, J. Towards real-time edge detection for event cameras based on lifetime and dynamic slicing. In *Advances in Intelligent Systems and Computing*; Springer International Publishing: Berlin/Heidelberg, Germany, 2020.
54. Lee, T.-J.; Yi, D.-H.; Cho, D.-I.D. A monocular vision sensor-based obstacle detection algorithm for autonomous robots. *Sensors* **2016**, *16*, 311. [CrossRef]
55. Haque, A.U.; Nejadpak, A. Obstacle avoidance using stereo camera. *arXiv* **2017**, arXiv:1705.04114.
56. Hartmann, W.; Tilch, S.; Eisenbeiss, H.; Schindler, K. Determination of the uav position by automatic processing of thermal images. In Proceedings of the ISPRS—International Archives of the Photogrammetry Remote Sensing and Spatial Information Sciences, Melbourne, Australia, 25 August–1 September 2012; pp. 111–116.
57. Gupta, A.; Padhi, R.; Indian Inst of Science Bangalore Dept of Aerospace Engineering. Reactive Collision Avoidance of Using Nonlinear Geometric and Differential Geometric Guidance. *J. Guid. Control Dyn.* **2010**, *34*, 303–310.
58. Khatib, O. Real-time obstacle avoidance for manipulators and mobile robots. In Proceedings of the 1985 IEEE International Conference on Robotics and Automation, St. Louis, MS, USA, 25–28 March 2005.
59. Oroko, J.; Nyakoe, G. Obstacle avoidance and path planning schemes for autonomous navigation of a mobile robot: A review. In Proceedings of the Sustainable Research and Innovation Conference, Delft, The Netherlands, 9 May 2014; pp. 314–318.
60. Choi, D.; Lee, K.; Kim, D. Enhanced Potential Field Based Collision Avoidance for Unmanned Aerial Vehicles in a Dynamic Environment. In Proceedings of the AIAA Scitech 2020 Forum, Orlando, FL, USA, 6–10 January 2020.

61. Grodzki, W.; Łukaszewicz, A. Design and manufacture of unmanned aerial vehicles (UAV) wing structure using composite materials. *Mater. Werkst.* **2015**, *46*, 269–278. [CrossRef]
62. Sun, J.; Tang, J.; Lao, S. Collision avoidance for cooperative UAVs with optimized artificial potential field algorithm. *Sens. IEEE Access Pract. Innov. Open Solut.* **2017**, *5*, 18382–18390. [CrossRef]
63. Wolf, M.T.; Burdick, J.W. Artificial potential functions for highway driving with collision avoidance. In Proceedings of the 2008 IEEE International Conference on Robotics and Automation, Pasadena, CA, USA, 19–23 May 2008.
64. Kim, C.Y.; Kim, Y.H.; Ra, W.-S. Modified 1D virtual force field approach to moving obstacle avoidance for autonomous ground vehicles. *J. Electr. Eng. Technol.* **2019**, *14*, 1367–1374. [CrossRef]
65. Yasin, J.N.; Mohamed, S.A.S.; Haghbayan, M.-H.; Heikkonen, J.; Tenhunen, H.; Plosila, J.M. Navigation of autonomous swarm of drones using translational coordinates. In *Advances in Practical Applications of Agents, Multi-Agent Systems, and Trustworthiness*; Springer International Publishing: Berlin/Heidelberg, Germany, 2020; pp. 353–362.
66. Yu, X.; Zhang, Y. Sense and avoid technologies with applications to unmanned aircraft systems: Review and prospects. *Prog. Aerosp. Sci.* **2015**, *74*, 152–166. [CrossRef]
67. Wang, M.; Voos, H.; Su, D. Robust online obstacle detection and tracking for collision-free navigation of multirotor UAVs in complex environments. In Proceedings of the 2018 15th International Conference on Control, Automation, Robotics and Vision (ICARCV), Singapore, 18–21 November 2018.
68. Sharma, S.U.; Shah, D.J. A practical animal detection and collision avoidance system using computer vision technique. *IEEE Access Pract. Innov. Open Solut.* **2017**, *5*, 347–358. [CrossRef]
69. De Simone, M.; Rivera, Z.; Guida, D. Obstacle avoidance system for unmanned ground vehicles by using ultrasonic sensors. *Machine* **2018**, *6*, 18; Version June 19, 2023 submitted to Journal Not Specified 12 of 12. [CrossRef]
70. Yu, Y.; Tingting, W.; Long, C.; Weiwei, Z. Stereo vision based obstacle avoidance strategy for quadcopter UAV. In Proceedings of the 2018 Chinese Control and Decision Conference (CCDC), Shenyang, China, 9–11 June 2018.
71. Bilimoria, K. A geometric optimization approach to aircraft conflict resolution. In Proceedings of the 18th Applied Aerodynamics Conference, Denver, CO, USA, 14–17 August 2000.
72. Goss, J.; Rajvanshi, R.; Subbarao, K. Aircraft conflict detection and resolution using mixed geometric and collision cone approaches. In Proceedings of the AIAA Guidance, Navigation, and Control Conference and Exhibit, San Francisco, CA, USA, 16–19 August 2004.
73. Seo, J.; Kim, Y.; Kim, S.; Tsourdos, A. Collision avoidance strategies for unmanned aerial vehicles in formation flight. *IEEE Trans. Aerosp. Electron. Syst.* **2017**, *53*, 2718–2734. [CrossRef]
74. Lin, Z.; Castano, L.; Mortimer, E.; Xu, H. Fast 3D collision avoidance algorithm for fixed wing UAS. *J. Intell. Robot. Syst.* **2020**, *97*, 577–604. [CrossRef]
75. Ha, L.N.N.T.; Bui, D.H.P.; Hong, S.K. Nonlinear control for autonomous trajectory tracking while considering collision avoidance of UAVs based on geometric relations. *Energies* **2019**, *12*, 1551. [CrossRef]
76. Pérez-Carabaza, S.; Scherer, J.; Rinner, B.; López-Orozco, J.A.; Besada-Portas, E. UAV trajectory optimization for Minimum Time Search with communication constraints and collision avoidance. *Appl. Artif. Intell.* **2019**, *85*, 357–371. [CrossRef]
77. Boivin, E.; Desbiens, A.; Gagnon, E. UAV collision avoidance using cooperative predictive control. In Proceedings of the 2008 16th Mediterranean Conference on Control and Automation, Ajaccio, France, 25–27 June 2008.
78. Biswas, S.; Anavatti, S.G.; Garratt, M.A. A particle swarm optimization based path planning method for autonomous systems in unknown terrain. In Proceedings of the 2019 IEEE International Conference on Industry 4.0, Artificial Intelligence, and Communications Technology (IAICT), Bali, Indonesia, 1–3 July 2019.
79. van den Berg, J.; Wilkie, D.; Guy, S.J.; Niethammer, M.; Manocha, D. LQG-obstacles: Feedback control with collision avoidance for mobile robots with motion and sensing uncertainty. In Proceedings of the 2012 IEEE International Conference on Robotics and Automation, St. Paul, MN, USA, 14–18 May 2012.
80. Zhu, H.; Alonso-Mora, J. Chance-constrained collision avoidance for MAVs in dynamic environments. *IEEE Robot. Autom. Lett.* **2019**, *4*, 776–783. [CrossRef]
81. Balestrieri, E.; Daponte, P.; De Vito, L.; Picariello, F.; Tudosa, I. Sensors and Measurements for UAV Safety: An Overview. *Sensors* **2021**, *21*, 8253. [CrossRef] [PubMed]
82. Pallottino, F.; Antonucci, F.; Costa, C.; Bisaglia, C.; Figorilli, S.; Menesatti, P. Optoelectronic proximal sensing vehicle-mounted technologies in precision agriculture: A review. *Comput. Electron. Agric.* **2019**, *62*, 859–873. [CrossRef]
83. Hu, W.; Li, X.; Hu, J.; Song, X.; Dong, X.; Kong, D.; Xu, Q.; Ren, C. A rear anti-collision decision-making methodology based on deep reinforcement learning for autonomous commercial vehicles. *IEEE Sens. J.* **2022**, *22*, 16370–16380. [CrossRef]
84. Chen, Y.; Cheng, C.; Zhang, Y.; Li, X.; Sun, L. A Neural Network-Based Navigation Approach for Autonomous Mobile Robot Systems. *Appl. Sci.* **2022**, *12*, 7796. [CrossRef]

sensors

MDPI

Article

UAV Trajectory Optimization in a Post-Disaster Area Using Dual Energy-Aware Bandits†

Amr Amrallah [1,2,*], Ehab Mahmoud Mohamed [3,4], Gia Khanh Tran [1,2] and Kei Sakaguchi [1,2]

1 Department of Electrical and Electronic Engineering, School of Engineering, Tokyo Institute of Technology, 2-12-1 Ookayama, Meguro-ku, Tokyo 152-8550, Japan
2 Academy for Super Smart Society, Tokyo Institute of Technology, 2-12-1 Ookayama, Meguro-ku, Tokyo 152-8550, Japan
3 Department of Electrical Engineering, College of Engineering in Wadi Addawasir, Prince Sattam Bin Abdulaziz University, Wadi Addawasir 11991, Saudi Arabia
4 Department of Electrical Engineering, Faculty of Engineering, Aswan University, Aswan 81542, Egypt
* Correspondence: amrallah@mobile.ee.titech.ac.jp

† This paper is an extended version of our paper published in Thirteenth International Conference on Ubiquitous and Future Networks (ICUFN), "Dual Energy-Aware based Trajectory Optimization for UAV Emergency Wireless Communication Network: A Multi-armed Bandit Approach", Barcelona, Spain, 5–8 July 2022.

Abstract: Over the past few years, with the rapid increase in the number of natural disasters, the need to provide smart emergency wireless communication services has become crucial. Unmanned aerial Vehicles (UAVs) have gained much attention as promising candidates due to their unprecedented capabilities and broad flexibility. In this paper, we investigate a UAV-based emergency wireless communication network for a post-disaster area. Our optimization problem aims to optimize the UAV's flight trajectory to maximize the number of visited ground users during the flight period. Then, a dual cost-aware multi-armed bandit algorithm is adopted to tackle this problem under the limited available energy for both the UAV and ground users. Simulation results show that the proposed algorithm could solve the optimization problem and maximize the achievable throughput under these energy constraints.

Keywords: unmanned aerial vehicle; trajectory optimization; reinforcement learning; multi-armed bandit; cost subsidy; post-disaster

1. Introduction

Across the globe, large-scale natural disasters are known for their severe casualties damage to property. Besides thousands of deaths and injuries resulting from various types of natural disasters around the world, there has been additional increase in material losses of about 100–150% [1]. The first few hours after a catastrophe are regarded as the "golden hours" of relief because rescue workers have a high probability of evacuating people from the damaged region during this period. Keep in mind that the wireless infrastructure in the disaster area might not be functional or even might be ravaged after the disaster. What makes the situation even more complicated is the paralysis of the power transmission lines after the disaster. The most powerful earthquake ever recorded in Japan, with a magnitude of 9.1, triggered a tsunami on the northeastern shore in March 2011. In the region of the catastrophe, around 6000 base stations (BSs) were wrecked, and the remaining BSs were highly overloaded with tremendous amounts of voice and data traffic. As a result of the high call block rate, communication services were suspended for roughly four days [2]. As a result, it is critical to develop an emergency wireless network that is completely independent of the conventional broadband network as soon as possible in order to preserve those valuable human lives. Unmanned aerial vehicles (UAVs) are well-known for their distinct characteristics, such as flexible deployment and rapid reaction. Thus, they can be deployed as temporary mobile BSs to establish this type of temporary emergency wireless network [3]. UAVs are now employed for a variety of emergency wireless communication applications, such as disaster management, surveillance, early warnings, post-disaster fusion centers, damage assessment, and supply-aid drop, in addition to temporary emergency wireless networks [4].

Notwithstanding the advantages of utilizing UAVs for establishing emergency wireless communication networks in a post-disaster area, there are a number of issues that need to be neutralized. In this tough environment induced by a natural disaster, the UAV must first design and optimize its flying route. This necessitates a quick online optimization procedure to accommodate the dramatic shift in the geographical field [5]. Secondly, the available energy for victims is ephemeral due to the limited battery capacity of their UEs and the destruction of the power supply infrastructure as a result of the natural disaster [6]. Thirdly, the UAV's operating duration is restricted by the onboard battery's capacity. The UAV should return to its base for recharging before it is completely depleted [7]. Therefore, while constructing an emergency wireless communication network, all of these concerns should be addressed. In addition, since this is a crucial mission, the UAV must assist as many people as possible in the disaster zone before its battery dies. Consequently, it is vital to seek out a robust mathematical tool capable of tackling such novel challenges.

Machine learning (ML) algorithms, and more precisely, reinforcement learning (RL) algorithms, are leveraged to tackle these kinds of optimization problems. Since RL algorithms are capable of achieving superb results in terms of efficiency and generalization, and due to their ability to deal with optimization problems with conflicting parameters, researchers have been inspired to utilize them in dealing with real-time issues in the field of wireless communications networks [8]. In this context, modern UAVs are equipped with wireless communications, ML, and image processing techniques. These techniques can support a UAV's trajectory optimization while avoiding obstacles and dealing with a limited battery capacity, which leads to serving more spots and enhancing the whole mission's energy efficiency. Recently, "follow me" drones have boomed in market value [9]. These drones are capable of filming a moving person with intelligent target-tracking and obstacle-avoidance algorithms, resulting in fabulous camera footage. Furthermore, novel UAV-related applications such as area surveillance, disaster relief, and traffic control are just a few applications that can be intelligently developed for future cities [10].

Multi-armed bandit (MAB) algorithms are considered one of the RL algorithms which are preferred in dealing with online optimization problems [11]. MAB algorithms can be defined as a set of arms, i.e., actions, of a bandit machine. At any given moment, pulling an arm leads to an instantaneous reward that is sampled from a certain distribution. A player wants to maximize his accumulated reward over the playing period by choosing an arm to pull during each moment of playing. Nevertheless, this player has no idea about the instantaneous reward behind each arm, since it will be revealed when the player decides to choose it. Therefore, some amount of the reward could be missed out due to this hidden setting. This loss is denoted by the term regret [12,13]. Thus, a player should develop a strategy to choose the arm that leads to the highest reward. On the other hand, this strategy should keep an eye on balancing between playing with the previously discovered arms that have high rewards or playing with the still-undiscovered ones that might have higher rewards. This is a common MAB dilemma, and it is called the exploration–exploitation trade-off [14,15]. Aiming to bolster disaster resilience, this paper describes a method of leveraging the latest advances in MAB algorithms and UAV wireless communications networks to improve the functionality of emergency wireless communication services for post-disaster response and assessment.

1.1. Prior Works and Motivations

One of the main benefits of deploying UAVs in emergency wireless communication networks is their capability of gathering extensive data from scattered ground devices, such as ground BSs, ground users, and even ground sensors [16]. The paper just cited gives a broad overview of different techniques but does not dive deeply in a specific direction. Furthermore, a UAV can operate as a flying edge server or a BS to support various traffic offloading scenarios [17], but it has a limited size of state action space. Due to its mobility, the planning and optimization of the UAV's trajectory and radio resource management of its wireless network are crucial issues. Researchers conducted many investigations on this topic during the past few years [18]. The UAV's speed and the location of its waypoints were used in [19] to design an optimal trajectory. However, the discussion was limited to cases where UAVs are used as relay stations in ad hoc networks. Minimizing the total energy consumption was studied in [20] using UAV speed control and a UAV data-scheduling-based heuristic algorithm, but it can be considered a theoretical approach only due to its large approximation factors. The authors of [21] considered UAVs with small cell capabilities to work as UAV-BSs. Particularly, the UAV movement, charging, and coverage action are considered in terms of jointly optimizing the energy and throughput through revenue and cost components. The UAV task scheduling was investigated in [22], where a mathematical framework for the optimization of UAV-aided video monitoring of a set of points of interest (PoI) distributed in a large urban area was proposed. Using this framework, which is based on mixed integer linear programming (MILP) techniques and real experimental data, particular energy-constrained UAVs are selected for recharging using public transportation buses, which also transfer the UAVs to desired PoIs in order to increase reliability and coverage.

UAV trajectory optimization may be carried out using traditional optimization approaches when realistic models of UAV wireless networks, including their flight dynamics, are available. Still, building these realistic network models is quite challenging; thus, model-free machine-learning methods can be used to manage the operation of UAVs that utilize wireless communication networks. By utilizing data gathered from prior experiences, machine learning algorithms are able to create autonomous control policies [23]. The authors of [24] studied the optimal deployment of UAVs equipped with directional antennas, using circle packing theory, where the 3D locations of the UAVs are determined such in a way that the total coverage area is maximized. The policy gradient approach for trajectory optimization used by the authors of [25] was able to maximize the overall distance covered by the UAV. However, this method took a lot of time and effort to find the best answer due to the large number of possible trajectories that the UAV must fly. The authors of [26] used the deep Q-learning method to optimize the UAV's flight path to maximize data rate during the flight period in an unknown environment. One major limitation of this proposed Q-learning approach for trajectory optimization is the long learning time, which makes it unfeasible even for moderate state spaces. By planning the UAV's flight trajectory, the authors of [27] were able to maximize the uplink transmission rate in a UAV cellular network. The deterministic policy gradient (DPG) approach was used to solve the optimization problem after it was converted into a Markov decision process (MDP). However, the characteristics of mmWave channels and beamforming were not taken into consideration during the optimization process.

Despite the existence of numerous excellent studies on UAV wireless communication networks, there are only a few works that focus on UAV-assisted emergency wireless communication networks. In our earlier studies [28,29], we investigated the radio resource allocation for a UAV emergency wireless communication network using a dynamic spectrum access system. The purpose of the deployment of UAVs as a cognitive radio network (CRN) was to maximize the downlink data rate in a post-disaster environment. Moreover, the limited transmission power of each UAV was used to control the constructed two multi-player MAB-based optimization problems called the power budget aware upper confidence bound (PBA-UCB) algorithm and the power budget aware Thompson sampling (PBA-TS) algorithm. The problem of gateway selection in a post-disaster area was addressed in [30], where a decentralized MAB algorithm was adapted to each UAV to let it maximize its data throughput by optimally choosing a suitable gateway. However, the optimization algorithm encountered some data loss due to not choosing the optimal strategy at the beginning of the optimization process. The authors of [31] built a system of a re-configurable intelligent surface (RIS) attached to a UAV. With the aid of a modified version of the MAB algorithm, the optimization problem aimed to find the optimum trajectory of the UAV that maximizes the total throughput while reducing the consumed flying power of the UAV. For a UAV with a limited battery capacity, the maximization problem for the number of served users was studied in [32] using two MAB algorithms called the ϵ-greedy algorithm and the D-UCB algorithm. The UAV trajectory optimization problem was studied in [33] to maximize the accumulated data volume from ground sensors under unknown network information. The optimization problem was transformed into a finite MDP and solved using two Q-learning-based UAV trajectory optimization frameworks called SUTOA and QUTOA. A Lyapunov-based deep Q-learning framed work called Safe-DQN was proposed in [34] to study the UAV trajectory optimization problem in a UAV-based emergency wireless communication network. The joint optimization problem aimed to maximize the total system rate under the constraints of the limited flight time of the UAV, the power capacity of the ground user, and the need to avoid obstacles in the disaster area. All the previous research was controlled by the limited capacity of the attached onboard battery for each UAV.

All of these studies on UAV emergency wireless communication networks focused on the optimization issue under a single power restriction, either a restricted UAV battery capacity or a limited amount of energy accessible to ground users (i.e., ground UE or ground sensors). We argue that these two elements together should be taken into account while constructing a UAV emergency wireless communication network. This is because the natural disaster destroys or at least renders the power supply network inoperable. Therefore, the goal of our suggested framework is to solve the UAV trajectory optimization problem under these two limited power constraints. In order to do this, our goal was to investigate a dual constraint optimization problem that might increase the UAV emergency wireless network's reliability in comparison to earlier studies. It should be noted that, to the best of our knowledge, our earlier work in [35] was the first study to investigate this sort of optimization issue with dual constrained energy capacity for both UAV and UEs at the same time. Furthermore, in the research, we extend our problem formulation by deeply evaluating the performance of our proposed framework against different benchmark methods. This evaluation was conducted in terms of the accumulated long-term uplink throughput of all UEs, the energy consumed by all UEs during the data-offloading process, and the energy efficiency of the UEs.

1.2. Contributions and Organization

According to the discussion in the preceding subsection, the majority of recent research on UAV emergency wireless communication networks concentrated primarily on the limited battery energy capacity of UAVs; just a small number of studies took into account the restricted energy capacity of ground users, i.e., ground users' equipment (UEs). We created a suggestion to fill this gap by examining an optimization scenario with constrained energy capacity for both UEs and UAVs. UAVs are seen as flying BSs that provide a wireless connection to ground UEs in the disaster-affected region from the sky. The information gathered from the UEs is deemed critical for estimating the status of the victims and assessing the damage in the post-disaster area. As a result, this critical data may be processed to help rescue crews save these precious lives. Our major goal is to acquire as much data from ground UEs as possible given the restricted power capacity of both the UAV and the ground UEs. However, since UAV coverage is somewhat limited in comparison to terrestrial BSs, our goal is to optimize the UAV flight trajectory to maximize the number of ground UEs visited before the battery runs out. Considering this limited battery capacity, another interesting idea is to have the UAVs maximize the scanned area while capturing photos to aid the rescue teams or to estimate the damage caused by the natural disaster. This goal was kept for our future work. The primary contributions of this work can be summarized as follows:

- In our situation, a UAV would gather user data in a disaster-affected region as part of a wireless emergency communication network. Ground BSs fail as a consequence of natural catastrophe damage, but ground UEs in the UAV coverage area may upload data using an alternate mode of connection from the sky thanks to the assistance of the UAV emergency wireless communication network. We propose an online optimization problem to optimize the uplink throughput for the UAV emergency wireless communication network by optimizing the flight trajectory of the UAV under these assumptions, taking into consideration the limited available energy for both the UAV and ground UEs in the post-disaster region.
- The optimization problem is adapted into a constrained MAB problem, with action, reward, and cost defined as the flight direction, uploaded data throughput, and dissipated energy for both the UAV and UEs, respectively.
- The numerical analysis of our proposed framework shows a considerable increase in long-term throughput and a slight increase in the energy consumption of the UEs in the post-disaster area, resulting in better energy efficiency for our proposed framework compared to other benchmark UAV trajectory optimization methods.

The rest of this paper is organized as follows. Section 2 presents the network architecture and formulates the online optimization problem for the long-term uplink throughput maximization problem. In Section 3, the general MAB framework is illustrated, followed by our proposed MAB-based framework for UAV trajectory optimization under dual energy constraints. Simulation results and numerical analysis are given in Section 4, and finally, the paper is concluded in Section 5.

2. Network Architecture and Problem Formulation

In this section, we discuss the architecture for the UAV-assisted emergency wireless communication network, including the flying model used for the UAV, the channel model for data uploading, and the optimization problem formulation.

2.1. UAV Flying Model

The system architecture for the UAV-assisted emergency wireless communication network is shown in Figure 1. In this scenario, a natural disaster, such as an earthquake or flood, strikes a specific location and causes the power grid and wireless network to fail. Our plan is to use the UAV to enable wireless access from the sky in this post-disaster area. In this approach, wireless connectivity may be enabled for victims, i.e., ground UEs, in this devastated region, allowing them to offload data that will be useful in guiding rescue crews and evaluating the damage. We assumed that there are M UEs trapped in this post-disaster area, denoted by $\mathcal{M} = \{1, \ldots, M\}$. Each of them has a fixed position designated by the following in Cartesian coordinates $l_m = (x_m, y_m)$. The UEs locations are supposed to be known to the UAV through self-reported global positioning system (GPS) coordinates. The discussion on how these data are transferred to the UAV is beyond the scope of this paper. It is assumed that the UAV will begin flying from the center of the post-disaster area, i.e., the center of the simulation area, which is denoted by $l_0 = (x_0, y_0)$. Additionally, it flies according to a constant speed of v and an altitude of H. We assume that this altitude is relatively high and that the data transmission duration is reasonably short and denoted by τ. As a result, the UAV is regarded immobile when uploading the UE data.

2.2. Wireless Communication Channel Model

For the convenience of designing an emergency wireless communication network, our designed system should utilize a channel in the unlicensed band, i.e., 2.4 GHz. In such a way, this system can be easily integrated with the hardware of modern UEs. Hence, the utilized channel model is expounded at [34], in accordance with the 3rd Generation Partnership Project (3GPP) specification in the technical report presented in [36]. This channel model represents the wireless communication link between the UAV and each of the served UEs into two components, i.e., the line-of-sight (LOS) component and the non-line-of-sight (NLOS) component, according to their corresponding probabilities, and can be calculated by (1).

$$L_m = \begin{cases} 30.9 + (22.25 - 0.5\log_{10} H)\log_{10} d_m + 20\log_{10} f, & \text{if LOS link} \\ \max\left(L_m^{\text{LOS}}, 32.4 + (43.2 - 7.6\log_{10} H)\log_{10} d_m + 20\log_{10} f\right), & \text{if NLOS link} \end{cases} \quad (1)$$

where H denotes the UAV flight altitude, f is the carrier frequency, and d_m is the distance between the UAV and any corresponding UE m, which can be calculated as follows:

$$d_m = \sqrt{H^2 + \|l_m - l_0\|^2},\ \forall m \in \mathcal{M} \quad (2)$$

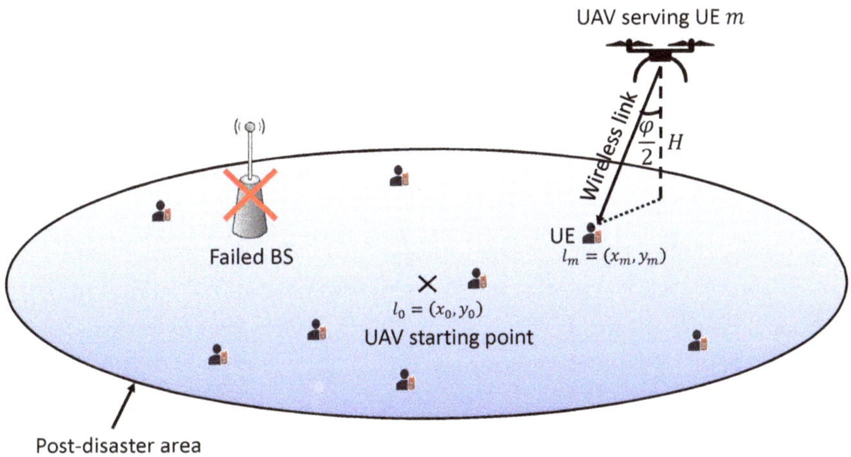

Figure 1. UAV emergency wireless communication network.

Since the calculation of path loss due to the NLOS component is a function of the path loss due to the LOS component L_m^{LOS}, the term L_m^{LOS} should be calculated prior to estimate the path loss of the NLOS component. The probability of the LOS link is denoted by $\mathcal{P}_m^{\text{LOS}}$ and given in (3).

$$\mathcal{P}_m^{\text{LOS}} = \begin{cases} 1, & \text{if} \sqrt{d_m^2 - H^2} \le d_0 \\ \frac{d_0}{\sqrt{d_m^2 - H^2}} + \exp\left\{\left(\frac{-\sqrt{d_m^2 - H^2}}{p_1}\right)\left(1 - \frac{d_0}{\sqrt{d_m^2 - H^2}}\right)\right\}, & \text{if} \sqrt{d_m^2 - H^2} > d_0 \end{cases} \quad (3)$$

$$d_0 = \max\left(294.05\log_{10} H - 432.94, 18\right) \quad (4)$$

$$p_1 = 233.98\log_{10} H - 0.95 \quad (5)$$

Furthermore, the probability of NLOS can be obtained naturally for the probability of LOS as follows:

$$\mathcal{P}_m^{\text{NLOS}} = 1 - \mathcal{P}_m^{\text{LOS}} \quad (6)$$

The channel gain between the UAV and any connected UE can be calculated as follows:

$$g_m = \mathcal{P}_m^{\text{LOS}}\left(10^{L_m^{\text{LOS}}/10}\right)^{-1} + \mathcal{P}_m^{\text{NLOS}}\left(10^{L_m^{\text{NLOS}}/10}\right)^{-1} \quad (7)$$

where L_m^{LOS} and L_m^{NLOS} are the path loss for the LOS and NLOS, respectively, and can be calculated from (1).

2.3. Data Transmission Model

For the sake of simplicity, we assumed that the UAV emergency wireless communication network can be established between the UAV and only one UE at any certain time. Hence, there are no simultaneous wireless connections from different UEs to the UAV. The effective radiation angle of the UAV antenna is denoted by φ, so the maximum distance between the UAV and any UE that permits the establishment of a wireless communication link is $H/\cos(\varphi)$. Additionally, it can be observed that the relationship between the channel gain g_m in (7) and the distance d_m in (2) is an inverse relationship. Therefore, our definition of the effective radiation angle φ is used as a parameter to make sure that this distance is suitable for establishing a wireless communication link. This can be done by evaluating the signal-to-noise ratio (SNR) value for a covered UE. When it reaches a certain threshold that permits the establishment of a wireless communication link, this covered UE can access the UAV to offload its data. Additionally, the value of φ can be chosen to be very narrow to shrink the UAV coverage. In such a way, the simultaneous transmission from different UEs can be easily eliminated. Hence, a UE can be within the UAV coverage if and only if it belongs to the following set:

$$\mathcal{M}_{\text{cov}} = \{m \in \mathcal{M} \, : \, d_m \leq H/\cos(\varphi)\} \tag{8}$$

It is assumed that each UE in the post-disaster area has an amount of data equal to Ψ bits. Then, a UE access indicator, denoted by α_m, is used to show whether the m-UE is connected to the UAV or not. This access indicator depends on two factors, i.e., the distance from the UAV, d_m, and the total uploaded bits from the m-UE to the UAV, Ω_m. Thus, α_m can be expressed as follows:

$$\alpha_m(t) = \begin{cases} 1, & \text{if } m \in \mathcal{M}_{\text{cov}}, \, \Omega_m(t) < \Psi \\ 0, & \text{otherwise} \end{cases} \tag{9}$$

where $t \in \mathcal{T}$, $\mathcal{T} = \{1, \ldots, T\}$ is the time elapsed while the UAV flies over the post-disaster area. The total uploaded bits from the m-UE to the UAV can be calculated as:

$$\Omega_m(t) = \sum_{i=1}^{t} \omega_m(i) \tag{10}$$

where ω_m is the instantaneous uploaded data size at time t and can be calculated as follows:

$$\omega_m(t) = R_m(t)\, \tau \tag{11}$$

where R_m is the transmission data rate from the m-UE to the UAV and can be calculated according to Shannon's theorem as follows:

$$R_m(t) = \alpha_m(t)\, B\, \log_2\left(1 + \frac{g_m\, P_m^{\text{Tx}}}{\sigma_0}\right) \tag{12}$$

where B is the available wireless channel bandwidth, P_m^{Tx} is the transmission power from m-UE, and σ_0 denotes the power of the additive white Gaussian noise (AWGN) at the UAV receiver.

2.4. Energy Model

From the perspective of the limited energy capacity, the consumed energy can be classified as follows: (1) the energy consumed by each m-UE while it is idle and during the data offloading period; (2) the energy consumed by the UAV while it is flying over the post-disaster area to provide the wireless connectivity for the trapped UEs. Thus, at any time t, these two consumed terms of energy can be denoted as follows:

$$e_m(t) = \begin{cases} \alpha_m(t)\ P_m^{\text{Tx}}\ \tau, & \text{if } m\text{-UE at Tx mode} \\ (1-\alpha_m(t))\ e^{\text{idle}}, & \text{if } m\text{-UE at idle mode} \end{cases} \quad (13)$$

$$E(t) = \Xi\ t \quad (14)$$

where e^{idle} is the energy consumed by each of m-UE during the idle mode, and Ξ is the UAV's flying power. Of course, there are many factors that control the UAV's energy consumption, such as the flying speed, acceleration, and mass of the UAV. However, we tried to simplify the notation of the energy consumption to be averaged per unit of time. In such a way, we can study the ability of our proposed solution to handle this dynamic energy consumption over time. Furthermore, the energy consumed by the UAV's receiver circuit and signal processing are relatively low compared to the energy consumed during flying, so it can be neglected. To expand this research to more detailed power consumption, the work presented in [3] is a straightforward extension, and it will be considered for our future work.

2.5. Problem Formulation

The ultimate goal of the post-disaster surveillance system is to improve the rescue success rate of victims and also to reduce casualties. This goal can be achieved by maximizing the data uploaded from the trapped victims in the post-disaster area over the UAV trajectory. At the same time, we must take into account the valuable limited energy of both UEs and the UAV. Mathematically speaking, our optimization problem can be expressed as follows:

$$\max_{m \in \mathcal{M}} \quad \frac{1}{T} \sum_{t=1}^{T} \sum_{m=1}^{M} \omega_m(t) \quad (15)$$

$$\text{s.t.} \ \sum_{t=1}^{T} e_m(t) \leq e_0, \ \forall m \in \mathcal{M} \quad (16)$$

$$\sum_{t=1}^{T} E(t) \leq E_0 \quad (17)$$

The optimization problem shown in (15) is considered an online optimization problem that aims to maximize the long-term throughput of the whole network by optimizing the UAV's flight trajectory. Since there is an unlimited number of routes that can be existed by changing the order of how the UAV serves the UEs, our optimization problem is an NP-hard problem. However, by considering energy constraints introduced in Equations (16) and (17), the optimization problem can be viewed as an NP-complete problem. The whole optimization process is done not only in an online manner but also in a decentralized way where there is no information exchange between different network elements. Furthermore, for any conventional programming solvers, all information should be gathered at one centralized entity to solve the optimization problem, which cannot be satisfied while designing an emergency wireless communications network for a post-disaster surveillance system. In such a case, we suggest using a reinforcement-learning-based algorithm to deal with this kind of online optimization problem.

The decision variables can be defined as the accumulated instantaneous throughput $\omega_m(t)$ for all the M UEs throughout the UAV's flight time T. The constraint (16) shows that the maximum energy available for each UE is limited by e_0, and the other constraint (17) limits the energy available for the UAV by E_0; both are considered the feasibility constraints of the optimization problem. Furthermore, the right-hand sides of constraints (16) and (17) are also long-term cumulative variables related to the UAV flight trajectory. Hence, the whole flight-trajectory process should be taken into account when solving the position of the UAV at any time t. Therefore, this optimization problem becomes difficult to figure out using conventional optimization methods. Additionally, sharing information on the remaining battery capacity for every UE in the post-disaster area is quite a changeling, especially when the commercial mobile network has malfunctioned. Therefore, for the sake of simplicity and without loss of generalization, our optimization problem was designed for the worst-case scenario for the available battery capacity for each UE. This value was chosen to be around 10% of modern UE's average total battery capacity [37]. In the next section, we introduce an MAB-based framework to tackle this issue.

3. Dual-Energy-Aware MAB-Based UAV Trajectory Optimization Approach

In this section, we explain the general MAB framework and then illustrate how the proposed dual-energy-aware MAB approach could address our previously described optimization problem.

3.1. General MAB Framework

Generally speaking, in any MAB-based framework, a player aims to maximize his long-term reward while playing with a set of arms of the bandit machine, $j \in \{1, \ldots, J\}$. This can be performed in a sequential way by selecting an arm at time t, i.e., $j(t)$, and observing their corresponding reward, i.e., $r_j(t)$. In the first few moments, the player tries to explore candidate arms as much as possible and observes their corresponding rewards. After that, the player exploits the arm with the highest reward, based on the gathered information from the already explored arms, to maximize the cumulative reward over the episode. This dilemma is quite well-known in the world of the MAB framework and is known as the exploration–exploitation trade-off [15]. The MAB framework can be classified as stochastic or adversarial based on the distribution of the rewards [14,15]. For the stochastic MAB framework, the rewards behind each arm are drawn from independent and identical distribution (i.i.d); however, for the adversarial MAB framework, rewards are selected arbitrarily with no prior distribution. For these two types of MAB frameworks, extensive research has been done to deal with these kinds of problems, resulting in the introduction of different algorithms, such as the ϵ-greedy algorithm [38], the upper confidence bound (UCB) algorithm [39], the Thompson sampling (TS) algorithm [40], and the exponential-weight algorithm for exploration and exploitation (EXP3) [41]. Furthermore, in real-world optimization problems, choosing an arm with a higher reward will have a high cost as well. Thus, cost-effective and budget-constrained MAB algorithms are introduced to deal with these kinds of scenarios [42,43].

3.2. The DEA-MAB Approach

To address the online optimization problem with the dynamic energy consumption over time that is given in (15), and which constrained by conditions (16) and (17), an MAB-based framework that is dual-energy-aware called DEA-MAB is proposed. Our DEA-MAB approach is inspired by the cost-subsidized explore-then-commit algorithm proposed in [43], where the chosen arm is accompanied by a certain cost. One of the traditional ways to optimize this reward/cost is to directly deduct the cost from the reward in the control formula. However, this is not usually meaningful in real-world problems, especially when the reward and the cost are defined in different quantities [43], such as the achievable throughput and the energy consumed, as illustrated in our problem formulation. Hence, it is necessary to find a better way to optimize for both the reward and the cost. In other words, the algorithm should avoid incurring an excessive cost for just a marginal increase in the reward. This may be done by building a feasible set of arms which is an estimate of all arms with a mean reward greater than the least tolerable value in each round, based on the upper confidence bound (UCB) and lower confidence bound (LCB) of the reward of each arm. Then, the arm with the lowest cost in this feasible set is selected to be played by it.

Though the cost-subsidized explore-then-commit algorithm is considered a good solution for separating the reward and the cost functions, it still needs some adaptation to tackle our optimization problem that is given in (15). Precisely speaking, our optimization problem considers two different energy costs, so the DEA-MAB algorithm adds a further step for checking the second cost. Thus, some controlling functions were added in the proposed algorithm to precisely address this issue.

Algorithm 1 summarizes how the DEA-MAB algorithm works. The DEA-MAB algorithm's input attributes are the state spaces of all available $\mathcal{M}$ UEs, including their corresponding locations $l_m \; \forall \; m \in \mathcal{M}$; the total flight time T; tuning parameters δ and ϵ; the available energy for each piece of UE e_0; and the total flight time of the UAV till its battery is completely depleted, T. At each time period t of the total flight time T, the UAV should select one of M UEs distributed in the post-disaster area via the DEA-MAB algorithm; then it will fly towards it to offload its data. In the beginning, the algorithm is initialized at $t = 0$ by setting the number of times each m-UE is selected, $Q_m(t)$, and their average achievable throughput, $\overline{\omega}_m(t)$, to 0. The DEA-MAB algorithm is divided into two phases, i.e., the pure exploration phase and the selection phase. During the exploration phase, the UAV randomly selects a UE to visit as follows:

$$m^*(t) = t \mod M \tag{18}$$

Then, the corresponding throughput $\omega_{m^*}(t)$ is observed, and the selection number, $Q_m(t)$, and the average throughput, $\overline{\omega}_m(t)$, are updated as in the following equations:

$$Q_{m^*}(t) = Q_{m^*}(t-1) + 1 \tag{19}$$

$$\overline{\omega}_{m^*}(t) = \frac{1}{Q_{m^*}(t)} \sum_{i=1}^{Q_{m^*}(t)} \omega_m(i) \tag{20}$$

The exploration phase is performed for a time period equal to $M\,\pi$, where $\pi = (T/M)^{2/3}$ is as given in [43]. After that, the DEA-MAB algorithm goes for the selection phase during each time $t \in [M\pi + 1, T]$, where both the UCB and LCB are calculated as follows:

$$\gamma_m^{\text{UCB}}(t) = \overline{\omega}_m(t) + \sqrt{2\ln(t)/Q_m(t)}, \; \forall \; m \in \mathcal{M} \tag{21}$$

$$\gamma_m^{\text{LCB}}(t) = \overline{\omega}_m(t) - \sqrt{2\ln(t)/Q_m(t)}, \; \forall \; m \in \mathcal{M} \tag{22}$$

Then, the UE index corresponding to the maximum value of the $\gamma_m^{\text{LCB}}(t)$ is calculated as follows:

$$\eta_t = \arg\max_m \gamma_m^{\text{LCB}}(t) \tag{23}$$

Afterwards, the feasibility region of all UEs having $\gamma_m^{\text{UCB}}(t) \geq (1-\delta)\gamma_{\eta_t}^{\text{LCB}}(t)$ is enumerated as follows:

$$F(t) = \left\{ m : \gamma_m^{\text{UCB}}(t) \geq (1-\delta)\gamma_{\eta_t}^{\text{LCB}}(t) \right\} \tag{24}$$

For this set of UEs, $F(t)$, the dissipated energy for each of the m-UE contained in this $F(t)$ list is obtained. Then, a control set, $C(t)$, is constructed out of all UEs in $F(t)$. A check is performed for the UEs' energy consumption; then priority is given to all UEs in the $F(t)$ list in case they exceed their energy consumption with a value of $1 - \varepsilon$ of the total available energy e_0. Otherwise, $C(t)$ is set to be equal to $F(t)$. This can be illustrated as follows:

$$C(t) = \begin{cases} m : \sum_{i=1}^{t} e_m(i) \geq (1-\varepsilon)e_0, & \sum_{i=1}^{t} e_m(i) \geq (1-\varepsilon)e_0 \\ F(t), & \text{otherwise} \end{cases} \tag{25}$$

Out of this list, $C(t)$, the UE corresponding to the minimum UAV energy cost, $E(T)$, is selected as a next-served UE for data offloading in the UAV flight trajectory as follows:

$$m^*(t) = \underset{m \in C(t)}{\arg\min}\, E(t) \tag{26}$$

Afterwards, values of the selection number, $Q_m(t)$, and the average throughput, $\overline{\omega}_{m^*}(t)$, are updated for the selected UE, $m^*(t)$, as given in Algorithm 1. Since the UAV should accomplish the whole data offloading task and ensure flying back to its base before the battery is used up, the UAV should confirm that there is enough remaining battery energy for returning. Otherwise, the UAV could be lost or damaged if it cannot arrive at its base before the battery becomes empty. Therefore, a checking step is provided to confirm this critical condition at each time before deciding to choose the next UE to be served. In this way, the DEA-MAB algorithm can optimize the UAV's flight trajectory considering limited energy of both the UAV and the UEs.

Algorithm 1: The proposed algorithm: DEA-MAB.

Output: $m^*(t)$
Input: $\mathcal{M}, l_m\ \forall\, m \in \mathcal{M}, T, \delta, \varepsilon, e_0, \nu, \Xi, E_0$
Initialization: at $t = 0$, Set $Q_m(0) = 0,\ \overline{\omega}_m(0) = 0,\ \forall m \in \mathcal{M}$
Exploration Phase:
Explore available UEs and calculated the corresponding throughput
for $t = 1$ **to** $M\pi$ **do**
  $m^*(t) = t \mod M$
  Fly towards a UE $m^*(t)$ and obtain $\omega_{m^*}(t)$
  $Q_{m^*}(t) = Q_{m^*}(t-1) + 1$
  $\overline{\omega}_{m^*}(t) = \frac{1}{Q_{m^*}(t)} \sum_{i=1}^{Q_{m^*}(t)} \omega_m(i)$
end for
Selection Phase:
for $t = M\pi + 1$ **to** T **do**
  $\gamma_m^{\mathrm{UCB}}(t) \leftarrow \overline{\omega}_m(t) + \sqrt{2\ln(t)/Q_m(t)},\ \forall\, m \in \mathcal{M}$
  $\gamma_m^{\mathrm{LCB}}(t) \leftarrow \overline{\omega}_m(t) - \sqrt{2\ln(t)/Q_m(t)},\ \forall\, m \in \mathcal{M}$
  $\eta_t = \underset{m}{\arg\max}\, \gamma_m^{\mathrm{LCB}}(t)$
  $F(t) = \left\{m : \gamma_m^{\mathrm{UCB}}(t) \geq (1-\delta)\gamma_{\eta_t}^{\mathrm{LCB}}(t)\right\}$
  Obtain $e_m\ \forall\, m \in F(t)$
  if $\sum_{i=1}^{t} e_m(i) \geq (1-\varepsilon)e_0$ **then**
    $C(t) = \left\{m : \sum_{i=1}^{t} e_m(i) \geq (1-\varepsilon)e_0\right\}$
  else
    $C(t) = F(t)$
  end if
  $m^*(t) = \underset{m \in C(t)}{\arg\min}\, E(t)$
  The UAV fly towards UE $m^*(t)$ and obtain $\omega_{m^*}(t)$
  $Q_{m^*}(t) = Q_{m^*}(t-1) + 1$
  $\overline{\omega}_{m^*}(t) = \frac{1}{Q_{m^*}(t)} \sum_{i=1}^{Q_{m^*}(t)} \omega_m(i)$
  if $E_0 - \sum_{i=0}^{t} E(i) < 2\,\Xi\,\sqrt{\|l_{m^*} - l_0\|^2}\,\nu^{-1}$ **then**
    Break the data offloading loop and the UAV returns to its base
  end if
end for

3.3. Complexity Analysis of the Proposed Approach

In the previous section, the task of the UAV finding the best trajectory in the post-disaster area is spotlighted. This is accomplished by finding the optimal policy to choose the next UE to be served through the learning process in Algorithm 1. In the beginning, the uplink throughput that can be achieved while the UAV connects to this UE is examined. Then, a higher priority is given to UEs whose batteries are nearly depleted. The consumed energy during UAV flying is also minimized. Moreover, it is assumed that the action space is deterministic; i.e., all actions are well-known to the UAV. Hence, the fundamental source of the computational complexity of the DEA-MAB algorithm comes from calculating both the UCB and the LCB. Then, other parameters are updated according to this selection. It should be mentioned that these parameters have the same computational complexity order as UCB or LCB. Hence, the overall computational complexity order of our proposed algorithm is a polynomial of $M + 1$, and can be expressed as $\mathcal{O}(M + 1)$ [43].

4. Simulation Results

In this section, the performance of the DEA-MAB algorithm is evaluated. In the simulation, it was assumed that the UAV will provide wireless connectivity for a previously allocated area where there are M trapped UEs which are randomly distributed. However, for a large post-disaster area, more than one UAV can be deployed to support the data offloading while considering the coordination between UAVs to facilitate rescue operations. This larger system is left for future work.

Table 1 shows the simulation parameters used in verifying our proposed algorithm. In order to investigate the effectiveness of our proposed framework, two trajectory optimization methods were used as benchmarks for the sake of comparison. These two methods can be described as follows:

1. The post-disaster area spiral scanning (PASS) method: This method is designed to scan the whole area using the spiral path where the UAV starts to fly from the center of the post-disaster area. With respect to the UAV antenna's radiation angle, a projected circle is created on the ground. This circle scans the whole post-disaster area from the center to the borders.
2. Shortest flight path (SFP) method: In this method, the UAV starts to fly from the center of the post-disaster area and then selects the UE with the shortest path. Then, the UAV flies toward this UE and hovers above it to offload its data. After that, the UAV searches for the next close UE and flies toward it. This operation is performed till the last UE.

In the following, the performance of the proposed framework is evaluated by comparing it with benchmark algorithms during the varying of both the number of trapped UEs in the post-disaster area and the UAV's battery capacity. For the sake of accuracy, and due to the randomness in UEs' distributions, all simulations were performed for a long enough time, i.e, 10^4 iterations. The average value of each case is provided for a better estimation of the result.

Figure 2 shows a sample of the UAV's flight trajectory in the post-disaster area. To visualize how our DEA-MAB algorithm could optimize the UAV's flight trajectory considering its available battery power, three different values were used, i.e., $E_0 = 20, 30, 40$ Wh, while keeping the number of UEs equal to 40. Obviously, increasing available UAV battery power increases the chance of serving more UEs in the post-disaster area.

Table 1. Simulation parameters.

Parameter	Value
Simulation area	500 m × 500 m
Number of UEs in the simulation area (M)	20, 30, 40, 50
UAV flight speed (v)	20 km/h
UAV flight altitude (H)	100 m
UAV antenna radiation angle (φ)	$\pi/8$ rad
Carrier frequency (f)	2.4 GHz
Channel bandwidth (B)	10 MHz
Data transmission duration (τ)	1 s
UE Transmission power (P_m^{Tx})	23 dBm
AWGN spectral density (σ_0)	−130 dBm/Hz
UAV battery capacity (E_0)	20, 30, 40 Wh
UAV flying power (Ξ)	120 W
UE battery capacity (e_0)	1 Wh
UE energy dissipation in idle mode (e^{idle})	0.01 J
Data rate feasibility region factor (δ)	0.6
Critical power feasibility region factor (ε)	0.5

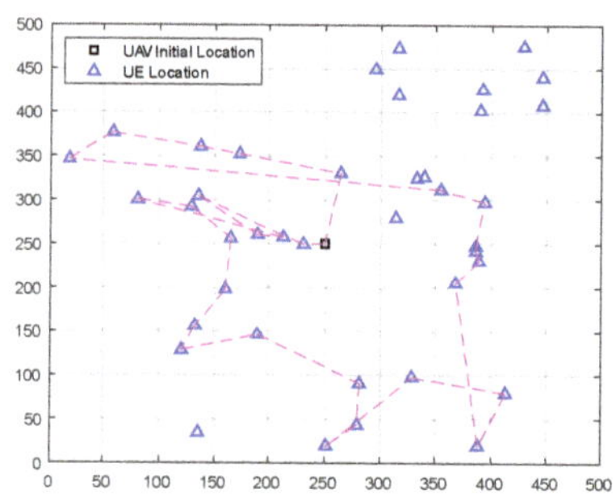

(**a**) $E_0 = 20$ Wh

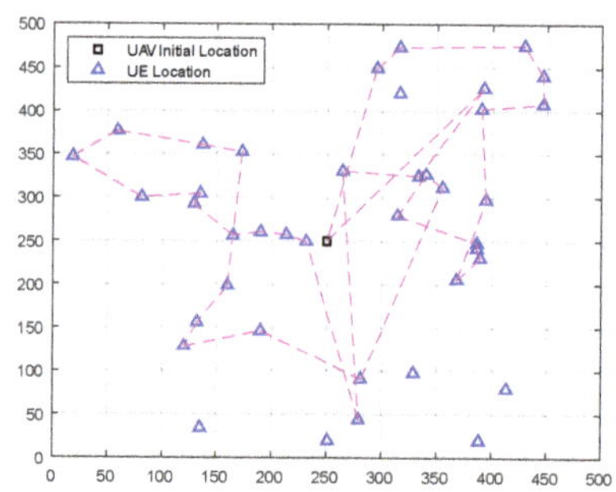

(**b**) $E_0 = 30$ Wh

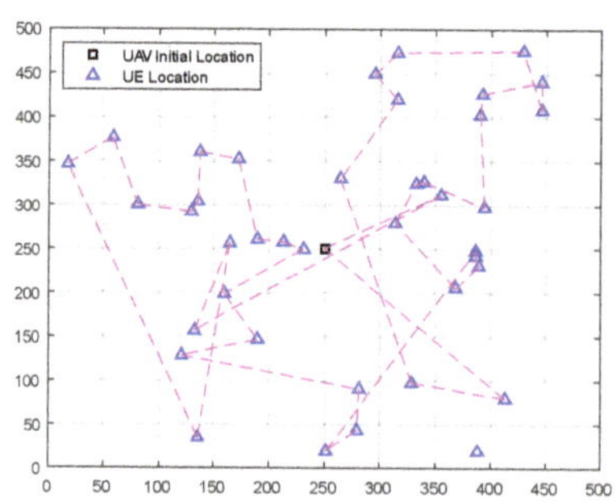

(**c**) $E_0 = 40$ Wh

Figure 2. A sample of the UAV flight trajectory using the DEA-MAB algorithm.

Figure 3 gives the long-term throughput for the data uploaded from UEs in the emergency wireless communication network. It is clearly visible that regardless of the value of the UAV's battery capacity or the algorithm used, as the number of UEs trapped in the post-disaster area increases, the uplink data throughput increases as well. Nevertheless, this upward trend gradually decreases, and all curves would saturate at a certain number of UEs. This is because the maximum capacity of a communication system with a fixed bandwidth is fixed. Hence, while the number of UEs increases, the accumulated uplink throughput of the emergency wireless network continues to approach this maximum capacity. When comparing the throughput performance of the DEA-MAB algorithm with other benchmark methods at various values of UAV battery capacity, it is clear that our proposed algorithm can achieve more uplink throughput than the PASS method, and much higher than the SFP method. For example, when (E_0 = 20 Wh, M = 30), (E_0 = 30 Wh, M = 40), and (E_0 = 40 Wh, M = 50), the DEA-MAB algorithm achieved higher throughput performance by 26%, 28%, and 24% compared to the PASS method, and high performance by 113%, 188%, and 184% than the SFP method, respectively.

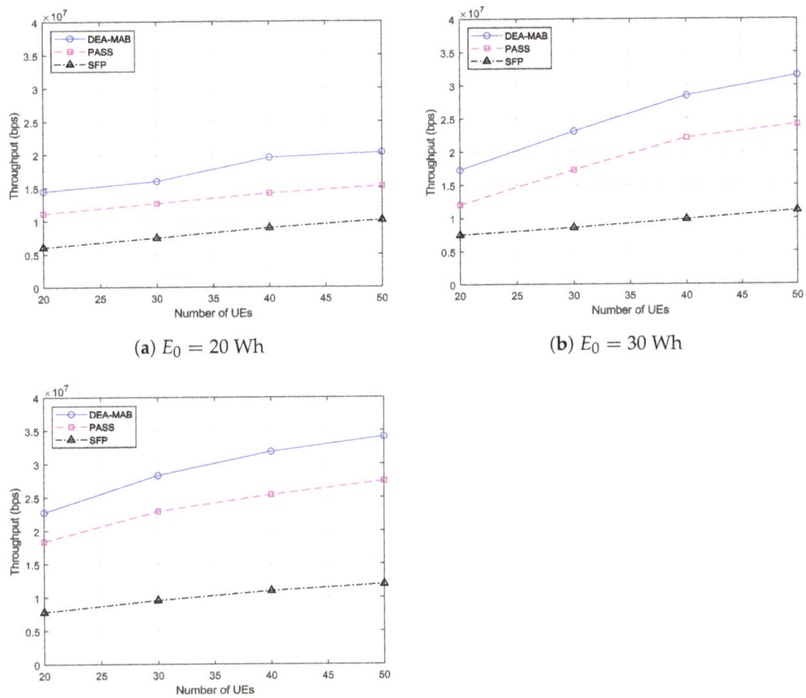

(**a**) E_0 = 20 Wh

(**b**) E_0 = 30 Wh

(**c**) E_0 = 40 Wh

Figure 3. The DEA-MAB algorithm's throughput versus the number of users.

In Figure 4, the normalized total energy consumption of all UEs trapped in the post-disaster area is compared among the three methods. It can be seen clearly that regardless of the used method, as the number of UEs increases, the total normalized energy consumption of UEs increases as well. Furthermore, for the same method with a certain number of UEs, the higher the UAV's battery capacity, the more energy consumed per UE. This can be justified, as when the UAV has a higher battery capacity, it can have a higher chance to offload data from a larger number of UEs before its battery becomes depleted. Additionally, since $P_m^{\mathrm{Tx}}\tau >> e^{\mathrm{idle}}$, more UEs tend to consume energy in the data-offloading process rather than just staying in idle mode. When comparing the normalized energy consumption performance of the DEA-MAB algorithm with other benchmark methods at the same values of UAV battery capacity, it can be shown that the DEA-MAB algorithm always has higher energy consumption than the PASS method, and much higher than the

SFP method. This can be explained by the overall system throughput being increased at the cost of more energy consumption by the UEs. For the sake of comparison, let us observe the same points at ($E_0 = 20$ Wh, $M = 30$), ($E_0 = 30$ Wh, $M = 40$), and ($E_0 = 40$ Wh, $M = 50$): the total energy consumption of all UEs using the DEA-MAB algorithm was increased by 11%, 24%, and 23% compared to the PASS method, and by 73%, 109%, and 169% compared to the SFP method.

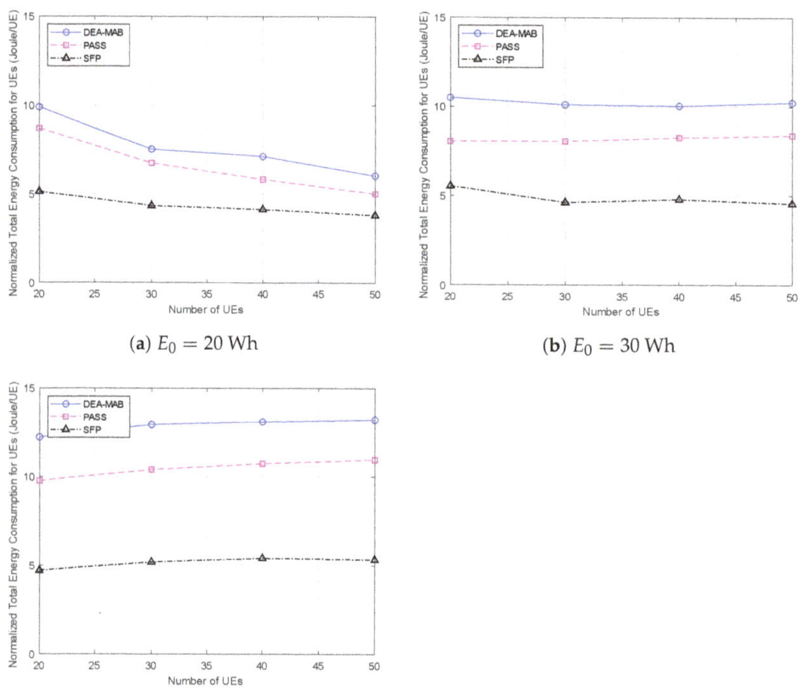

(**a**) $E_0 = 20$ Wh (**b**) $E_0 = 30$ Wh (**c**) $E_0 = 40$ Wh

Figure 4. Normalized total energy consumption versus the number of users.

As observed from the analysis of results in Figures 3 and 4, it can be concluded that the DEA-MAB algorithm can achieve a considerable increase in the uplink throughput of UEs with a reasonable increase in the UEs energy consumption. Hence, for a better understanding of the advantages of using the DEA-MAB algorithm, the UEs' energy efficiency (μ) is compared using our proposed algorithm against benchmark methods. μ can be defined as the ratio of the long-term UEs uplink throughput over the total UEs energy consumption in bit/Joule as follows:

$$\mu = \frac{\sum_{t=1}^{T} \sum_{m=1}^{M} \omega_m(t)}{\sum_{t=1}^{T} \sum_{m=1}^{M} e_m(t)} \quad (27)$$

In the energy efficiency performance shown in Figure 5, it is observed clearly that whatever the UAV's battery capacity or the number of UEs trapped in the post-disaster area, the DEA-MAB algorithm can surpass benchmark methods in terms of energy efficiency, which, of course, means enhancing the overall performance of the emergency wireless communication network. It should be mentioned that, when increasing the UAV's battery capacity to 40 Wh, as in Figure 5c, the PASS method achieved a performance that is very close to that of the DEA-MAB algorithm. This can be justified, as the UAV's battery at this point becomes quite enough to accomplish the spiral scanning for a major part of the post-disaster area.

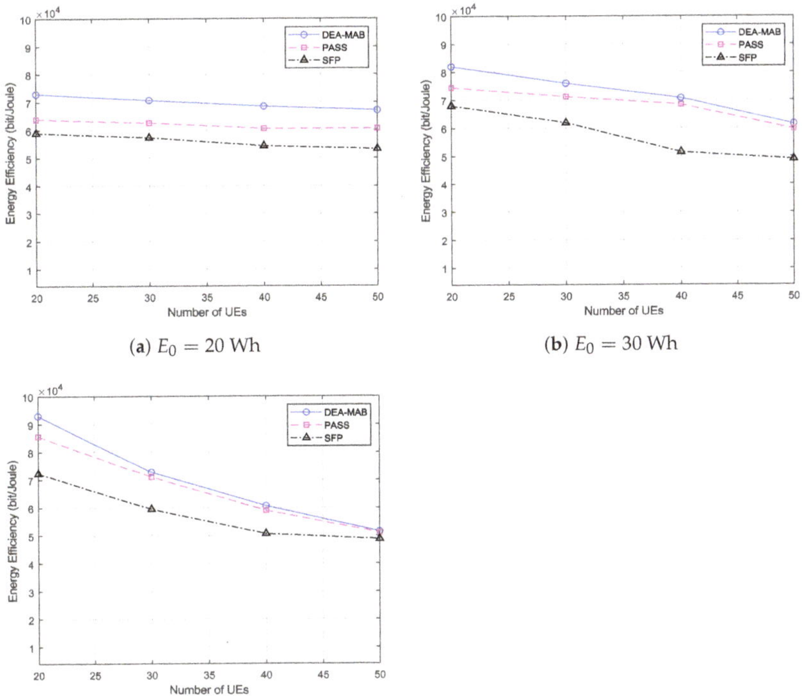

(a) $E_0 = 20$ Wh

(b) $E_0 = 30$ Wh

(c) $E_0 = 40$ Wh

Figure 5. UEs' energy efficiency versus the number of users.

5. Conclusions

In this paper, the trajectory optimization for a UAV-assisted emergency wireless communication network was investigated. The UAV is deployed as a temporary BS to provide wireless connectivity from the sky for trapped UEs in a post-disaster area where all BSs are damaged or have malfunctioned due to a natural disaster. The UAV's target is to optimize its flying trajectory to maximize the long-term uplink throughput from UEs. However, due to the malfunctioning of the power supplies in the disaster area as well, this optimization problem is performed with limited battery capacity of not only the UAV but also UEs in the post-disaster area. We proposed an MAB-based algorithm constrained with these two energy limitations to address this optimization problem. The proposed algorithm can solve the trajectory optimization problem with respect to this dynamic energy consumption over time. Simulation results showed that our algorithm outperforms benchmark methods in terms of long-term uplink throughput and energy efficiency. Furthermore, it could increase the energy consumption of the UEs during the data offloading process, which reflects success in maximizing the UEs served in a post-disaster area and accomplishing the task of information collection in the post-disaster area. A straightforward extension could be to expand the simulation area to be served with more than one UAV. In such a case, each UAV would have to develop a strategy to not only maximize the objective function but also to avoid collisions with other UAVs. One of these strategies would be to keep a certain operating distance between each pair of UAVs. This distance could be designed using optical sensors attached to the UAV to recognize the surrounding UAVs, or by detecting a low-power beacon signal transmitted from each operating UAV. A detailed system design was kept for our future work. Additionally, for a more realistic scenario, UEs might be considered as moving objects, and the UAV should consider an accurate methodology for estimating the location of each UE that should be served.

Author Contributions: Conceptualization, A.A., E.M.M. and G.K.T.; methodology, A.A. and G.K.T.; software, A.A. and G.K.T.; validation, A.A. and G.K.T.; formal analysis, A.A., E.M.M. and G.K.T.; investigation, A.A. and G.K.T.; resources, A.A., E.M.M. and G.K.T.; data curation, A.A. and G.K.T.; writing—original draft preparation, A.A.; writing—review and editing, A.A., E.M.M., G.K.T., and K.S.; visualization, A.A.; supervision, E.M.M., G.K.T. and K.S.; project administration, G.K.T. and K.S.; funding acquisition, G.K.T. and K.S. All authors have read and agreed to the published version of the manuscript.

Funding: This research received no external funding.

Institutional Review Board Statement: Not applicable.

Informed Consent Statement: Not applicable.

Data Availability Statement: Not applicable.

Acknowledgments: This work is partially supported by the Telecommunications Advancement Foundation, Japan.

Conflicts of Interest: The authors declare no conflict of interest.

Abbreviations

The following abbreviations are used in this manuscript:

BS	Base station
UAV	Unmanned aerial vehicle
ML	Machine learning
RL	Reinforcement learning
MAB	Multi-armed bandit
DPG	Deterministic policy gradient
MDP	Markov decision process
CRN	Cognitive radio network
UCB	Upper confidence bound
TS	Thompson sampling
RIS	Re-configurable intelligent surface
SUTOA	state-action-reward-state-action based UAV-trajectory optimization algorithm
QUTOA	Q-learning based UAV-trajectory optimization algorithm
UE	User equipment
GPS	Global positioning system
3GPP	3rd generation partnership project
LOS	Line-of-sight
NLOS	Non-line-of-sight
SNR	Signal-to-noise ratio
AWGN	Additive white Gaussian noise
EXP3	The exponential-weight algorithm for exploration and exploitation
LCB	Lower confidence bound
DEA	Dual-energy aware
PASS	Post-disaster area spiral scanning
SFP	Shortest flight path
PoI	Points of interest
MILP	Mixed Integer Linear Programming

References

1. Hoeppe, P. Trends in weather related disasters–Consequences for insurers and society. *Weather Clim. Extrem.* **2016**, *11*, 70–79. [CrossRef]
2. Deepak, G.; Ladas, A.; Sambo, Y.A.; Pervaiz, H.; Politis, C.; Imran, M.A. An overview of post-disaster emergency communication systems in the future networks. *IEEE Wirel. Commun.* **2019**, *26*, 132–139.
3. Tran, G.K.; Ozasa, M.; Nakazato, J. NFV/SDN as an Enabler for Dynamic Placement Method of mmWave Embedded UAV Access Base Stations. *Network* **2022**, *2*, 479–499. [CrossRef]
4. Saad, W.; Bennis, M.; Mozaffari, M.; Lin, X. *Wireless Communications and Networking for Unmanned Aerial Vehicles*; Cambridge University Press: Cambridge, UK, 2020.
5. Erdelj, M.; Natalizio, E.; Chowdhury, K.R.; Akyildiz, I.F. Help from the sky: Leveraging UAVs for disaster management. *IEEE Pervasive Comput.* **2017**, *16*, 24–32. [CrossRef]
6. Kwasinski, A.; Weaver, W.W.; Chapman, P.L.; Krein, P.T. Telecommunications power plant damage assessment for hurricane katrina–site survey and follow-up results. *IEEE Syst. J.* **2009**, *3*, 277–287. [CrossRef]
7. Fotouhi, A.; Qiang, H.; Ding, M.; Hassan, M.; Giordano, L.G.; Garcia-Rodriguez, A.; Yuan, J. Survey on UAV cellular communications: Practical aspects, standardization advancements, regulation, and security challenges. *IEEE Commun. Surv. Tutor.* **2019**, *21*, 3417–3442. [CrossRef]
8. Sutton, R.S.; Barto, A.G. *Reinforcement Learning: An Introduction*; MIT Press: Cambridge, MA, USA, 2018.
9. Federal Aviation Administration. FAA Aerospace Forecast: Fiscal Years 2019–2039; U.S. Department of Transportation: Washington, DC, USA, 2019; 43p. Available online: https://www.faa.gov/data_research/aviation/aerospace_forecasts/media/fy2019-39_faa_aerospace_forecast.pdf (accessed on 19 December 2022).
10. Hashesh, A.O.; Hashima, S.; Zaki, R.M.; Fouda, M.M.; Hatano, K.; Eldien, A.S.T. AI-Enabled UAV Communications: Challenges and Future Directions. *IEEE Access* **2022**, *10*, 92048–92066. [CrossRef]
11. Chen, X.; Nie, Y.; Li, N. Online Residential Demand Response via Contextual Multi-Armed Bandits. *IEEE Control Syst. Lett.* **2021**, *5*, 433–438. [CrossRef]
12. Katehakis, M.N.; Veinott Jr, A.F. The multi-armed bandit problem: Decomposition and computation. *Math. Oper. Res.* **1987**, *12*, 262–268. [CrossRef]
13. Bubeck, S.; Cesa-Bianchi, N. Regret analysis of stochastic and nonstochastic multi-armed bandit problems. *arXiv* **2012**, arXiv:1204.5721.
14. Auer, P.; Cesa-Bianchi, N.; Fischer, P. Finite-time analysis of the multiarmed bandit problem. *Mach. Learn.* **2002**, *47*, 235–256. [CrossRef]
15. Audibert, J.Y.; Munos, R.; Szepesvári, C. Exploration–exploitation tradeoff using variance estimates in multi-armed bandits. *Theor. Comput. Sci.* **2009**, *410*, 1876–1902. [CrossRef]
16. Lahmeri, M.A.; Kishk, M.A.; Alouini, M.S. Artificial intelligence for UAV-enabled wireless networks: A survey. *IEEE Open J. Commun. Soc.* **2021**, *2*, 1015–1040. [CrossRef]
17. Mamaghani, M.T.; Hong, Y. Intelligent trajectory design for secure full-duplex MIMO-UAV relaying against active eavesdroppers: A model-free reinforcement learning approach. *IEEE Access* **2020**, *9*, 4447–4465. [CrossRef]
18. Han, S.I. Survey on UAV Deployment and Trajectory in Wireless Communication Networks: Applications and Challenges. *Information* **2022**, *13*, 389. [CrossRef]
19. Zeng, Y.; Xu, X.; Zhang, R. Trajectory design for completion time minimization in UAV-enabled multicasting. *IEEE Trans. Wirel. Commun.* **2018**, *17*, 2233–2246. [CrossRef]
20. Sugihara, R.; Gupta, R.K. Speed control and scheduling of data mules in sensor networks. *ACM Trans. Sens. Netw. (TOSN)* **2010**, *7*, 4. [CrossRef]
21. Chiaraviglio, L.; D'Andreagiovanni, F.; Liu, W.; Gutierrez, J.A.; Blefari-Melazzi, N.; Choo, K.K.R.; Alouini, M.S. Multi-area throughput and energy optimization of UAV-aided cellular networks powered by solar panels and grid. *IEEE Trans. Mob. Comput.* **2020**, *20*, 2427–2444. [CrossRef]
22. Trotta, A.; Andreagiovanni, F.D.; Di Felice, M.; Natalizio, E.; Chowdhury, K.R. When UAVs ride a bus: Towards energy-efficient city-scale video surveillance. In Proceedings of the IEEE Infocom 2018-IEEE Conference on Computer Communications, Honolulu, HI, USA, 16–19 April 2018; pp. 1043–1051.
23. Chen, M.; Challita, U.; Saad, W.; Yin, C.; Debbah, M. Artificial neural networks-based machine learning for wireless networks: A tutorial. *IEEE Commun. Surv. Tutors* **2019**, *21*, 3039–3071. [CrossRef]
24. Mozaffari, M.; Saad, W.; Bennis, M.; Debbah, M. Efficient deployment of multiple unmanned aerial vehicles for optimal wireless coverage. *IEEE Commun. Lett.* **2016**, *20*, 1647–1650. [CrossRef]
25. Pearre, B.; Brown, T.X. Model-free trajectory optimisation for unmanned aircraft serving as data ferries for widespread sensors. *Remote Sens.* **2012**, *4*, 2971–3005. [CrossRef]
26. Bayerlein, H.; De Kerret, P.; Gesbert, D. Trajectory optimization for autonomous flying base station via reinforcement learning. In Proceedings of the 2018 IEEE 19th International Workshop on Signal Processing Advances in Wireless Communications (SPAWC), Kalamata, Greece, 25–28 June 2018; pp. 1–5.
27. Yin, S.; Zhao, S.; Zhao, Y.; Yu, F.R. Intelligent trajectory design in UAV-aided communications with reinforcement learning. *IEEE Trans. Veh. Technol.* **2019**, *68*, 8227–8231. [CrossRef]

28. Amrallah, A.; Mohamed, E.M.; Tran, G.K.; Sakaguchi, K. Enhanced dynamic spectrum access in UAV wireless networks for post-disaster area surveillance system: A multi-player multi-armed bandit approach. *Sensors* **2021**, *21*, 7855. [CrossRef]
29. Amrallah, A.; Mohamed, E.M.; Tran, G.K.; Sakaguchi, K. Radio Resource Management Aided Multi-Armed Bandits for Disaster Surveillance System. In Proceedings of the 2020 International Conference on Emerging Technologies for Communications (ICETC2020), Virtual, 2–4 December 2020.
30. Mohamed, E.M.; Hashima, S.; Aldosary, A.; Hatano, K.; Abdelghany, M.A. Gateway selection in millimeter wave UAV wireless networks using multi-player multi-armed bandit. *Sensors* **2020**, *20*, 3947. [CrossRef] [PubMed]
31. Mohamed, E.M.; Hashima, S.; Hatano, K. Energy Aware Multi-Armed Bandit for Millimeter Wave Based UAV Mounted RIS Networks. *IEEE Wirel. Commun. Lett.* **2022**, *11*, 1293–1297. [CrossRef]
32. Lin, Y.; Wang, T.; Wang, S. UAV-assisted emergency communications: An extended multi-armed bandit perspective. *IEEE Commun. Lett.* **2019**, *23*, 938–941. [CrossRef]
33. Cui, J.; Ding, Z.; Deng, Y.; Nallanathan, A.; Hanzo, L. Adaptive UAV-trajectory optimization under quality of service constraints: A model-free solution. *IEEE Access* **2020**, *8*, 112253–112265. [CrossRef]
34. Zhang, T.; Lei, J.; Liu, Y.; Feng, C.; Nallanathan, A. Trajectory optimization for UAV emergency communication with limited user equipment energy: A safe-DQN approach. *IEEE Trans. Green Commun. Netw.* **2021**, *5*, 1236–1247. [CrossRef]
35. Amrallah, A.; Mohamed, E.M.; Tran, G.K.; Sakaguchi, K. Dual Energy-Aware based Trajectory Optimization for UAV Emergency Wireless Communication Network: A Multi-armed Bandit Approach. In Proceedings of the 2022 Thirteenth International Conference on Ubiquitous and Future Networks (ICUFN), Barcelona, Spain, 5–8 July 2022; pp. 43–48.
36. 3GPP. *Study on Enhanced LTE Support for Aerial Vehicles (Release 15)*; The 3rd Generation Partnership Project (3GPP): Valbonne, France, 2017. Available online: https://portal.3gpp.org/desktopmodules/Specifications/SpecificationDetails.aspx?specificationId=3231 (accessed on 19 December 2022).
37. González-Cañete, F.J.; Casilari, E. Consumption analysis of smartphone based fall detection systems with multiple external wireless sensors. *Sensors* **2020**, *20*, 622. [CrossRef] [PubMed]
38. Vermorel, J.; Mohri, M. Multi-armed bandit algorithms and empirical evaluation. In Proceedings of the European Conference on Machine Learning, Porto, Portugal, 3–7 October 2005; Springer: Berlin/Heidelberg, Germany, 2005; pp. 437–448.
39. Agrawal, R. Sample mean based index policies by o (log n) regret for the multi-armed bandit problem. *Adv. Appl. Probab.* **1995**, *27*, 1054–1078. [CrossRef]
40. Scott, S.L. A modern Bayesian look at the multi-armed bandit. *Appl. Stoch. Model. Bus. Ind.* **2010**, *26*, 639–658. [CrossRef]
41. Auer, P.; Cesa-Bianchi, N.; Freund, Y.; Schapire, R.E. The nonstochastic multiarmed bandit problem. *SIAM J. Comput.* **2002**, *32*, 48–77. [CrossRef]
42. Ding, W.; Qin, T.; Zhang, X.D.; Liu, T.Y. Multi-armed bandit with budget constraint and variable costs. In Proceedings of the Twenty-Seventh AAAI Conference on Artificial Intelligence, Bellevue, WA, USA, 14–18 July 2013.
43. Sinha, D.; Sankararaman, K.A.; Kazerouni, A.; Avadhanula, V. Multi-armed bandits with cost subsidy. In Proceedings of the International Conference on Artificial Intelligence and Statistics, PMLR, Virtual Event, 13–15 April 2021; pp. 3016–3024.

sensors

MDPI

Article

Computational Study of a Motion Sensor to Simultaneously Measure Two Physical Quantities in All Three Directions for a UAV

Kamran Siddique * and Yoshifumi Ogami *

Department of Mechanical Engineering, College of Science and Engineering, Ritsumeikan University, 1-1-1 Noji-Higashi, Kusatsu 525-8577, Japan
* Correspondence: 14bemeksiddique@smme.edu.pk (K.S.); ogami@se.ritsumei.ac.jp (Y.O.); Tel.: +81-80-4499-9876 (K.S.)

Abstract: Cross-axis sensitivity is generally undesirable, and lower values are required for the accurate performance of a thermal accelerometer. In this study, errors in devices are utilized to simultaneously measure two physical quantities of an unmanned aerial vehicle (UAV) in the X-, Y-, and Z-directions, i.e., where three accelerations and three rotations can also be simultaneously measured using a single motion sensor. The 3D structures of thermal accelerometers were designed and simulated in a FEM simulator using commercially available FLUENT 18.2 software Obtained temperature responses were correlated with input physical quantities, and a graphical relationship was created between peak temperature values and input accelerations and rotations. Using this graphical representation, any values of acceleration from 1*g* to 4*g* and rotational speed from 200 to 1000°/s can be simultaneously measured in all three directions.

Keywords: thermal accelerators; unmanned aerial vehicle; motion sensor; cross-axis sensitivity; microelectromechanical system

1. Introduction

Unmanned aerial vehicles (UAVs) offer a cost-effective and time-saving method for performing various functions, providing safety and convenience compared to traditional methods. They have a wide range of applications and are crucial to multiple industries. They can be useful in search and rescue operations [1], including disaster relief and emergency response. They can be used to deliver packages and goods to remote areas or places where road access is impossible. Additionally, UAVs can be used to inspect infrastructures such as buildings and bridges, provide high-quality images for making the right monitoring decisions [2], and monitor landslides [3].

Microelectromechanical system (MEMS)-based sensor–actuator applications continue to develop in various industries due to their improved sensitivity, accuracy, and reliability of operations, as well as their low power consumption [4]. The use of MEMS sensors for vehicular sensing is gradually increasing [5]. MEMS accelerometers have been extensively utilized in UAVs owing to their compactness, lightweight, low power consumption, and high sensitivity. Piezoelectric, capacitive, and thermal accelerometers are among the most commonly used MEMS accelerometer types in UAVs. Stephan [6] conducted a performance analysis of different accelerometer types using data sheets of 118 accelerometers from 27 different manufacturers. It was found that piezoelectric accelerometers have the highest measurement range, which can be greater than 10,000*g*. Thermal accelerometers, however, have outstanding shock limits of 50,000*g* with a low measurement range of 5*g*. Jiang [7] provides an extensive overview of piezoelectric accelerometers and mentions how these are effective for high-temperature applications, such as those encountered in aerospace, aircraft, automotive and energy industries. Thermal accelerometers, however, do not have a solid-proof mass; hence, their fabrication is simpler, and the integration of the sensor with a signal conditioning circuit is easier, resulting in improved device durability and measurement consistency [8].

UAV states are generally estimated via fusing data from accelerometers, gyroscopes, and global navigation satellite systems (GNSS) to determine pitch, yaw, and roll, as shown in Figure 1. The common types of MEMS gyroscopes used in UAVs include ring lasers, fiber optics, MEMS vibrating structures, and Coriolis vibratory types. Environmental thermal fluctuations are major issues in such devices. Several researchers have presented a temperature compensation method that changes the structure of a gyroscope to decrease frequency variation under different temperatures [9,10]. This limitation of MEMS gyroscopes remains a challenge. In this study, a novel concept is proposed that involves the measurement of rotational speed in addition to acceleration, achieved by modifying a conventional thermal accelerometer that can only measure acceleration.

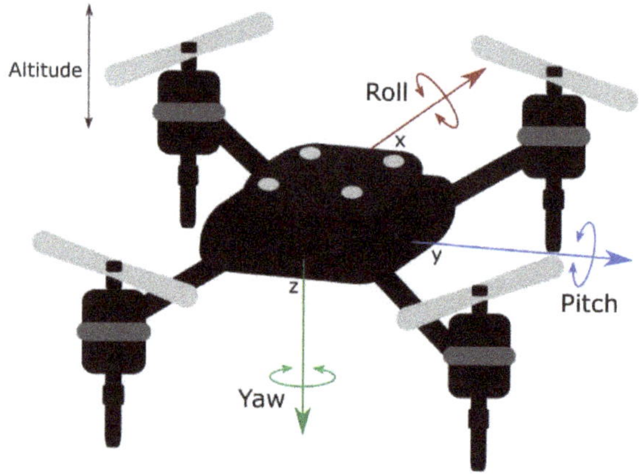

Figure 1. A drone's pitch, roll, and yaw.

Thermal-based accelerometers offer several advantages over conventional proof-mass accelerometers. They exhibit no measurable resonance, delivering immunity to vibration; no temperature hysteresis; and excellent zero g offset stability with the added shock resistance, increasing their reliability [11]. This enhances the sensing range of the device while avoiding failure or vibrational limitations. The working principle is that a heating source creates a thermal profile inside a cavity, and a set of temperature sensors placed equidistant from the heater measures the temperature. When acceleration occurs, the sensors' temperature values change, corresponding to the applied acceleration. This phenomenon is illustrated in Figure 2. As shown in the figure, sets of isotherms were created around the heating source, and isotherms near the heater had higher temperatures. A consistent temperature profile was obtained without applying any motion, as shown in Figure 2(left). However, when acceleration was applied in the right direction, the sets of isotherms shifted to the right (Figure 2(right)). This temperature change was detected by sensors placed on both sides of the heater and correlated with applied acceleration.

Different numerical and experimental studies have been conducted to optimize multiple-axis thermal accelerometers. Novel triple-axis thermal accelerometers were introduced by [12] and improved by [13]. However, their sensitivities were still insufficient and required further improvement. Mukherjee [14] modified the cavity structure of the device to achieve better sensitivity. Jiang [15] showed that increasing heater power enhanced the sensitivities along the X-, Y-, and Z-axes. Wang [16] achieved improved sensitivity and a wider ambient temperature measurement range in a recent study. This design can precisely detect a range of temperatures in various environments. In this study, we aimed to select the parameters with high sensitivity and resolution, as reported in previous studies.

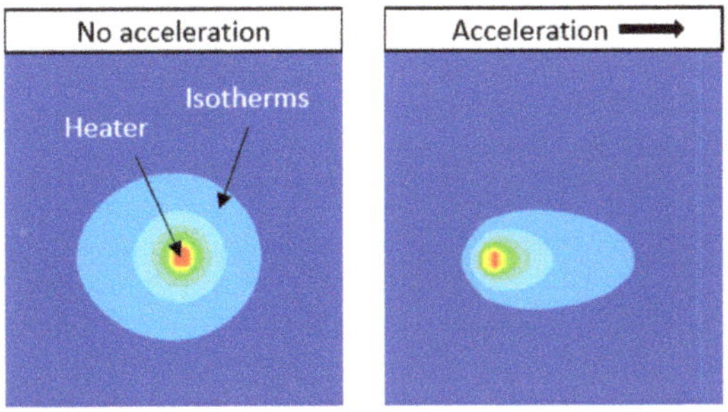

Figure 2. Changes in isotherms with no acceleration (**left**) and acceleration applied to the right side (**right**).

For a dual-axis thermal accelerometer, two sets of temperature sensors are required around the two axes. In contrast, three sets of temperature sensors are required to detect temperatures around the X-, Y-, and Z-axes for a triple-axis thermal accelerometer. To reduce installation and maintenance costs, Ogami's concept [17], which uses cross-axis sensitivity (CAS), can be used. Due to CAS, when motion is applied to a single axis, a correlating temperature change is observed on the other two perpendicular axes. According to Farahani [18], CAS is a good measure of sensor performance in response to external factors. Therefore, for high-accuracy applications, a lower CAS is expected. However, a computational study presented by Siddique [19] showed that multiple physical quantities could be measured using CAS. In this study, using X- and Y-sensor outputs, acceleration in the X-direction and rotational speed on the Z-axis were measured.

With the development of technology, multiple sensors have been incorporated into engineering devices. In micro- and insect-scaled UAVs, multiple sensors from accelerometers to magnetic sensors have been installed [20]. The installation and maintenance of a considerable number of sensors can be expensive and time-consuming. Therefore, this study introduces a method for simultaneously measuring three accelerations and three rotations in UAVs using a single-motion sensor. The measurement of rotational speed, including pitch, roll, and yaw, is crucial for controlling the flight of the device. Using fluid flow and thermodynamic principles and leveraging CAS, we derived three inverse functions from a set of axial accelerations and the rotational speeds perpendicular to them. The inverse functions could then be implemented in the computing unit of a real motion sensor.

2. Materials and Methods

In this study, ANSYS FLUENT 18.2 was employed to perform computational analyses. The temperature response was obtained via the simultaneous application of acceleration and rotation around the axis perpendicular to the other axis, on which the direction of acceleration was applied to the computational model. The design of a triple-axis thermal accelerometer was considered, and heating sources were incorporated into the model using C-programmed user-defined functions (UDFs). Additionally, the positions of the temperature sensors were determined using UDFs by tracking their cell IDs, which are unique to every mesh element.

The thermal accelerometer is based on heat transfer and especially on free convection heat transfer in a closed chamber. This phenomenon is governed by a Navier–Stokes equation based on the principle of conserving mass, momentum, and energy, as follows:

$$\frac{\partial \rho}{\partial t} + \nabla(\rho \boldsymbol{u}) = 0$$

$$\rho\left[\frac{\partial \boldsymbol{u}}{\partial t} + (\boldsymbol{u} \cdot \nabla)\boldsymbol{u}\right] = -\nabla p + \nabla I + f$$

$$\rho C_p \left(\frac{\partial T}{\partial t} + \boldsymbol{u} \cdot \nabla T \right) = k \nabla^2 T$$

where $\boldsymbol{u}$ is the flow velocity vector field, ∇ is the spatial divergence operator, p is the pressure, I is the total stress tensor, and f is the body forces acting on the fluid. The parameters C_p, ρ, and k are the specific heat, density, and thermal conductivity of the fluid in the cavity, respectively.

In FLUENT, a pressure-based transient solver was used along with an energy model because the flow characteristics were not highly compressible. The DEFINE_CG_MOTION UDF was applied to define linear and rotational motions. Carbon dioxide (CO_2) was selected as a gas medium because of its high density and low kinematic viscosity. The low viscosity of CO_2 enables a more efficient flow and results in greater sensitivity than gases with higher viscosities because high viscosity impedes gas flow [21].

The steps involved in measuring multiple physical quantities using a motion sensor are shown in Figure 3. The first step was to generate a cylindrical model of the motion sensor with a diameter of 2 cm and depth of 1.3 cm, in which heaters and sensors were placed. Next, a mesh that served as an input to ANSYS FLUENT was developed. The next step was to input different values of acceleration and rotation and the temperature–time curves were generated. Analysis of these curves revealed the maximum and minimum temperature values, which were then linked to input acceleration and rotation values. Finally, a three-dimensional graph was drawn to visually represent the relationship between acceleration, rotation, and extreme temperature values (T_{max} and T_{min}).

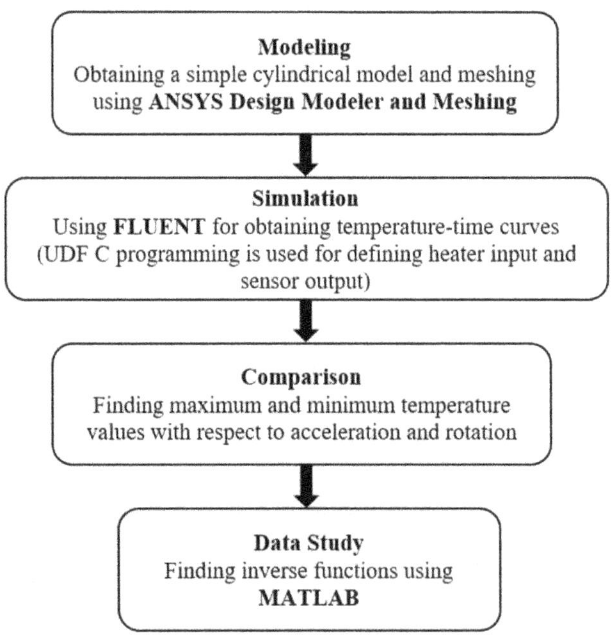

Figure 3. Methodology flow.

In thermal accelerometers, a heating source that generates temperature contours must be defined. The temperature response is generally greater when the heating power is high and the ambient temperature is low. Therefore, the differences in peak values increase. However, this results in higher temperatures inside the cavity, which can lead to the heating of the motion sensor walls. In our previous study [19], a heating power of 70 mW was considered for applications in small-scale UAVs and robots. In this study, we reduced the heating power from 70 to 40 mW, so that it can be applied to significantly smaller UAVs.

For the simulations, a time step size of 0.0042 s was considered with regard to increment velocities. The acceleration and rotation ranges were set to 1–4g and 250–1000°/s, respectively. For the case of acceleration, the maximum velocity for a flow time of 3 s was 117.72 m/s with an increment velocity of 0.164808 m/s at 4g. In contrast, for rotational velocities, a maximum linear speed of 0.174533 m/s was practically the same as the maximum increment velocity at 4g. The values are listed in Table 1 for comparison.

Table 1. Maximum and increment velocities at different accelerations and rotations.

		Maximum Velocity at 3 s (m/s)	Increment Velocity (m/s)
Acceleration	1g	29.43	0.0412
	2g	58.86	0.0824
	3g	88.29	0.1236
	4g	117.72	0.1648
Rotation (°/s)	250	0.0436	0.0436
	500	0.0873	0.0873
	750	0.1309	0.1309
	1000	0.1745	0.1745

To validate the reliability and independence of computational simulation results, a grid-independence test was conducted because no experimental or theoretical models were available for comparison. The velocity, pressure (at sensor X11), and temperature variables were evaluated at 500°/s around the Z-axis, and 2g was applied to the device in the X-direction using five different meshes with varying numbers of elements. The velocity and pressure distributions (viewed from the top plane of the motion sensor) of a mesh with 166,675 elements are shown in Figure 4. It can be observed that the velocity of CO_2 molecules was practically the same throughout the motion sensor cavity with a maximum velocity of 49.23 m/s^2 at the instant of t = 2.5 s. Furthermore, the temperature change with respect to time for all five meshes is shown in Figure 5. The maximum values in this graph were extracted and used for comparison.

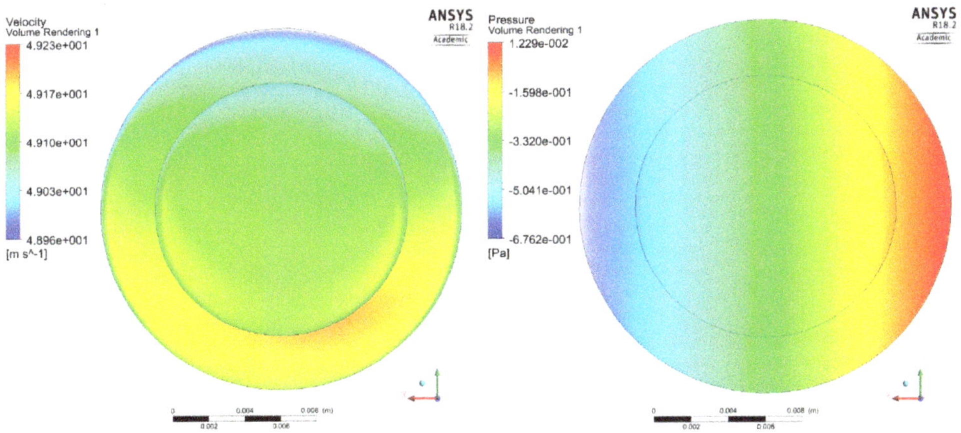

Figure 4. Velocity (**left**) and pressure distribution (**right**) at t = 2.5 s for the mesh with 166,675 elements.

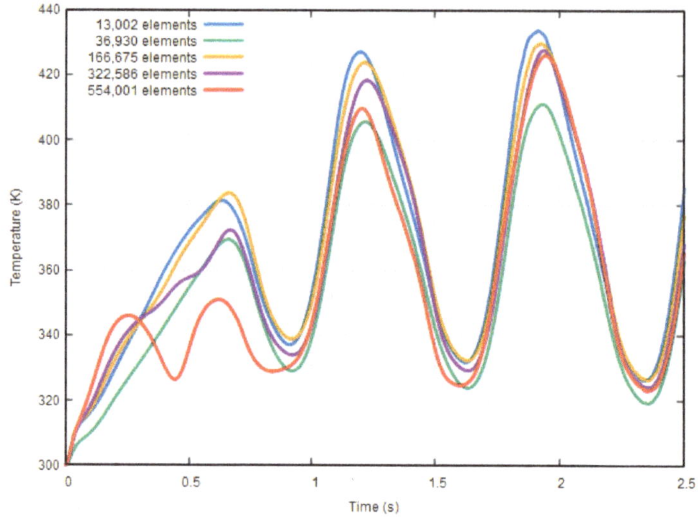

Figure 5. Temperature response at 500°/s around the Z-axis and 2*g* applied in the X-direction for different mesh sizes.

The results of the grid-independence test, which compare the T_{max} values with respect to the number of mesh elements, are presented in Table 2 and illustrated in Figure 6.

Table 2. Changes in variables with different numbers of mesh elements.

S	No. of Mesh Elements	V_{max} (m/s)	P_X11$_{min}$ (Pa)	P_X11$_{max}$ (Pa)	T_{max} (K)
1	13,002	49.23	−0.497392	−0.168213	433.7
2	36,930	49.23	−0.509235	−0.167053	411.0
3	166,675	49.235	−0.508066	−0.181941	429.8
4	322,586	49.235	−0.513125	−0.183977	427.5
5	554,001	49.24	−0.518651	−0.185735	426.0
6	752,760	49.16	−0.523673	−0.186014	421.6

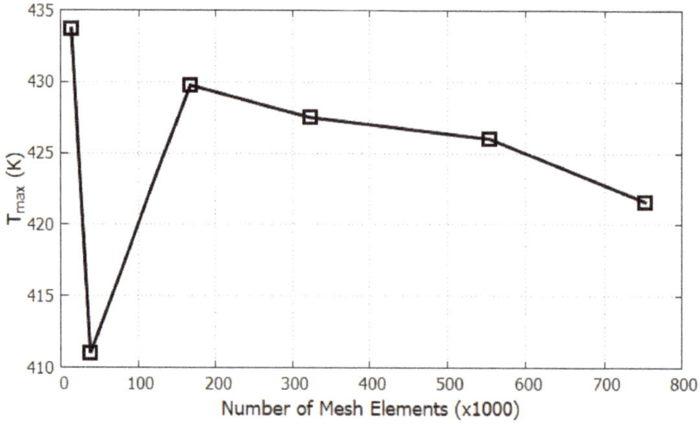

Figure 6. T_{max} vs. number of mesh elements.

As shown in Figure 6, a significant increase of 4.6% in the T_{max} value can be observed between the second and third meshes. However, from the third mesh onwards, the difference in T_{max} values between the meshes was approximately 1%, which was minimal compared to the differences between the first three meshes. Therefore, a mesh with 166,675 elements was selected for future calculations because it provides a reliable solution with minimal computational requirements such as CPU time. The placement of the heaters and sensors is shown in Figure 7, and the meshing structure used in the computational study is shown in Figure 8. Four heaters (H1–H4) were placed on all four axes, 40 mm from the center of the cavity. Heaters H1 and H3 on the X-axis are surrounded by pairs of X-sensors: X21 and X22, and X11 and X12, respectively. Similarly, H2 and H4 on the Y-axis are bounded by Y-sensor pairs: Y21 and Y22, and Y11 and Y12, respectively. On the Z-axis, each heater constitutes a temperature sensor located 10 mm away from the heater in the Z-direction. The notation of the Z-sensors is such that the number denotes the heater number (e.g., Z1 around H1).

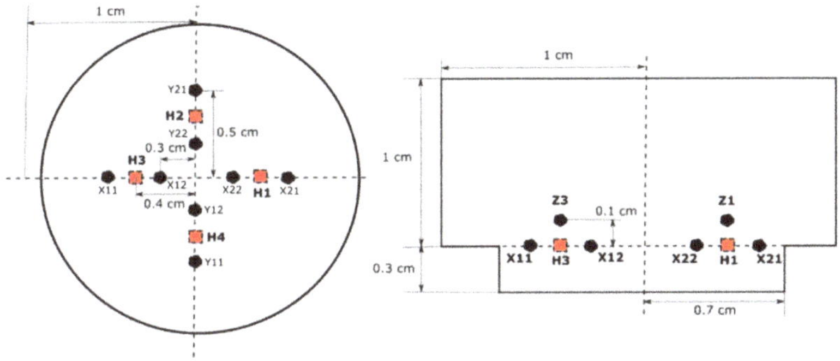

Figure 7. Position of heaters and sensors; cross-sectional (**left**) and side views (**right**).

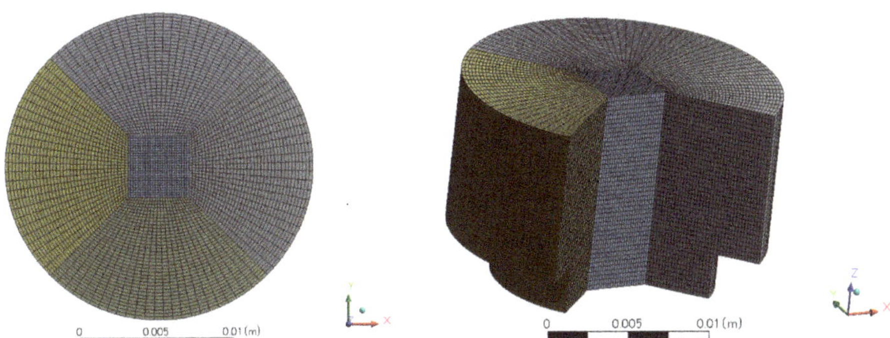

Figure 8. Computational mesh; top view (**left**) and isometric view of the inside of the mesh (**right**).

To define the relationship between two input physical quantities (PQ) and two output variables, the following equation was established:

$$(\text{Output 1}, \text{Output 2}) = (\text{PQ 1}, \text{PQ 2})$$

Once this relationship was established, the next step was to find the inverse function of this relationship, which is represented as:

$$(\text{PQ 1}, \text{PQ 2}) = {}^{-1}(\text{Output 1}, \text{Output 2})$$

Using the above equation, any two PQs can be measured simultaneously. In this study, this relationship is calculated for two quantities in three directions. Therefore, three inverse functions must be obtained, and the data can be installed in the computing unit of the motion sensor. This is illustrated in Figure 9.

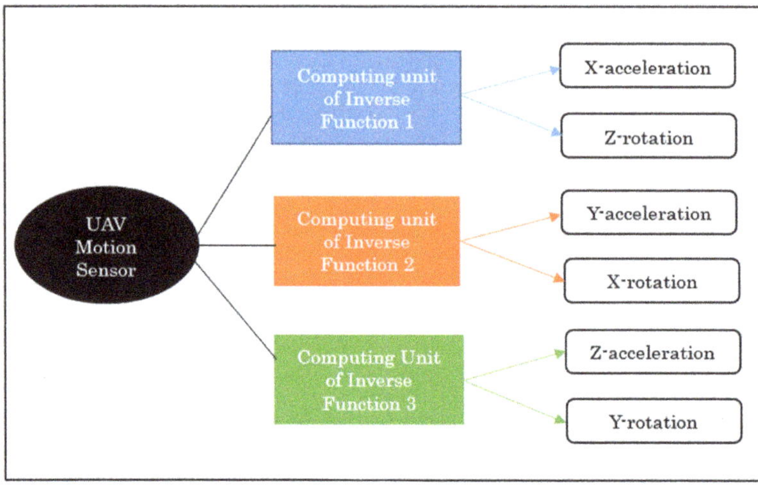

Figure 9. Schematic of obtaining three inverse functions for the measurement of acceleration and rotation in all three directions.

3. Results and Discussions

A numerical study was conducted using computational simulations to measure two physical quantities (acceleration and rotation) in all three directions. As described in Section 2, three inverse functions were obtained via computational fluid dynamics (CFD) simulations using the ANSYS Fluent 18.2 software, and each inverse function measured one acceleration (a) and one rotation (ω) perpendicular to a particular acceleration value. The inverse functions can be described as follows:

$$(a_x, \omega_z) = f^{-1}(\mathrm{T}_{x\mathrm{max}}, \mathrm{T}_{x\mathrm{min}})$$

$$(a_y, \omega_x) = f^{-1}(\mathrm{T}_{y\mathrm{max}}, \mathrm{T}_{y\mathrm{min}})$$

$$(a_z, \omega_y) = f^{-1}(\mathrm{T}_{z\mathrm{max}}, \mathrm{T}_{z\mathrm{min}})$$

Cross-axis sensitivity (CAS) is a significant issue in thermal devices, including those examined in this study. To measure acceleration and rotation simultaneously, we can evaluate the peak temperature values at the axis where acceleration is applied. A previous study [12] demonstrated that temperature data from both the X- and Y-axes can be simultaneously utilized to measure the X-acceleration and Z-rotation. Therefore, we can extract data from the axis where only acceleration is applied by accounting for the CAS influence.

As indicated in Figure 3 of Section 2, temperature–time curves were generated via simulations using FLUENT software. Acceleration and rotation were simultaneously applied to the cylindrical model, and the temperature graphs were analyzed to extract maximum and minimum values. To illustrate this process, Figure 10 shows the data acquired by the Y12 sensor at a rotational velocity of 500°/s and 1–4g accelerations. The identified peak values are listed with respect to physical input quantities (Tables 3–5).

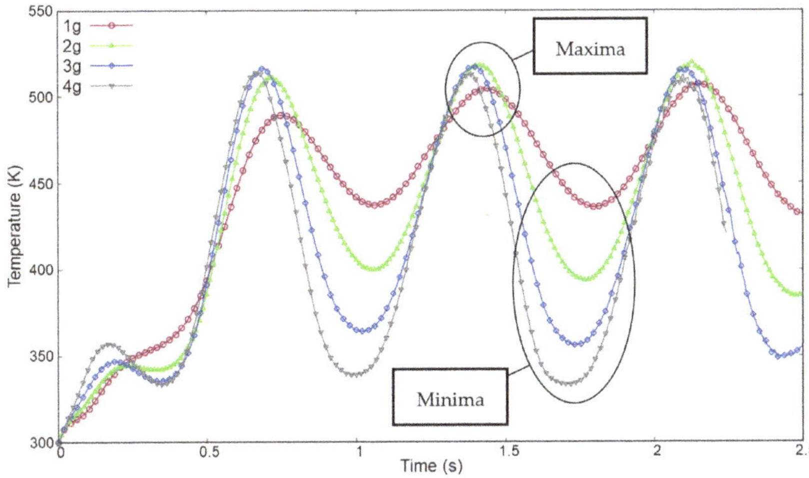

Figure 10. Temperature–time curve for Y12 sensor at 500°/s with varying accelerations from 1 to 4*g*.

Table 3. Data for T_X11$_{max}$ and T_X11$_{min}$.

	a \ ω	**250**	**500**	**750**	**1000**
T_X11$_{max}$	1*g*	477.9	488.9	488.3	489.7
	2*g*	495.0	502.6	501.1	500.7
	3*g*	500.9	504.8	501.4	501.4
	4*g*	498.4	503.3	495.5	496.7
	a \ ω	**250**	**500**	**750**	**1000**
T_X11$_{max}$	1*g*	407.1	415.7	417.3	415.7
	2*g*	359.1	373.2	379.5	380.1
	3*g*	328.3	338.5	346.1	348.9
	4*g*	312.7	316.4	320.3	322.1

Table 4. Data for T_Y12$_{max}$ and T_Y12$_{min}$.

	a \ ω	**250**	**500**	**750**	**1000**
T_X11$_{max}$	1*g*	497.3	507.0	512.6	514.3
	2*g*	499.4	519.3	529.5	536.2
	3*g*	495.6	516.1	527.1	534.8
	4*g*	496.1	512.2	523.5	533.2
	a \ ω	**250**	**500**	**750**	**1000**
T_X11$_{max}$	1*g*	425.9	435.8	440.6	440.0
	2*g*	381.3	393.7	404.0	408.1
	3*g*	353.7	355.9	361.3	361.0
	4*g*	337.6	333.4	332.3	329.3

Table 5. Data for T_YZ4_{max} and T_Z4_{min}.

	a \ ω	250	500	750	1000
T_X11_{max}	1*g*	483.1	490.3	497.3	500.2
	2*g*	471.4	475.5	486.3	494.3
	3*g*	452.8	454.8	471.9	486.9
	4*g*	438.9	426.8	444.1	457.3
	a \ ω	**250**	**500**	**750**	**1000**
T_X11_{max}	1*g*	442.4	440.6	438.4	444.8
	2*g*	388.2	398.6	401.7	399.7
	3*g*	350.7	354.8	360.0	360.2
	4*g*	332.2	329.4	333.2	334.0

This research study explores accelerations ranging from 1 to 4*g* (9.81–39.24 m/s^2) and rotations ranging from 200 to 1000°/s in all three directions. As described in Section 2 and Figure 7, four heaters and four pairs of temperature sensors were positioned in all three directions. Owing to the symmetry of the structure, identical results were obtained for X11 and X21, and X12 and X22. Similarly, Y11 and Y21, and Z1 and Z3 had similarly extreme temperature values. Therefore, to obtain these extreme values, the maxima and minima of X11, Y12, and Z4 were considered. These values were then recorded for all accelerations and rotations in all three directions and correlated with the applied physical quantities of acceleration and rotation.

Because we obtained results from the computational simulation for only four data points of acceleration and rotation, we had to employ the interpolation technique to obtain more data between the upper and lower limits of input physical quantities. This was performed in MATLAB using a cubic interpolation technique, which required four data points to compute a polynomial. In this method, no constraints were present in the derivatives, unlike other interpolation techniques such as spline interpolation.

The T_{max} and T_{min} values for all three axes are listed in Tables 3–5, and the inverse functions used to obtain acceleration and rotational speed in all three directions corresponding to the measured maximum and minimum temperature values are shown in Figures 11–13. Node values indicate each data point. These inverse functions were installed in the computing unit of a real thermal motion sensor.

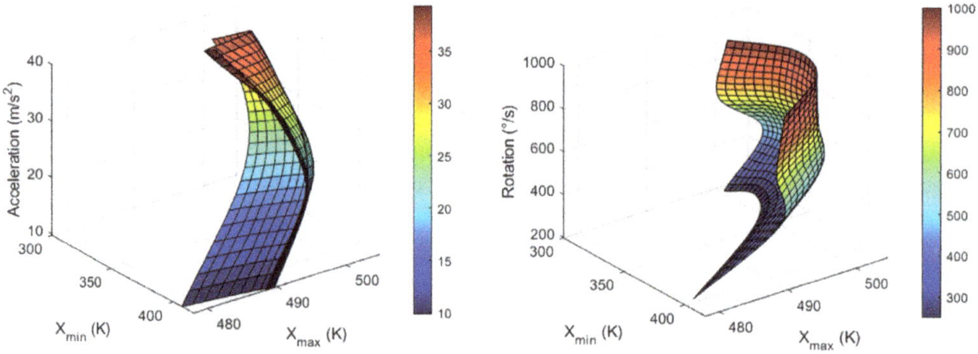

Figure 11. Graphs for X-acceleration (**left**) and Z-rotation (**right**) values from X_{min} and X_{max} measured around heater 1.

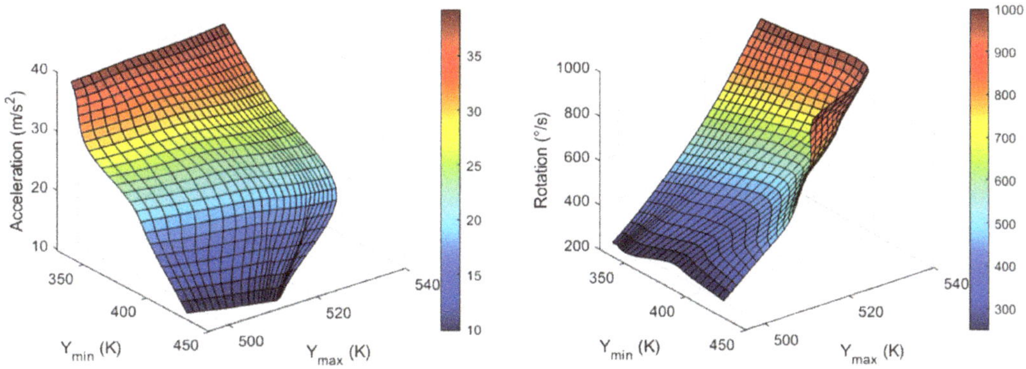

Figure 12. Graphs for Y-acceleration (**left**) and X-rotation (**right**) values from Y_{min} and Y_{max} measured around heater 2.

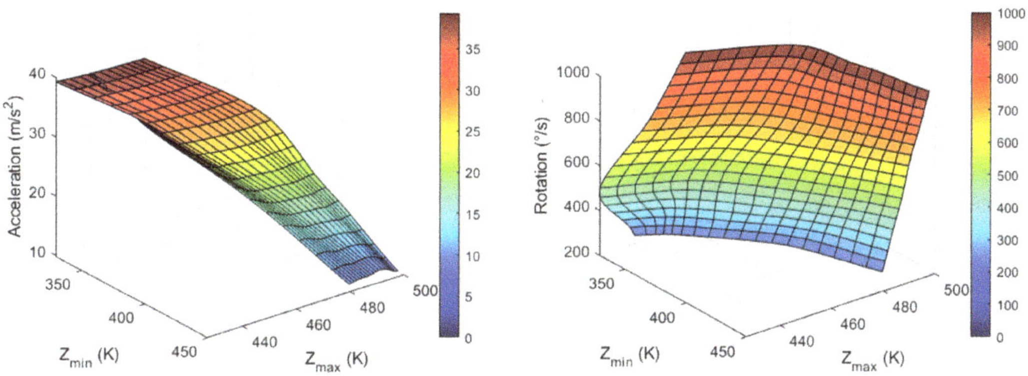

Figure 13. Graphs for Z-acceleration (**left**) and X-rotation (**right**) values from Z_{min} and Z_{max} measured around heater 4.

Using the data from Figures 11–13, a real thermal motion sensor can simultaneously measure all three accelerations at any value between 1 and 4g and rotation speeds between 200 and 1000°/s. The range of acceleration and rotational speeds can be further increased by conducting additional simulations. Furthermore, to obtain more accurate results, more data should be extracted using simulations rather than relying on interpolation techniques.

The main issue encountered in the aforementioned inverse functions is in the region of multiple solutions. In this region, two identical combinations of maximum and minimum temperature values generated different input physical quantities. This yielded inaccurate results. This region can be viewed by drawing a vertical line from the XY-plane parallel to the input physical quantity axis. The Y-acceleration was measured using Y_{max} and Y_{min}, and this region is represented by an ellipse in Figure 14. Therefore, the results should be verified, and the parameters that generate unique solutions should be determined. This problem can be reduced by either altering the cavity shape of the sensor or changing the positions of heaters and sensors.

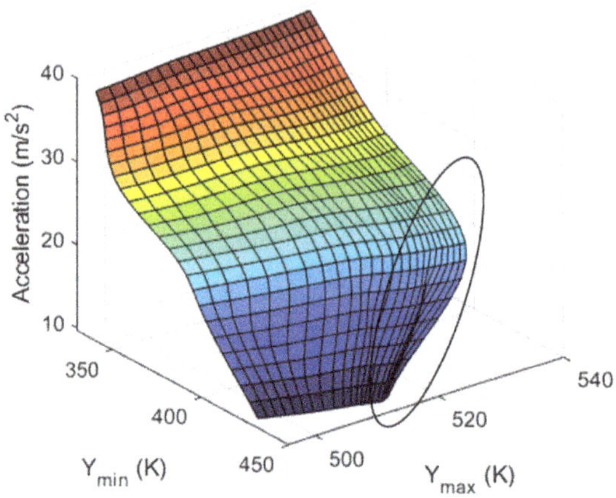

Figure 14. Region of multiple solutions indicated on the graph to obtain Y-acceleration using Y_{max} and Y_{min}.

As mentioned in Section 2, no experimental or theoretical model is presented in this study to support or validate numerical simulations. Our future research will involve the manufacturing of a real thermal device based on the results presented in this study. However, the technique presented in this study may provide a cost-effective and time-saving method in the field of UAV sensor technology.

This method has the disadvantage of requiring both acceleration and rotation. However, considering both rotary-wing and fixed-wing UAVs, the pitch motion along the transverse or lateral axis is effective throughout the flight of the device. Furthermore, study in which more than two quantities can be measured using a single inverse function must be conducted. The idea for this is proposed below and will be explored in future research.

In future studies, a mapping method (2D or 3D) could be utilized to measure multiple physical quantities without using an inverse function. In our study, taking the examples of Y-acceleration and X-rotation for the 2D method, input data, i.e., acceleration and rotational speeds, were first plotted in a graph, as shown in the mapping of input data in Figure 15a (red lines). Similarly, Figure 15b (red lines) shows the output data (T_{max} and T_{min}) extracted from Table 4. For real measurements, if the output (measured) values of the sensor can be obtained (plotted in Figure 15b (purple circle)), the input values for the output values can be approximately and geometrically calculated without using an inverse function, as shown in the blue circle in Figure 15a. If quadrilateral shapes overlap with each other for the output data, it will indicate multiple solutions. This technique can also be utilized in 3D to simultaneously measure three physical quantities.

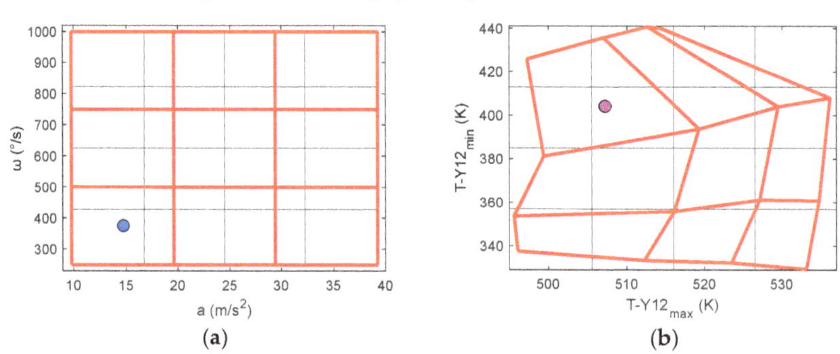

Figure 15. A 2D mapping of input data (**a**) and output data (**b**).

4. Conclusions

In this study, we propose a concept using a computational simulation in which acceleration and rotational speeds of unmanned aerial vehicles (UAVs) can be simultaneously measured in all three directions using a single device. Cross-axis sensitivity, which is an error in accelerometers and gyroscopes, was utilized to achieve this goal. For X-acceleration and Z-rotation, maximum and minimum temperature values from X-sensor data were extracted and from these output values, and input quantities were measured. In a similar manner, Y-axis output data were considered for Y-acceleration and X-rotation, and Z-sensor data were considered for Z-acceleration and Y-rotation. Six graphical plots are presented for each quantity, by which any acceleration from 1 to 4g and rotational speed between 200 and 1000°/s can be measured in the X-, Y-, and Z-directions. The inverse function plots can then be installed in the computing unit of a real thermal motion sensor to measure the quantities.

We also proposed a new technique of 2D or 3D mapping, which we can use to effectively measure multiple physical quantities numerically from the plots of output values corresponding to their input values. This may be a better approach than using inverse function and will be a focus of our future research.

Validation using theoretical and experimental models is absent in this study and will also be a focus of our future research. A microelectromechanical system (MEMS)-based motion sensor, as described in this study, will be manufactured, and using the same flow conditions and parameters, simulation results will be validated in the future. Overall, this methodology provides a robust framework for measuring multiple physical quantities and enables researchers to gain deeper insights into their experimental data. This study provides an excellent solution for challenges faced by sensor technology in UAVs.

Author Contributions: CFD simulations and data analysis was conducted by K.S. under the supervision of Y.O.; the original draft was prepared by K.S.; reviewed and edited by Y.O. All authors have read and agreed to the published version of the manuscript.

Funding: This work was supported by JSPS KAKENHI Grant Numbers 18K04915.

Institutional Review Board Statement: Not applicable.

Informed Consent Statement: Not applicable.

Conflicts of Interest: The authors declare no conflict of interest.

References

1. Iob, P.; Frau, L.; Danieli, P.; Olivieri, L.; Bettanini, C. Avalanche rescue with autonomous drones. In Proceedings of the IEEE 7th International Workshop on Metrology for AeroSpace (MetroAeroSpace), Pisa, Italy, 22–24 June 2020; pp. 319–324. [CrossRef]
2. Nawaz, H.; Ali, H.M.; Massan, S. Applications of unmanned aerial vehicles: A review. *Tecnol. Glosas InnovaciÓN Apl. Pyme* **2019**, *2019*, 85–105. [CrossRef]
3. Yaprak, S.; Yildirim, Ö.; Susam, T. UAV based landslide monitoring. Tema. *J. Land Use Mobil. Environ.* **2017**, *10*, 325–338. [CrossRef]
4. Mishra, M.K.; Dubey, V.; Mishra, P.M.; Khan, I. MEMS technology: A review. *J. Eng. Res. Rep.* **2019**, *4*, 1–24. [CrossRef]
5. Fanse, T.S. Micro-Electro-Mechanical System (MEMS) Application and Prospects in Automobile. *IOSR J. Mech. Civ. Eng.* **2022**, *19*, 17–21.
6. Elies, S. Performance analysis of commercial accelerometers: A parameter review. *Sens. Transducers* **2015**, *193*, 179–190.
7. Jiang, X.; Kim, K.; Zhang, S.; Johnson, J.; Salazar, G. High-Temperature Piezoelectric Sensing. *Sensors* **2014**, *14*, 144–169. [CrossRef] [PubMed]
8. Mukherjee, R.; Basu, J.; Mandal, P.; Guha, P.K. A review of micromachined thermal accelerometers. *J. Micromech. Microeng.* **2017**, *27*, 123002. [CrossRef]
9. Wang, W.; Zhang, T.; Fan, D.; Xing, C. Study on frequency stability of a linear-vibration MEMS gyroscope. *Microsyst. Technol.* **2014**, *20*, 2147–2155. [CrossRef]
10. Chang, H.; Gong, X.; Wang, S.; Zhou, P.; Yuan, W. On improving the performance of a triaxis vortex convective gyroscope through suspended silicon thermistors. *IEEE Sens. J.* **2014**, *15*, 946–955. [CrossRef]

11. Fennelly, J. Capacitive versus Thermal MEMS for High-Vibration Applications. MEMSIC White Pap. 2016. Available online: https://www.automation.com/en-us/articles/2016-1/capacitive-vs-thermal-mems-for-high-vibration-appl (accessed on 3 May 2023).
12. Tsang, S.H.; Ma, A.H.; Karim, K.S.; Parameswaran, M.; Leung, A.M. Monolithically fabricated polymermems 3-axis thermal accelerometers designed for automated wirebonder assembly. In Proceedings of the IEEE 21st International Conference on Micro Electro Mechanical Systems (MEMS), Tucson, AZ, USA, 13–17 January 2008; pp. 880–883. [CrossRef]
13. Ma, A.H.; Leung, A.M. Three-axis thermal accelerometer based on buckled cantilever microstructure. In Proceedings of the IEEE Sensors Conference, Lecce, Italy, 26–29 October 2008; pp. 1492–1495. [CrossRef]
14. Mukherjee, R.; Mandal, P.; Guha, P.R. Sensitivity improvement of a dual axis thermal accelerometer with modified cavity structure. *Microsyst. Technol.* **2017**, *23*, 5357–5363. [CrossRef]
15. Jiang, L.; Cai, Y.; Liu, H.; Zhao, Y. A micromachined monolithic 3 axis accelerometer based on convection heat transfer. In Proceedings of the 8th Annual IEEE International Conference on Nano/Micro Engineered and Molecular Systems, Suzhou, China, 7–10 April 2013; pp. 248–251. [CrossRef]
16. Wang, X.; Xu, W.; Izhar; Lee, Y.K. Theoretical and experimental study and compensation for temperature drifts of micro thermal convective accelerometer. *J. Microelectromech. Syst.* **2020**, *29*, 277–284. [CrossRef]
17. Ogami, Y.; Murakita, Y.; Fukudome, K. Computational experiments on the Step and Frequency Responses of a three-axis thermal accelerometer. *Sensors* **2017**, *17*, 2618. [CrossRef] [PubMed]
18. Farahani, H.; Mills, J.K.; Cleghorn, W.L. Design, fabrication and analysis of micromachined high sensitivity and 0% cross-axis sensitivity capacitive accelerometers. *Microsyst. Technol.* **2009**, *15*, 1815–1826. [CrossRef]
19. Siddique, K.; Ogami, Y. Computational study on thermal motion sensors that can measure acceleration and rotation simultaneously. *Sensors* **2022**, *22*, 6744. [CrossRef]
20. Brzozowski, B.; Rochala, Z.; Wojtowicz, K. Overview of the research on state-of-the-art measurement sensors for UAV navigation. In Proceedings of the 4th IEEE International Workshop on Metrology for AeroSpace (MetroAeroSpace), Padua, Italy, 21–23 June 2017; pp. 565–570. [CrossRef]
21. Han, M.; Kim, J.K.; Park, J.H.; Kim, W.; Kang, S.W.; Kong, S.H.; Jung, D. Sensitivity and frequency-response improvement of a thermal convection–based accelerometer. *Sensors* **2017**, *17*, 1765. [CrossRef] [PubMed]

sensors

MDPI

Article

Resilient Multi-Sensor UAV Navigation with a Hybrid Federated Fusion Architecture

Sorin Andrei Negru *, Patrick Geragersian, Ivan Petrunin and Weisi Guo

School of Aerospace, Transport, and Manufacturing, Cranfield University, Cranfield MK43 0AL, UK; patrick.geragersian@cranfield.ac.uk (P.G.); i.petrunin@cranfield.ac.uk (I.P.); weisi.guo@cranfield.ac.uk (W.G.)
* Correspondence: s.negru.965@cranfield.ac.uk

Abstract: Future UAV (unmanned aerial vehicle) operations in urban environments demand a PNT (position, navigation, and timing) solution that is both robust and resilient. While a GNSS (global navigation satellite system) can provide an accurate position under open-sky assumptions, the complexity of urban operations leads to NLOS (non-line-of-sight) and multipath effects, which in turn impact the accuracy of the PNT data. A key research question within the research community pertains to determining the appropriate hybrid fusion architecture that can ensure the resilience and continuity of UAV operations in urban environments, minimizing significant degradations of PNT data. In this context, we present a novel federated fusion architecture that integrates data from the GNSS, the IMU (inertial measurement unit), a monocular camera, and a barometer to cope with the GNSS multipath and positioning performance degradation. Within the federated fusion architecture, local filters are implemented using EKFs (extended Kalman filters), while a master filter is used in the form of a GRU (gated recurrent unit) block. Data collection is performed by setting up a virtual environment in AirSim for the visual odometry aid and barometer data, while Spirent GSS7000 hardware is used to collect the GNSS and IMU data. The hybrid fusion architecture is compared to a classic federated architecture (formed only by EKFs) and tested under different light and weather conditions to assess its resilience, including multipath and GNSS outages. The proposed solution demonstrates improved resilience and robustness in a range of degraded conditions while maintaining a good level of positioning performance with a 95th percentile error of 0.54 m for the square scenario and 1.72 m for the survey scenario.

Keywords: UAV; urban air mobility; computer vision; multipath; resilient navigation; hybrid fusion; GRU; EKF

1. Introduction

The emergence of the UAM (urban air mobility) concept necessitates more stringent requirements and regulations to ensure safe operations between manned and unmanned vehicles within the same airspace. Authorities such as EASA (European Union Aviation Safety Agency), CAA (Civil Aviation Authority), and FAA (Federal Aviation Administration) have already established specific requirements for regulating the air traffic in urban, semi-urban, and rural environments for UAVs, as specified in [1–3]. In this context, to ensure the safety of operations in urban environments, the PNT solutions provided by UAVs operating in the proximity of buildings and obstacles must be continuous, robust, and resilient. In addition, as it can be seen from [4], UAVs can play a key role in multi-spectral mapping applications and other civil applications [5], where the need for a stable PNT system is crucial to fulfilling all the mission requirements.

Given that the GNSS receivers serve as the primary source of PNT data for UAVs, which can provide good accuracy in open-sky conditions as presented in [6], where an RTK (real-time kinematics) system is implemented, external disturbances can quickly degrade their accuracy. Due to the nature of urban and semi-urban environments, NLOS (non-line-of-sight) and multipath signal propagation can decrease the quality of the GNSS signals, leading to erroneous localization. As the GNSS receivers are low-powered, low-cost jamming devices can easily emit electromagnetic interference over the same frequencies used by the GNSS receivers, resulting in an untrustworthy PNT solution. Spoofing is another threat affecting the PNT integrity, where false GNSS signals are broadcasted to deliberately degrade the PNT data. Thus, considering all the potential threats to the GNSS receivers, A-PNT (alternative position, navigation, and timing) sensors should be used to achieve better navigation performance, even when the GNSS is not able to provide a reliable PNT solution.

An IMU serves as an A-PNT sensor, typically formed by three accelerometers and three gyroscopes, providing data regarding the linear acceleration and angular velocity of the carrier in each direction of the body frame. Usually, UAVs are equipped with MEMS (microelectromechanical system) IMU sensors to derive the position and attitude using an INS (inertial navigation system) mechanization process. Unfortunately, the INS mechanization leads to positioning drift over time as specified in [7,8], making the IMU unreliable when used in a standalone mode for long flight operations.

A-PNT sources, including sensors such as stereo or monocular optical cameras, are alternative methods of improving positioning accuracy and precision in situations when the GNSS is unavailable. Motion estimation for optical A-PNT sources can be classified into two categories: RVL (relative visual localization) or AVL (absolute visual localization). RVL is computed through the application of VO (visual odometry) [9,10] and SLAM (simultaneous localization and mapping) algorithms, as presented in [11–13]. VO methods involve the analysis of the frames captured by an optical sensor to estimate its motion through the environment. Instead, the SLAM (simultaneous localization and mapping) approach represents a more intricate navigation algorithm capable of estimating the relative motion of the UAV while simultaneously building the surroundings on the map. Therefore, VO can be regarded as a subset of SLAM-based navigation. However, it is important to note that in challenging environments characterized by low light conditions or scarcity of distinctive features, both VO and SLAM algorithms can cause divergence in their motion estimation due to drift. Instead, for estimating the UAV's absolute position, the VPS (visual positioning system) algorithm can be used, as presented in [14,15]. To successfully implement the VPS, it is essential to have a proper dataset with georeferenced aerial images that covers the AoI (area of interest). Additionally, the tilt angle of the camera during the flight should align with the used dataset in order to enhance its overall performance. Positioning accuracy can be affected by additional factors, such as seasonal changes and the ongoing construction of new buildings and roads, potentially causing mismatches with the dataset in use. Therefore, updated and recurrent datasets are required.

Although it is possible to extract PNT information from various sensors, a fusion approach is required in order to combine all the advantages offered by each A-PNT sensor. Multi-sensor fusion frameworks can be categorized into either CF (centralized fusion) or DF (decentralized fusion) frameworks. Even if a CF architecture can provide a reliable PNT solution, its high computational cost can lead to the so-called 'computational disaster' effect, as described in [16]. To mitigate computational costs and enhance the robustness of the fusion framework, a DF approach can be used. Instead of using only one filter, as in the CF architecture, the DF framework implements multiple local filters in parallel, fusing their output into a final master filter. Thus, the computational cost can be divided among all the local filters, and multiple A-PNT sensors can be added easily as subsystems. This method results in a federated DF framework suitable for real-time applications, as described in [17], where KFs (Kalman filters) were implemented. While the FF (federated fusion) architecture with KFs demonstrates good performances, it is important to note that in the real world, A-PNT sensors are susceptible to external noise, as discussed in [18], which can negatively impact the accuracy of KFs. In addressing non-linear systems, various solutions such as EKFs (extended Kalman filters), UKFs (unscented Kalman filters), and PFs (particle filters) have been widely used before, as specified in [19]. Their effectiveness is contingent upon prior knowledge of the measurement noise and processing noise.

Instead, fusion frameworks based on RNNs (recurrent neuronal networks) have demonstrated good performance in modeling and predicting the behavior of A-PNT sensors in real-world testing scenarios, as presented in [20], by fusing the INS and GNSS to cope during GNSS outages. This approach has demonstrated a notable 60% improvement against a traditional EKF (extended Kalman filter). However, there are certain drawbacks to using RNNs, including their high computational cost and the challenge of long-term data storage, which can introduce errors over extended periods of time. Moreover, if the weights are too small, the learning rate becomes slow, and managing data over time can decrease the performance of the RNN, leading to the so-called 'vanishing gradient' effect. Conversely, if the weight is too large, the output can diverge, leading to an 'exploding gradient' effect. Hence, to improve the performance of RNNs, LSTM (long short-term memory) and GRUs (gated recurrent units) introduce gates that aid in the longer-term memory capability of the RNN. As it can be seen in [21,22], LSTM models are used to enhance positioning accuracy in urban environments. Even better performances were obtained by implementing a GRU model to cope with GNSS outages, as presented in [23,24].

Thus, to assess the performance of combining GRUs with traditional fusion methods such as EKFs, this paper introduces a hybrid federated fusion architecture for 3D positioning. The federated architecture uses two EKFs as local filters and a GRU model as a master filter to predict the position of the UAV during a flight mission performed in an urban environment. The system gathers data from various A-PNT sensors, including a GNSS receiver, a MEMS IMU sensor, a monocular camera, and a MEMS barometer. To enhance the realism of the data collected, a HIL (hardware in the loop) set-up is used, which involves using Spirent's GSS7000 simulator tools (SimGEN and SimSENSOR) along with OKTAL-SE (Sim3D) to gather GNSS data with multipath and MEMS IMU data. At the same time, a virtual environment in Unreal Engine is used to integrate a monocular camera to be used by a VO algorithm in order to estimate the UAV's ego motion through the urban environment. In addition, the MEMS barometer readings from the virtual UAV in Unreal Engine are integrated into the federated fusion framework to cope with the instabilities introduced by the VO on the z axis during the flight mission. The paper's key contributions can be summarized as follows:

1. The research introduces, in a three-dimensional scenario, a hybrid fusion architecture that integrates GRU (gated recurrent unit) and EKF (extended Kalman filter) systems. This study offers a detailed comparison of the new hybrid approach against the traditional FF (federated fusion) architecture.
2. To evaluate the performance of the proposed hybrid FF architecture using a range of realistic trajectories with the aim of mimicking real-world UAV operations, including multipath and GNSS outages.
3. To assess the influence of the optical part of the fusion algorithm by introducing various weather conditions, including dust and fog. In addition, the VO algorithm was tested during different light intensities, both in the afternoon and in the evening, using realistic photogrammetry data.

The reminder of the paper is structured as follows: In Section 3, the proposed hybrid federated fusion architecture is presented; in Section 4, the HIL configuration is detailed; in Section 5, the trajectories, the camera calibration steps, and the performances of the hybrid federated fusion architecture are discussed; and in Section 6, the conclusions and future work are given.

2. Related Works

Multi-sensor fusion frameworks have been widely used to assure a robust and resilient position and navigation for autonomous systems. With the advent of ML, hybrid fusion frameworks have been increasingly adopted in recent times by combining KFs and ML models. Standalone ML fusion models can be used without the integration of any KF, but their performance is limited when used with MEMS IMU sensors, as presented in [25]. The main disadvantage of implementing standalone ML models used to fuse GNSS and MEMS IMU sensors without KFs is represented by the absence of feedback to update the measurement model of the MEMS IMU inertial sensor, which is crucial due to its rapid change in dynamics over time. Thus, hybrid fusion methods can combine the advantages of KFs and ML models, which can be divided into three categories, as specified in [26].

In the first category, ML models are used as aids to tune KFs, as presented in [27], where a RBFN (radial basis function network) and a PSO (particle swarm optimization) are used as aids to cope with the non-linearities of the system. On the other hand, in [28], the authors developed a NN (neuronal network) as an aid to an AKF (adaptive Kalman filter) to adjust the system noise parameters.

Instead, in the second category, hybrid fusion methods are used in combination with ML models to predict INS errors, while GNSS signals are not available. As presented in [29], an UKF is used with a BP (back propagation) neuronal network to cope with GNSS outages. When the GNSS receiver is not affected by external disturbances, the BP model is trained using the position errors provided by the UKF as input, and when there are GNSS outages, the BP-trained model is used to enhance the positioning output by correcting the INS data. Although the solution proposed by the authors improves the position output during GNSS outages, the BP model has inferior performance compared to the UKF model during normal operations when the GNSS is available. Meanwhile, better results are presented in [30], where the authors implemented a GRU model along with an AKF to cope with GNSS outages. The GRU model is trained with GNSS data when available and used to predict GNSS position measurements during GNSS disturbances, measurements that are used as input for an AKF with INS data. Results showed a reduction in root mean square error of 83.03% and 75.39% during the 180 and 120 seconds of GNSS outages, respectively, proving the efficiency of the GRU-trained model in combination with an AKF. As a drawback, the solution presented by the authors considers only one scenario, and further data collection is required to better evaluate the presented hybrid fusion framework.

In the third category, ML models can be used to enhance fusion methods in combination with fault detection approaches for real-time applications and in complex environments. As presented in [31], a RBFNN (radial basis function neuronal network) is used to predict pseudo-GNSS measurements when faulty GNSS data is detected, aiming to improve fault isolation and system reconfiguration in a tightly coupled approach. The main challenge in the solution proposed by the authors is to optimally tune the POP (precision of positioning) and RDOP (relative differential precision of positioning) thresholds, which define the filter precision.

Hybrid fusion methods have been widely used to predict errors related to GNSS/IMU fusion configurations. Considering the complexity of urban environments, it is unlikely that UAVs will rely solely on GNSS and MEMS IMU sensors to cope with all the external disturbances. Hence, the proposed fusion framework is investigating the performance of a hybrid federated fusion framework in a complex urban environment, relying solely on a GNSS, a MEMS IMU, an optical camera, and a MEMS barometer.

3. Proposed Hybrid Federated Fusion Solution

The proposed hybrid federated fusion architecture involves the fusion of four different sensors, formed by a GNSS receiver, a MEMS IMU sensor, a monocular camera, and a MEMS barometer sensor, to enhance the PNT solution, even in proximity to urban and sub-urban areas. The proposed fusion architecture adopts a hybrid approach, combining machine learning techniques using a GRU with traditional fusion architectures such as the EKF, as can be seen in Figure 1. The hybrid approach is used to improve the accuracy of the final positioning output, particularly when dealing with GNSS data affected by multipaths and outages.

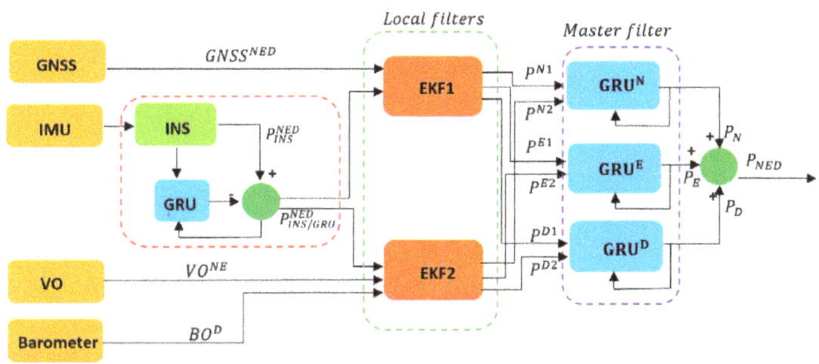

Figure 1. Proposed hybrid federated multi-sensor fusion architecture.

3.1. Trained INS GRU

To cope with the INS drift over time, which leads to an erroneous position estimation, a GRU model is used to enhance the INS positioning output by predicting the INS behavior in time, as can be seen in the left part of Figure 2. During the training part, the GRU gathers raw MEMS IMU data as the input, formed by readings from the MEMS accelerometer and the MEMS gyroscope. As depicted in the right part of Figure 2, the input layer of the GRU block, which takes data from the provided dataset, is formed by the following components: a representing the linear acceleration and ω representing the angular velocity. For the training part, 80% of the dataset was used, while the other 20% was used for the testing part. As output, the GRU model provides INS corrections by comparing the estimated INS corrections against the ground truth, where δP_{INS}^{N}, δP_{INS}^{E}, and δP_{INS}^{D} represent the position error in the NED frame without using the input block called 'time since last GNSS', as presented in [24].

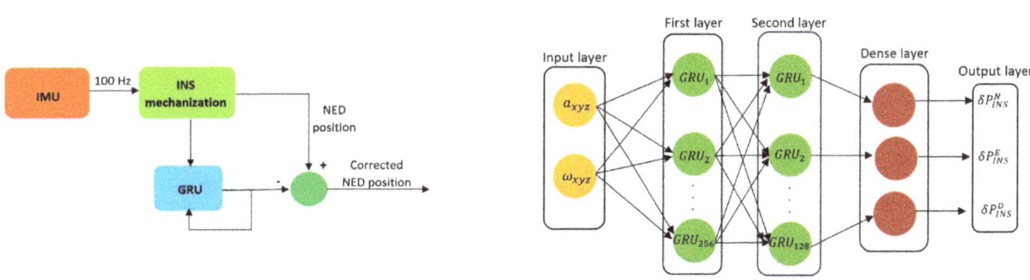

Figure 2. INS/GRU corrections—**left**; GRU diagram—**right**.

Once the GRU model has been trained, the estimated INS data are utilized to provide input to the federated fusion architecture, feeding the output into the two local filters, as can be seen in Equation (1). As a result, the impact of INS drift on the local filters diminishes over time.

$$P_{INS/GRU}^{NED} = P_{INS}^{NED} - \delta P_{INS}^{NED} \tag{1}$$

3.2. GNSS/INS EKF

The first local filter fuses data from the GRU block, presented in the previous section, and the GNSS in a loosely coupled approach. To fuse the output from the two sensors, both have to share the same navigation frame, as defined in Appendix B. Thus, GNSS data must be converted from the LLA (latitude, longitude, and altitude) frame to a NED coordinated frame as it can be seen in Figure 3. To set the conversion, an LLA reference base is defined as:

$$\varphi_{REF} = 43.604441^\circ \quad \lambda_{REF} = 1.4427133^\circ \quad h_{REF} = 0\,\text{m} \tag{2}$$

where φ_{REF}, λ_{REF}, and h_{REF} are the initial altitude, longitude, and altitude, respectively. With the initial starting point coordinates, it is possible to realize the conversion from geodetic coordinates to geocentric coordinates. To begin, it is necessary to convert the data from the WGS84 (World Geodetic System 1984) to an ECEF (Earth-centered, Earth-fixed) coordinate system, as presented in [32]. After that, the ECEF position is converted to a NED coordinate frame, as specified in [33] and in Appendix B. Once the GNSS data is converted to NED coordinates, it can be fused with the GRU output. The state vector for the EKF is defined as follows:

$$x_k = \begin{bmatrix} P_{xN} \\ P_{yE} \\ P_{zD} \end{bmatrix} \tag{3}$$

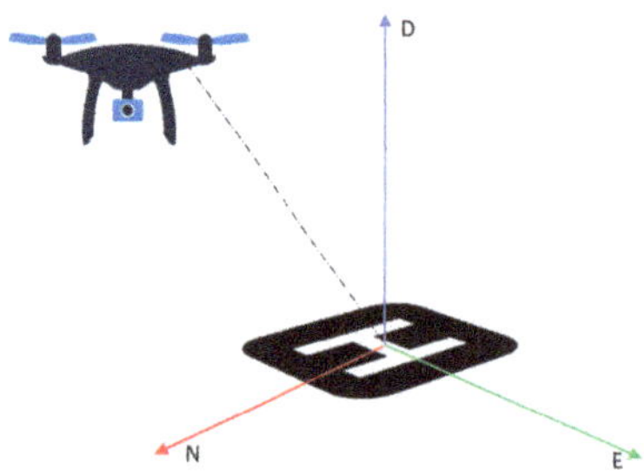

Figure 3. UAV in a NED frame.

The initialization step of the EKF is given by the UAV base position in the NED frame along the initial covariance matrix. Furthermore, the prediction step has the aim of estimating future states and is defined as follows:

$$\begin{gathered} \hat{x}_k^- = g\left(\hat{x}_{k-1}^-, u_{k-1}\right) \\ \hat{x}_k^- = \hat{x}_{k-1}^- F_{k-1} \\ P_k^- = F_{k-1} P_{k-1} F_{k-1}^T + G_{k-1} Q_{k-1} G_{k-1}^T \end{gathered} \tag{4}$$

where $\hat{x}_{k-1}^-$ is the initial UAV position, u_{k-1} is the control input given by the INS GRU source, F_{k-1} is the dynamic covariance matrix, P_k^- is the priori covariance matrix, G_{k-1} is the noise covariance matrix, and Q_{k-1} is the process noise covariance matrix. The last step is represented by the update step (correction step). The number of states dictates the number of columns of the measurement matrix, and the number of measurements dictates the size of the rows. Once the measurement matrix is defined, the measurement residual, or innovation, can be defined as:

$$y_k = z - h\left(\hat{x}_k^-\right) \tag{5}$$

where y_k is the measurement residual, z is the observation vector, and $h\left(\hat{x}_k^-\right)$ is the measurement equation based on the predicted states. Once the measurement residual is obtained, the Kalman gain can be calculated, as defined in Equation (6). If its value is low (closer to 0), the predicted values are closer to the actual states; otherwise, the value will be closer to 1, meaning that the predicted values have more errors.

$$K_k = P_k^- H_k^T \left(H_k P_k^- H_k^T + R_{k-1}\right)^{-1} \tag{6}$$

The measurement update from the GNSS is defined as:

$$\begin{gathered}z_{GNSS_k}=\begin{bmatrix}1&0&0\\0&1&0\\0&0&1\end{bmatrix}\begin{bmatrix}P_{xN}\\P_{yE}\\P_{zD}\end{bmatrix}+\vartheta_{1,k}\\ z_{GNSS_k}=\begin{bmatrix}P_{xN\_GNSS}\\P_{yE\_GNSS}\\P_{zD\_GNSS}\end{bmatrix}\end{gathered} \tag{7}$$

where z_{GNSS_k} is the measurement vector, H_{GNSS} is the measurement matrix for the GNSS measurements, and $\vartheta_{1,k}$ is the Gaussian noise related to the measurements formed by a covariance matrix defined as R_k^{GNSS}. Finally, the predicted state vector and the predicted matrix are defined as:

$$\begin{gathered}x_k=\hat{x}_k^-+K_k(z-h(\hat{x}_k^-))\\ P_k=(I-K_kH_k)P_k^-\end{gathered} \tag{8}$$

3.3. INS/VO/Barometer EKF

The second local filter relies on positioning data from the trained INS GRU block and the positioning data generated by the VO algorithm along the MEMS barometer sensor. The VO algorithm is a visual-based technique widely implemented in robotics that can be used to estimate user motion from a sequence of images, especially when a GNSS solution cannot be provided. VO algorithms can be divided into two main categories: appearance-based and feature-based.

The appearance-based approach estimates the robot's motion by analyzing pixel intensity information obtained from the output of an optical camera, as defined in [34]. Based on this approach, it is possible to derive two additional methods. The first method consists of using a template matching method, which can provide a motion solution by aligning two consecutive frames and measuring local unchanged similarities. The second method implements an optical flow algorithm that directly analyzes the changing intensity of pixels in two consecutive frames, computing a field of vectors from which motion can be estimated.

The second category is formed by feature-based methods, which do not track all the data from two consecutive frames but only key features such as lines or corners, which are effective in environments rich in details. From a computational point of view, the feature-based methods are more effective than the appearance-based methods.

Considering that urban and sub-urban environments are characterized mainly by a multitude of details, the authors implemented a feature-based approach.

Different feature-based algorithms can be implemented, each having different performances, such as the Harris-Corner detector [35], Shi-Tomasi corners [36], FAST (features from accelerated segment test) corners [37], SURF (speeded-up robust features) features [38], SIFT (scale invariant feature transform) features [39], and ORB (oriented FAST and rotated BRIEF) features [40]. In a more detailed analysis, as specified in [41], ORB shows the best performance in terms of computational load; thus, an ORB approach is chosen to deal with real-time missions. Furthermore, the ORB feature detector algorithm can be utilized without the requirement for a license. In contrast, other methods such as SIFT and SURF are subject to patents and, as a result, entail associated costs for usage.

The first step in using the ORB algorithm is to detect features using the FAST corners approach. The FAST algorithm begins by selecting a reference pixel to serve as the center and then considers all the pixels within a radius. After that, a threshold based on pixel intensity is calculated, and the position of features can be determined. Although the features can be detected, the FAST approach does not provide any direction information. Thus, as specified in [40], the intensity centroid (IC) approach is used to find and define the orientation vector, as specified in [42]. This solution increases the robustness of the detected features during rotatory movements, as specified in [40]. Furthermore, a steered BRIEF (binary robust independent elementary feature) descriptor is used, as defined in [40].

To match the descriptors between two frames, a matcher algorithm is needed. Thus, the FLANN (Fast Library for Approximate Nearest Neighbors) [43] is used for its real-time features and matching performance when many features occur against the BFmatcher algorithm.

After the feature matching step, a motion estimation method is needed to compute the ego motion of the camera, which is rigidly attached to the UAV. There are mainly three methods to estimate the ego motion of the camera, as follows [44]:

- 2D to 2D (both features are specified in 2D image coordinates between two frames)
- 3D to 3D (both features are specified in 3D image coordinates between two frames)
- 3D to 2D (previous features are specified in 3D coordinates and the current features in 2D image coordinates)

Considering that a single monocular camera is used, a 2D–2D method is adopted. Furthermore, the essential matrix is required in order to extract the ego motion of the camera, defined in [9] as follows:

$$E = [t]_x R \tag{9}$$

where R is the rotational matrix and t is the translational vector. From the estimated essential matrix, it is possible to extract the rotational matrix and the translation vector. Usually, four solutions are provided, but with triangulation, one single solution is extracting the ego motion of the monocular camera. In addition, to increase the accuracy of the estimated trajectory, the authors implemented the RANSAC (random sample consensus) algorithm, as specified in [45]. To utilize the VO data accurately, it is necessary to execute a conversion from the camera frame to the navigation frame, as specified in [46,47] and in Equation (A1).

$$\begin{bmatrix} p^n_{t,N} \\ p^n_{t,E} \\ 1 \end{bmatrix} = \lambda K^{-1} \left[T^{NED}_{camera}; \quad T^{NED}_{camera} r^n_{nc} \right]^{-1} \begin{bmatrix} u \\ v \\ 1 \end{bmatrix} \tag{10}$$

where T^{NED}_{camera} is the transformation from camera to the navigation frame, K is the intrinsic camera matrix, r^n_{nc} is the position of the camera in the navigation frame, and λ is the scale factor. In comparison to the previous EKF presented in Section 3.2, a barometer is used to provide altitude information to cope with the instabilities provided by the monocular camera on the z axis. Thus, the N and E positioning coordinates are provided by the monocular camera using the VO algorithm, and the D positioning coordinate is provided by the barometer. Furthermore, the output from the VO and barometer is fused with the NED positioning output from the GRU model, which provides an enhancement of the INS output, as presented in Section 3.1.

3.4. Master Filter

The final section of the fusion framework is represented by three GRU models, used to enhance the output from the two EKFs as depicted in Figure 4. Since only the N, E, and D positions are considered, the first GRU model is used to process only the N position data, while the second and third GRU models are used to process only the E and D position data, respectively.

Each GRU model is formed by a layer consisting of 128 GRUs, a RELU activation layer, and a dense layer, followed by the final output layer. Each GRU model was trained using 80% of the dataset and tested using the remaining 20% of the dataset. After the training phase, the final output is defined as follows:

$$P_{NED} = \hat{P}_N + \hat{P}_E + \hat{P}_D \tag{11}$$

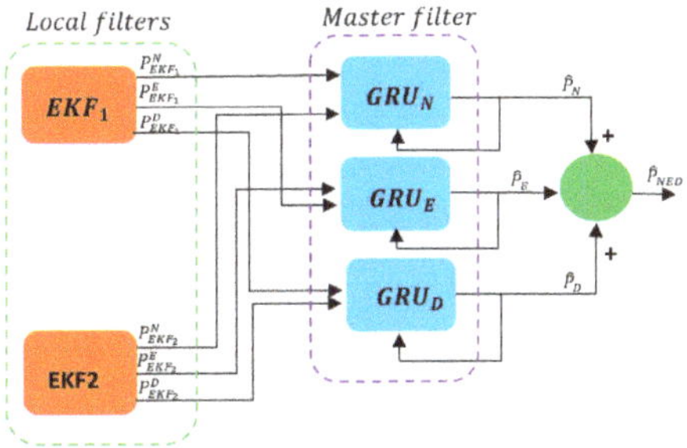

Figure 4. Master filter architecture.

4. Hardware in the Loop Configuration

To enhance the realism of the dataset used by the hybrid sensor fusion architecture, a HIL configuration is established, as can be seen in Figure 5. For the HIL set-up, a Pixhawk 2.4.8 board is used and configured in the HIL mode by running the px4fmu-v2_default firmware, which represents the main FCU (flight control unit) of the UAV responsible for the navigation, control, and stability of the flying device.

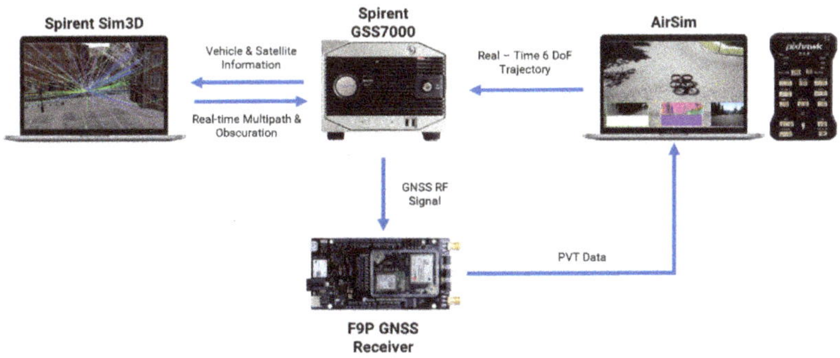

Figure 5. HIL set-up.

Furthermore, once the connection is established with the hosting computer, the Pixhawk is calibrated properly by using the QGroundControl interface. After the calibration step, on the same hosting computer, Unreal Engine 4.27.2, Cesium v2.0.0, and AirSim 1.7.0 are launched with the HIL configuration file (required by AirSim), starting the HIL simulation. Unreal Engine is a 3D graphics interface that can be used to model specific simulation environments, such as urban, semi-urban, and rural environments. The hosting computer is equipped with an Intel Keon CPU E5-1650 v4, 32 GB of RAM, and an NVIDIA GeForce GTX 1080Ti 11 GB GPU (Lenovo, Bratislava, Slovakia). In addition, with the aid of Google Earth and Cesium, photogrammetry data can be easily imported into the UE interface, and the UAV dynamics and sensors are included using the AirSim plugin.

Once the set-up is finalized, a Python v3.6.0 file is used to establish a UDP connection between the Spirent GSS7000 hardware (Spirent PLC, Paignton, UK), and the hosting computer. Thus, a link between AirSim and the GSS7000 allows the recording of IMU data using the SimGEN v7.02 software. At the same time, RF signals are generated and sent further to the C009-F9P Ublox board (u-blox AG, Zürcherstrasse, Switzerland), which is responsible for processing the GNSS signals. By using OKTAL-SE Sim3D v4.7 in conjunction with Spirent's GSS7000 hardware, the generated RF GNSS signals also include multipath effects. More details regarding the IMU and U-Blox F9P GNSS receiver can be viewed in Table 1.

Table 1. INS and U-Blox specifications.

Sensor Specifications					
Accelerometer		**Gyroscope**		**U-Blox F9P GNSS Receiver Specification**	
Scaling factor (ppm)	500	Scaling factor (ppm)	500	Pseudo-range accuracy (m)	3
Bias (mg)	0.1	Bias (deg/h)	0.001	Pseudo-range rate accuracy (m/s)	0.5
ARW (m/s/sqrt(h))	0.003	GRW (deg/sqrt(h))	0.003	Update rate (Hz)	1
Update rate (Hz)		100		-	

5. Evaluation

5.1. Scenario Definition

To evaluate the hybrid FF architecture, Unreal Engine is implemented along with AirSim, as specified in Section 3. For this specific simulation, real photogrammetry data from the city center of Toulouse is integrated into Unreal Engine to replicate an urban environment. To further assess the fusion framework, two trajectories are used, as can be seen in Figure 6. The first trajectory is formed by different waypoints covering a larger area, while the second trajectory is limited to a 150-meter square area. Both trajectories are replicating a survey mission conducted in an urban environment. The simulation aims to test the VO algorithm while assessing the influence of multipath on the GNSS and the drift introduced by the MEMS IMU over time.

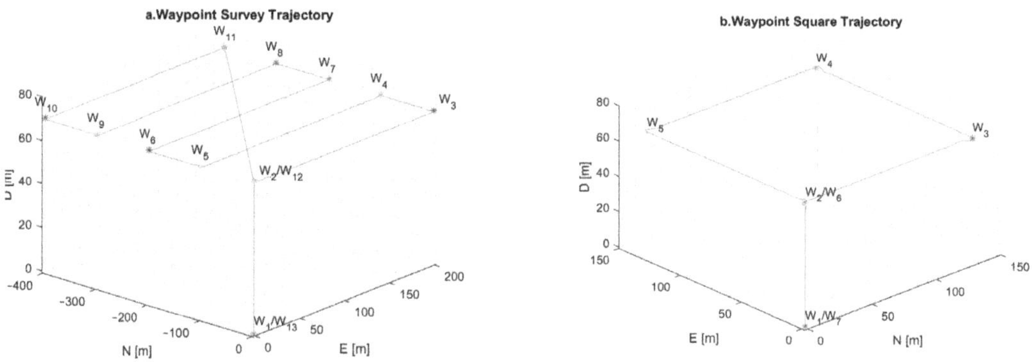

Figure 6. Waypoint survey trajectory—**left** (**a**); waypoint square trajectory—**right** (**b**).

5.2. Light and Weather Evaluation

In addition, both fusion architectures are tested under different light and weather conditions. The first scenario simulates the flight of the UAV during normal daylight conditions at 14:51 p.m. local time in Unreal Engine, while the second flight is at 18:00 p.m., as can be seen in Figure 7. Additionally, to further evaluate the accuracy of the VO algorithm, two more scenarios are considered, performing the two trajectories under fog and dust conditions, as can be seen in Figure 8.

5.3. Camera Calibration Set-Up

Considering the realism introduced into the simulation, before evaluating both fusion architectures, the monocular camera, rigidly fixed on the UAV and pointing downward, is configured in AirSim with an image width of 752 pixels and an image height of 480 pixels.

Figure 7. Toulouse in UE during the afternoon—**left**; Toulouse in UE during the evening—**right**.

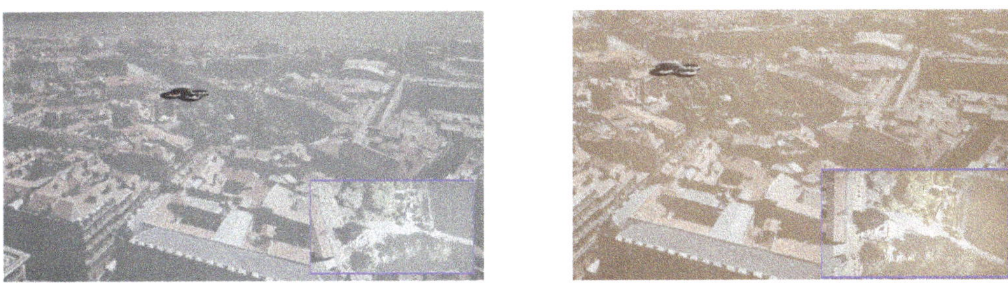

Figure 8. The UAV during fog operations—**left**; the UAV during dust operations—**right**.

Before using the monocular camera with the VO algorithm, a chessboard is introduced into Unreal Engine in order to calibrate the camera by finding the intrinsic matrix, defined as K, formed by the focal lengths f_x and f_y and by the optical centers c_x and c_y. The chessboard is characterized by 10 rows and 10 columns, featuring alternating white and black squares, as can be seen in Figure 9. Each block has a dimension of 2 m in both length and width on the defined chessboard. Furthermore, in Unreal Engine, the UAV performed a small square trajectory over the chessboard at a cruise altitude of 70 m while recording all the camera frames. Then, the 220 frames extracted from Unreal Engine are processed using the MATLAB 2023b 'Camera Calibrator' toll, obtaining the intrinsic matrix, as defined in Equation (A3). Thus, the intrinsic matrix is used to enhance the realism of the VO algorithm.

$$K = \begin{bmatrix} f_x & 0 & c_x \\ 0 & f_y & c_y \\ 0 & 0 & 1 \end{bmatrix} = \begin{bmatrix} 378.6062 & 0 & 376.2584 \\ 0 & 378.6560 & 240.4231 \\ 0 & 0 & 1 \end{bmatrix} \tag{12}$$

Figure 9. Calibration process for the monocular camera used in Unreal Engine.

5.4. Conventional Federated Filter Architecture

To compare the advantages of the proposed hybrid FF architecture, a conventional FF architecture formed only by EKFs is used, as can be seen in Figure 10, in a loosely coupled approach.

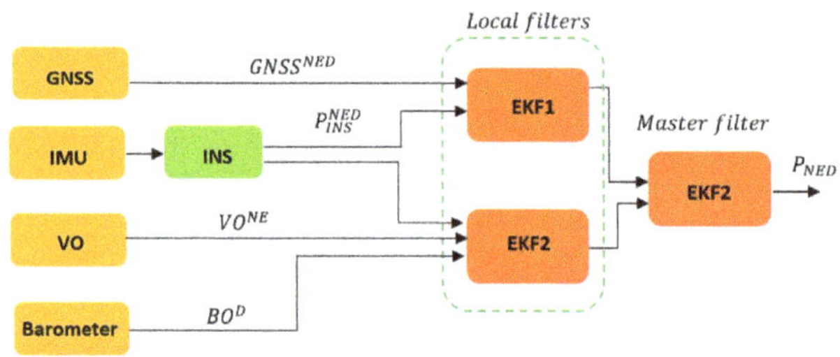

Figure 10. FF architecture with EKFs.

Thus, the INS block does not have any GRU model that predicts the IMU drift over time, and the master filter is only formed by a conventional EKF that fuses the positioning output from the two local filters. To further evaluate the FF, the same architecture is used, but while using the trained GRU model to enhance the INS position output.

5.5. Evaluation of the Square Trajectory

To evaluate the performance of both fusion architectures, various metrics are implemented, including 3D positioning, horizontal and vertical error, and RMSE (root mean square error) on each axis, as specified in Appendix B.

Although the pseudo-range accuracy of the GNSS receiver is 3 m as it can be seen in Table 1, the multipath introduced is consistently affecting the position accuracy of the first local filter when no GRU correction are used. As it can be seen in Table 2, for the square trajectory, significant positioning improvements can be observed in the first EKF by fusing the output from the GRU aid, which enhances the INS position output, with GNSS data, obtaining an equivalent horizontal error of 0.59 m (95th percentile). In contrast, the EKF without the GRU model shows a higher horizontal error of 9.76 m (95th percentile). Analysing the vertical error in both local filters, it can be observed that the filter without GRU corrections has the worst performance against the local filter with GRU corrections.

Table 2. Positioning performance for Toulouse under different light conditions—square trajectory.

Toulouse—Afternoon							
Position Source	**3D Position Error (95th Percentile) [m]**	**Horizontal Error (95th Percentile) [m]**	**Vertical Error (95th Percentile) [m]**	**RMSE N [m]**	**RMSE E [m]**	**RMSE D [m]**	**RMSE NED [m]**
EKF1 IMU/GNSS (no GRU aid)	10.40	9.76	3.5	4.49	1.35	3.36	5.77
EKF2 IMU/VO/BO (no GRU aid)	9.09	9.03	1.58	4.17	1.60	0.78	4.54
Master EKF filter (no GRU aid)	9.09	9.03	1.50	4.18	1.35	0.77	4.46
EKF1 IMU/GNSS (with GRU aid)	0.64	0.59	0.29	0.20	0.17	0.13	0.30
EKF2 IMU/VO/BO (with GRU aid)	4.57	4.57	1.48	1.32	1.45	0.76	2.11
Master GRU filter (with GRU aid)	0.58	0.54	0.27	0.16	0.14	0.12	0.25
Master EKF filter (with GRU aid)	0.76	0.73	0.29	0.21	0.21	0.13	0.33
Toulouse—Evening							
EKF2 IMU/VO/BO (no GRU aid)	9.16	9.00	2.54	4.17	1.64	1.40	4.68
Master EKF filter (no GRU aid)	9.13	8.95	2.44	4.15	1.63	1.37	4.58
EKF2 IMU/VO/BO (with GRU aid)	5.07	4.92	1.63	1.52	1.58	1.00	2.35
Master GRU filter (with GRU aid)	0.59	0.54	0.27	0.16	0.14	0.12	0.25
Master EKF filter (with GRU aid)	0.77	0.75	0.29	0.22	0.21	0.13	0.34

On the other hand, the second local filter, which fuses the output from the IMU with the VO and barometer data, shows slightly better performance with a horizontal error of 9.03 m (95th percentile) without relying on any corrections from the GRU model. Instead, when the IMU/GRU corrections are implemented, the horizontal error tends to achieve an equivalent positioning output of 4.57 m (95th percentile), with an improvement in the N, E, and D coordinates with an equivalent RMSE of 1.32 m, 1.45 m, and 0.76 m, respectively. The overall RMSE, considering all the NED coordinates, equals 2.11 m against the filter without GRU corrections, which equals 4.54 m. By analyzing the output of the master filters, it is possible to notice an enhancement in positioning, shifting from a horizontal error of 9.03 m (95th percentile) to 0.54 m (95th percentile) when employing the GRU model as the master filter instead of the master EKF. If a master EKF is considered with a GRU aid, slightly worse performances can be observed, maintaining a sub-meter horizontal error. By changing the light conditions from a daylight flight to an evening flight, in Table 2, it is possible to notice the influence of the VO over the output of the EKF. Despite the change in light conditions while executing the same trajectory, comparable performances are attained, as can be seen in Figure 11. Although comparable positioning performances are achieved during both afternoon and evening flights, a degradation in positioning is evident when weather conditions change from clear-sky conditions to the presence of fog and dust effects.

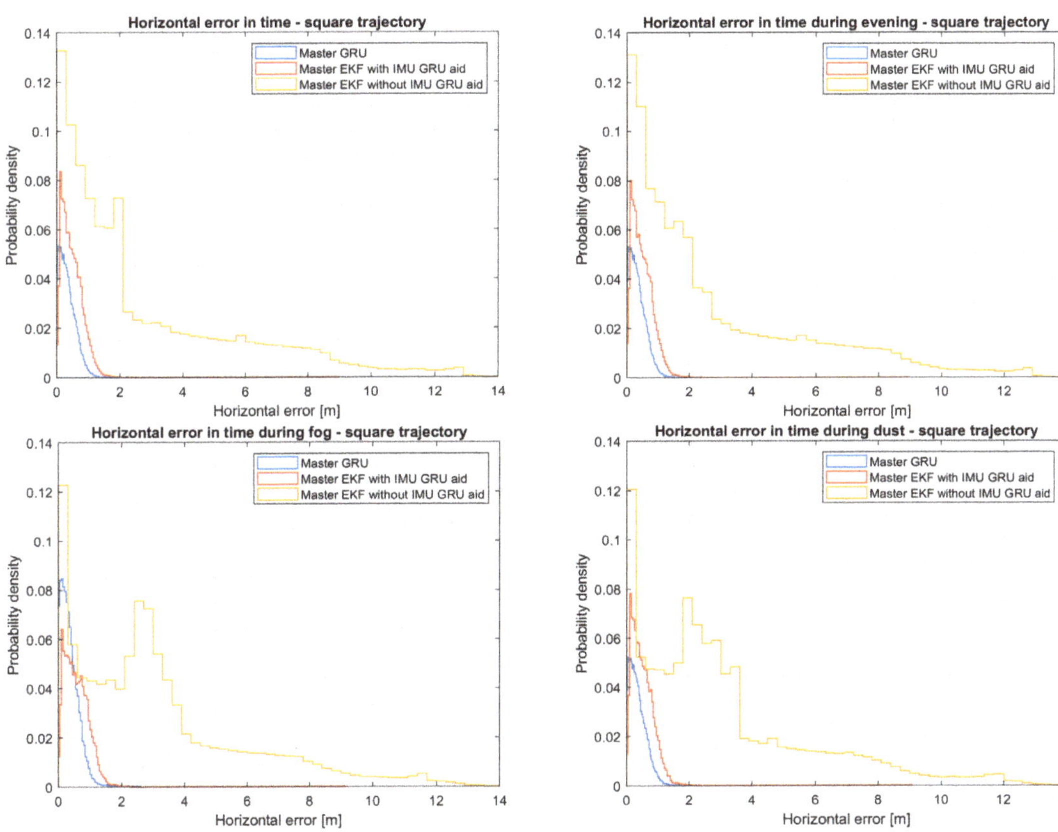

Figure 11. Horizontal error comparisons during different light conditions considering a square trajectory.

From Figure 11, it is possible to notice that the distribution of the horizontal error under fog and dust conditions is greater compared to the horizontal error observed in flights conducted during the afternoon and evening conditions. By analyzing the outputs of the second local EKF, which uses VO for positioning, from Table 3, it is possible to highlight that fog, in comparison to dust, degrades the EKF positioning output more, leading to a horizontal error of 14.25 m (95th percentile). In comparison, the dust effect leads to a horizontal error of 13.62 m (95th percentile). On the one hand, processing both fog and dust outputs with a master EKF leads to similar results due to the fusion with the output from the first local filter. On the other hand, the aid of the master GRU filter substantially improves the positioning output, decreasing the horizontal error to 0.57 m during fog conditions and 0.55 m during dust conditions. On the one hand, the master GRU model does cope with all the additional instabilities introduced by the VO during adverse and challenging weather conditions. In contrast, if a master EKF is used with a GRU aid, comparable results are obtained, as in the previous cases, during afternoon and evening flights. Thus, the VO algorithm can increase the overall position of an UAV in an urban environment during normal conditions, up to good light and weather conditions.

Table 3. Positioning performance for Toulouse under different weather conditions—square trajectory.

			Toulouse—Fog				
Position Source	**3D Positioning Error (95th Percentile) [m]**	**Horizontal Error (95th Percentile) [m]**	**Vertical Error (95th Percentile) [m]**	**RMSE N [m]**	**RMSE E [m]**	**RMSE D [m]**	**RMSE NED [m]**
EKF2 IMU/VO/BO (no GRU aid)	14.63	14.25	3.00	6.34	5.07	1.60	8.28
Master EKF filter (no GRU aid)	9.44	8.96	2.98	4.12	1.71	1.60	4.74
EKF2 IMU/VO/BO (with GRU aid)	8.63	8.63	3.25	2.80	2.53	1.48	4.07
Master GRU filter (with GRU aid)	0.61	0.57	0.28	0.17	0.17	2.30	0.27
Master EKF filter (with GRU aid)	0.89	0.87	0.29	0.25	0.24	0.13	0.37
			Toulouse—Dust				
EKF2 IMU/VO/BO (no GRU aid)	13.74	13.62	4.90	6.54	3.90	2.54	8.03
Master EKF filter (no GRU aid)	9.65	8.82	4.85	4.09	1.57	2.50	5.06
EKF2 IMU/VO/BO (with GRU aid)	6.00	5.51	2.50	1.77	1.61	1.45	2.80
Master GRU filter (with GRU aid)	0.60	0.55	0.27	0.16	3.07	0.12	0.25
Master EKF filter (with GRU aid)	0.79	0.77	0.29	0.22	0.22	0.13	0.34

5.6. Evaluation of the Survey Trajectory

If a more complex trajectory is considered, it can be observed from Table 4 that the second local filter introduces more errors into the fusion system. This correlates to the drift introduced into the second EKF filter over time by the VO algorithm and INS. Considering that the survey trajectory covers a larger area, green areas such as parks decrease the efficiency of the VO algorithm due to the lack of features. Although slightly better performances are achieved when the VO output is fused with the output from the MEMS barometer and MEMS IMU with GRU corrections, the horizontal error is higher in comparison to the values presented in both Tables 2 and 3. Although the perturbances of the VO algorithm are higher, it can be observed that both the master EKF and master GRU models substantially reduce the final positioning error.

As can be seen in Figure 12, the master EKF has better performance during the afternoon flight, while the master EKF with data collected during evening conditions has more errors. In contrast, the master GRU model shows better performance, boasting a horizontal error of 1.72 m (95th percentile) and a vertical error of 0.28 m (95th percentile). Considering the effects of weather on the VO algorithm, as presented for the square trajectory, it can be seen from Figure 12 and Table 5 that more errors are introduced into the fusion framework during foggy conditions. However, in both situations, the trained GRU model, by filtering both fusion outputs and considering all the NED coordinates, achieves an RMSE of 0.83 m.

Table 4. Positioning performance for Toulouse under different light conditions—survey trajectory.

Position Source	3D Positioning Error (95th Percentile) [m]	Horizontal Error (95th Percentile) [m]	Vertical Error (95th Percentile) [m]	RMSE N [m]	RMSE E [m]	RMSE D [m]	RMSE NED [m]
Toulouse—Afternoon							
EKF1 IMU/GNSS (no GRU aid)	11.37	10.19	4.05	1.77	4.52	3.77	6.15
EKF2 IMU/VO (no GRU aid)	22.92	22.90	5.00	9.38	7.36	1.60	12.03
Master EKF filter (no GRU aid)	10.44	10.19	2.11	1.94	4.48	1.80	5.21
EKF1 IMU/GNSS (with GRU aid)	1.86	1.85	0.26	0.76	0.49	0.16	0.92
EKF2 IMU/VO (with GRU aid)	11.74	11.74	4.10	4.79	4.63	1.27	6.79
Master GRU filter (with GRU aid)	1.72	1.72	0.28	0.67	0.44	0.12	0.81
Master EKF filter (with GRU aid)	1.93	1.93	0.26	0.77	0.53	0.16	0.95
Toulouse—Evening							
EKF2 IMU/VO (no GRU aid)	23.43	23.43	5.53	10.00	7.55	1.64	12.64
Master EKF filter (no GRU aid)	10.57	10.40	5.31	2.71	4.48	1.57	5.47
EKF2 IMU/VO (with GRU aid)	13.38	13.32	3.99	5.27	5.46	1.73	7.79
Master GRU filter (with GRU aid)	1.73	1.73	0.28	0.67	0.16	0.12	0.82
Master EKF filter (with GRU aid)	1.97	1.96	0.26	0.78	0.54	0.16	0.97

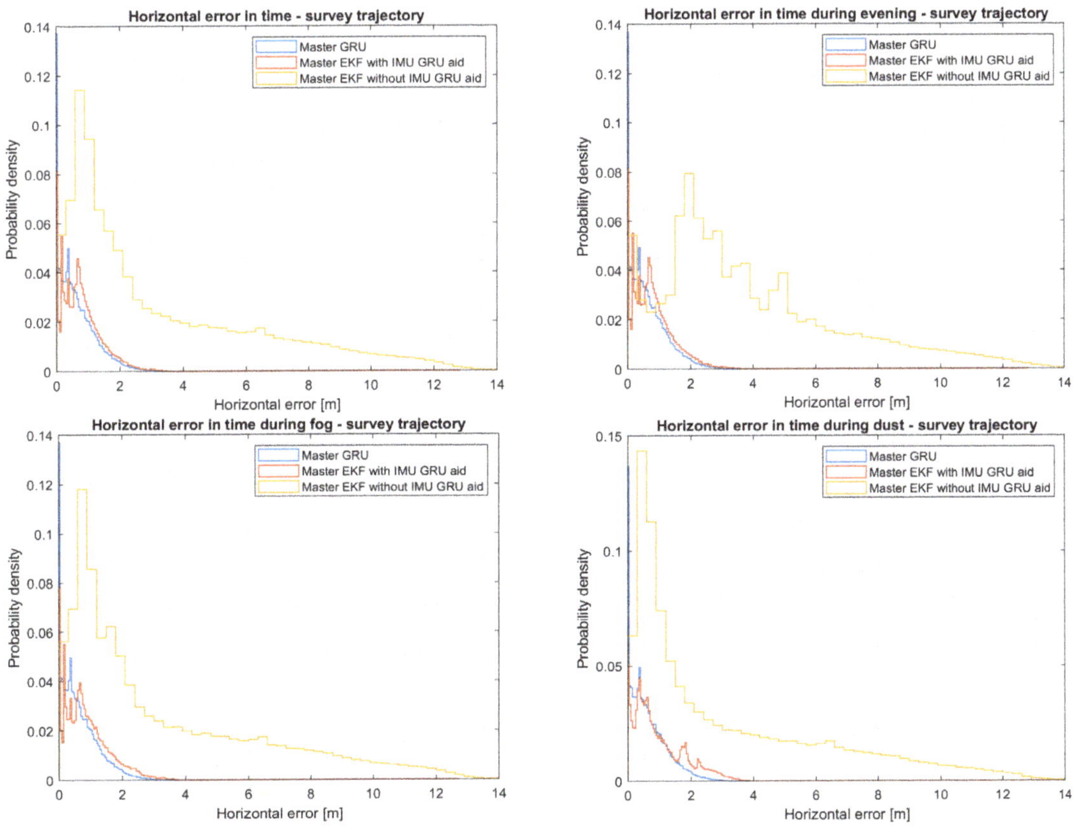

Figure 12. Horizontal error comparisons during different light and weather conditions considering a survey trajectory.

Table 5. Positioning performance for Toulouse under different weather conditions—survey trajectory.

Position Source	3D Positioning Error (95th Percentile) [m]	Horizontal Error (95th Percentile) [m]	Vertical Error (95th Percentile) [m]	RMSE N [m]	RMSE E [m]	RMSE D [m]	RMSE NED [m]
Toulouse—Fog							
EKF2 IMU/VO (no GRU aid)	23.76	23.76	10.44	9.10	7.95	3.01	12.50
Master EKF filter (no GRU aid)	11.36	10.21	10.31	1.80	4.52	3.10	5.77
EKF2 IMU/VO (with GRU aid)	25.22	25.19	8.34	10.39	9.92	2.80	14.64
Master GRU filter (with GRU aid)	1.74	1.73	0.29	0.67	0.46	0.13	0.83
Master EKF filter (with GRU aid)	2.22	2.21	0.26	1.07	0.87	0.60	1.07
Toulouse—Dust							
EKF2 IMU/VO (no GRU aid)	36.78	36.75	8.04	14.83	10.24	2.77	18.24
Master EKF filter (no GRU aid)	11.32	10.18	3.97	1.84	4.5	3.71	6.12
EKF2 IMU/VO (with GRU aid)	22.51	22.12	5.95	8.73	9.12	4.08	13.27
Master GRU filter (with GRU aid)	1.74	1.73	0.29	0.67	0.46	0.13	0.83
Master EKF filter (with GRU aid)	2.53	2.53	0.26	10.03	0.56	0.16	1.18

5.7. Evaluation of the Square and Survey Trajectories Considering GNSS Outages

To further investigate the performance of the proposed hybrid fusion architecture, GNSS outages are introduced into the simulation. For the square trajectory, two small outages lasting 15 seconds each are considered, along with the introduction of a more extended outage lasting 50 seconds. As can be seen from Table 6, the local EKF fusing data from the GNSS and IMU have an equivalent horizontal error of 12.16 m (95th percentile), which is higher in comparison to the previous scenario where outages were not considered, as can be seen in Figure 13. On the one hand, when the GNSS data are fused with the second local filter and the master EKF, an improvement in position can be seen, leading to a horizontal error of 9.44 m (95th percentile). Although the GNSS data is affected by multipaths and outages, it can be observed that the VO algorithm along the barometer from the second local EKF contains the errors introduced by GNSS outages. On the other hand, the GRU model used to predict INS errors substantially reduces the GNSS errors, leading to a horizontal error of 0.82 m (95th percentile), while the master GRU filter further reduces the horizontal error, improving it by 32%. Although similar performances are obtained during the evening flight scenario, it can be observed that due to the VO degradation in foggy and dusty conditions, the master EKF leads to a higher position error. In contrast, the trained master GRU filter achieves a sub-meter position error, despite the disturbances introduced during the simulation. Similar results are obtained using a master EKF with GRU corrections, leading to sub-meter accuracy with a horizontal error of 0.75 m (95th percentile) and a vertical error of 0.31 m (95th percentile).

Instead, for the survey trajectory, two short GNSS outages lasting 15 seconds and two extended GNSS outages lasting 50 seconds are introduced into the simulation, as can be seen from Figure 14. The horizontal and vertical errors are degraded in comparison to the scenario where outages are not considered. At the same time, in the square scenario, the IMU, monocular camera, and MEMS barometer sensor reduce the errors introduced by the GNSS trying to cope with all the outages. It can be observed from Table 7 that even if a federated multi-sensor fusion framework is implemented, the master EKF cannot guarantee a reliable and stable flight in a multipath environment with outages. However, if a master GRU filter is used, the horizontal error tends to be within 2 m (95th percentile) despite all the external instabilities. Instead, if a master EKF filter is implemented with INS GRU corrections, the horizontal error tends to be within 4 m (95th percentile).

Table 6. Positioning performance for Toulouse under different light conditions with GNSS outages—square trajectory.

Position Source	3D Position Error (95th Percentile) [m]	Horizontal Error (95th Percentile) [m]	Vertical Error (95th Percentile) [m]	RMSE N [m]	RMSE E [m]	RMSE D [m]	RMSE NED [m]
Toulouse—Afternoon							
EKF1 IMU/GNSS (no GRU aid)	15.81	12.16	11.73	5.19	5.51	7.22	13.92
Master EKF filter (no GRU aid)	9.48	9.44	1.52	4.03	2.23	0.82	6.01
EKF1 IMU/GNSS (with GRU aid)	0.98	0.82	0.59	0.34	0.33	0.34	1.01
Master GRU filter (with GRU aid)	0.60	0.55	0.27	0.17	0.18	0.12	0.28
Master EKF filter (with GRU aid)	0.80	0.75	0.31	0.23	0.23	0.17	0.37
Toulouse—Evening							
Master EKF filter (no GRU aid)	9.81	9.59	2.74	4.43	4.04	1.49	6.18
Master GRU filter (with GRU aid)	0.60	0.56	0.28	0.18	0.17	0.15	0.29
Master EKF filter (with GRU aid)	0.81	0.76	0.31	0.23	0.23	0.17	0.37
Toulouse—Fog							
Master EKF filter (no GRU aid)	19.66	19.44	2.94	8.29	7.59	1.65	11.36
Master GRU filter (with GRU aid)	0.86	0.81	0.43	0.28	0.17	0.16	0.37
Master EKF filter (with GRU aid)	0.92	0.88	0.31	0.25	0.26	0.17	0.40
Toulouse—Dust							
Master EKF filter (no GRU aid)	19.29	19.07	2.94	8.14	7.40	1.65	11.12
Master GRU filter (with GRU aid)	0.62	0.58	0.28	0.19	0.17	0.15	0.30
Master EKF filter (with GRU aid)	0.73	0.83	0.31	0.23	0.23	0.17	0.37

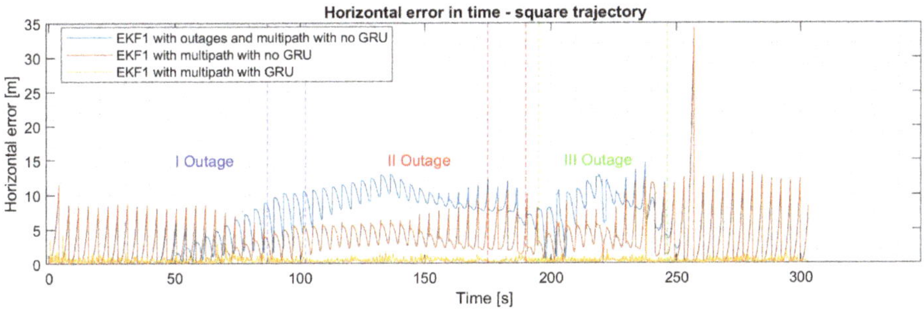

Figure 13. Horizontal error in time for the survey trajectory with multipath and outages over the EKF1.

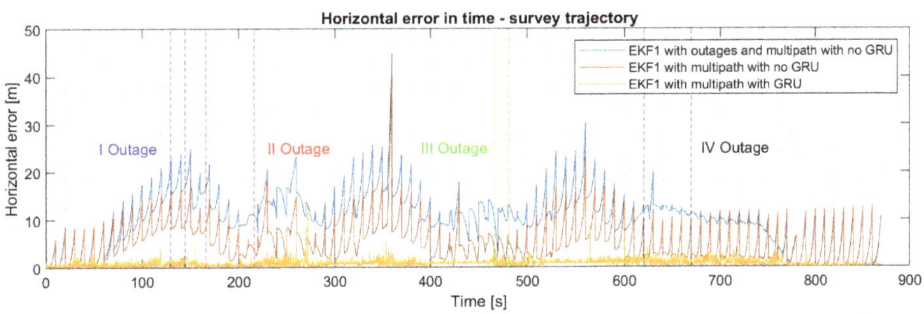

Figure 14. Horizontal error in time for the square trajectory with multipath and outages over the EKF1.

Table 7. Positioning performance for Toulouse under different light conditions with GNSS outages—survey trajectory.

Position Source	3D Position Error (95th Percentile) [m]	Horizontal Error (95th Percentile) [m]	Vertical Error (95th Percentile) [m]	RMSE N [m]	RMSE E [m]	RMSE D [m]	RMSE NED [m]
Toulouse—Afternoon							
EKF1 IMU/GNSS (no GRU aid)	17.83	17.83	13.16	6.34	8.91	10.85	15.41
Master EKF filter (no GRU aid)	17.82	17.78	4.99	6.35	8.88	1.59	11.03
EKF1 IMU/GNSS with GRU aid	3.39	3.35	0.60	1.71	0.83	0.49	1.98
Master GRU filter (with GRU aid)	1.77	1.77	0.27	0.49	0.69	0.13	0.86
Master EKF filter (with GRU aid)	3.42	3.37	0.60	1.68	0.87	0.49	1.96
Toulouse—Evening							
Master EKF filter (no GRU aid)	17.79	17.75	5.51	6.38	8.85	1.63	11.03
Master GRU filter (with GRU aid)	1.77	1.76	0.27	0.49	0.69	0.13	0.86
Master EKF filter (with GRU aid)	3.43	3.39	0.60	1.69	0.87	0.49	1.96
Toulouse—Fog							
Master EKF filter (no GRU aid)	17.87	17.74	10.64	6.35	8.85	3.20	11.36
Master GRU filter (with GRU aid)	1.76	1.76	0.27	0.49	0.69	0.13	0.86
Master EKF filter (with GRU aid)	3.50	3.46	0.60	1.69	0.89	0.49	1.98
Toulouse—Dust							
Master EKF filter (no GRU aid)	18.01	17.81	8.02	6.50	8.89	2.77	11.36
Master GRU filter (with GRU aid)	1.76	1.76	0.27	0.49	0.69	0.13	0.86
Master EKF filter (with GRU aid)	3.48	3.44	0.60	1.69	0.89	0.49	1.97

5.8. Performance Comparison

To evaluate the results obtained for the second local filter, which fuses the position output from the VO algorithm, the altitude from the MEMS barometer, and the position from the MEMS IMU, the following paper [48] is used as a benchmark. The paper implements a VINS-mono [49] algorithm that gathers real data from a monocular camera mounted on a UAV pointing downward. It can be observed that the output from the second local filter shows significant improvements at 60 m while considering both the square and survey trajectories, with an equivalent improvement of 61.78% for the square scenario and 17.89% for the survey trajectory. It can be observed that with longer trajectories, the errors introduced by the VO accumulate over time, decreasing the position accuracy.

Instead, to compare the overall performances obtained by the proposed hybrid federated fusion architecture, the solution presented in [50] is used as a comparison. The paper presents a robust adaptive Kalman filter for gathering data from a GNSS sensor, an IMU MEMS sensor, and an optical camera. As observed in Table 8, the proposed solution with an ML aid shows better performances considering both scenarios during normal VO operations and with multipaths.

Table 8. Benchmark analysis.

Algorithm	MAE N [m]	MAE E [m]	MAE D [m]	RMSE NED [m]	Square Scenario		Survey Scenario	
Mono-VIO [48]	-	-	-	16.54	61.78%		17.89%	
Robust Adaptive Kalman filter [50]	0.06	0.07	0.06	-	N	193%	N	188%
					E	180%	E	192%
					D	187%	D	199%

6. Conclusions

In this study, the authors presented and demonstrated the following:

- A novel hybrid sensor fusion framework based on a federated approach was developed and tested in a loosely coupled set-up, integrating data from diverse sources, including a GNSS receiver, a MEMS IMU sensor, a monocular camera, and a MEMS barometer.
- A virtual environment was developed in UE, along with AirSim, Cesium, and photogrammetry data imported from Google Earth, allowing the authors to test and validate the effects of the VO algorithm over the hybrid fusion framework under different light and weather conditions. To further validate the framework, the hybrid FF architecture was compared to a classic FF framework. GNSS data were enhanced using the Spirent GSS7000 simulator with the OKTAL-SE Sim 3D software stack, introducing multipath during the data collection phase, and collected using a C009-F9P Ublox board. At the same time, IMU data were gathered using the Spirent GSS7000. In addition, GNSS outages were considered for both scenarios.
- Based on the performance metrics presented in Tables 2 and 3 for a square trajectory and Tables 4 and 5 for a survey trajectory, it is evident that the corrections offered by the master GRU model surpass those of the master EKF filter. The master GRU model demonstrates the capability to achieve a sub-meter positioning error in terms of horizontal and vertical error for the square trajectory and below 2 m for the survey trajectory, under different weather and light conditions.
- The presented feature-based VO algorithm does improve the position accuracy of the UAV, as can be seen in Tables 2 and 3 under good weather and light conditions. If more complex and longer missions are considered, the VO algorithm does not provide major position correction.

Author Contributions: Conceptualization, S.A.N. and P.G.; methodology, S.A.N.; software, S.A.N. and P.G.; validation, S.A.N. and P.G.; formal analysis, S.A.N.; investigation, S.A.N.; data curation, S.A.N.; writing—original draft preparation, S.A.N.; writing—review and editing, S.A.N., I.P., P.G., and W.G.; visualization, S.A.N.; supervision, I.P. and W.G.; All authors have read and agreed to the published version of the manuscript.

Funding: This research received no external funding.

Institutional Review Board Statement: Not applicable.

Informed Consent Statement: Not applicable.

Data Availability Statement: Data are contained within the article.

Conflicts of Interest: The authors declare no conflict of interest.

Appendix A. Coordinate Systems Definition and Transformations

Appendix A.1. Body Frame

The first coordinate system is defined by the body frame having its origin in the UAV's center of gravity, where the x axis points ahead, the z axis downwards, and the y axis on the right side of the flying vehicle, while rotation on the x axis is defined as φ (roll), rotation on the y axis as θ (pitch), and rotation on the z axis as ψ (yaw), as can be seen in the figure below.

Appendix A.2. ECEF (Earth-Centered Earth-Fixed Frame)

The Earth-Centered Earth-Fixed Frame has its origin at the Earth's center, and it is fixed and rotates with the Earth. The x axis points to the prime meridian, the z axis points towards the North Pole, and the y axis completes the right-hand frame.

Appendix A.3. Inertial Frame

The inertial frame aligns with the Earth's center, similar to the ECEF frame, except its rotation is independent. The x axis points towards the mean vertical equinox, the z axis is parallel with the Earth's rotation, and the y axis completes the right-handed frame, similar to the ECEF frame.

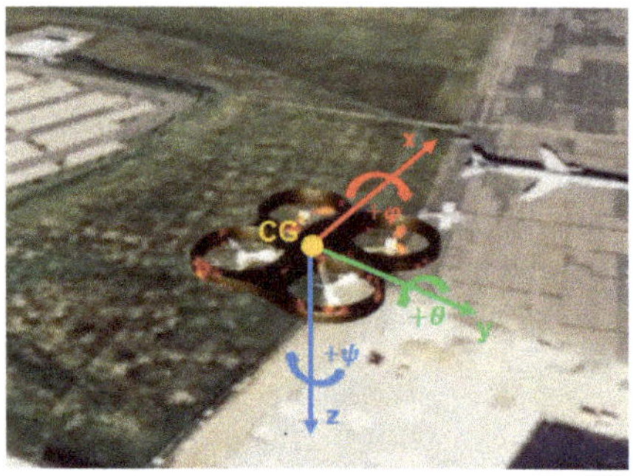

Figure A1. UAV's body frame definition in AirSim.

Appendix A.4. Navigation Frame

A NED (North, East, and Down) [51] navigation coordinate frame attached to the vehicle is adopted by the author. Thus, the x axis points towards north (N), the z axis points downwards (D), and the y axis points east (E), considering a WGS84 ellipsoid Earth model.

Appendix A.5. Conversion from WGS84 to ECEF

$$\begin{aligned} x_{ECEF} &= (N+h)\cos\varphi\cos\lambda \\ y_{ECEF} &= (N+h)\cos\varphi\sin\lambda \\ z_{ECEF} &= \left[N\left(1-e^2\right)+h\right]\sin\varphi \end{aligned} \tag{A1}$$

$$N = \frac{a}{\sqrt{1-e^2\sin\varphi^2}}$$
$$e^2 = 2f - f^2$$
$$f = \frac{a-b}{a}$$
$$P_{ECEF} = \begin{bmatrix} x_{ECEF} \\ y_{ECEF} \\ z_{ECEF} \end{bmatrix}$$

where x_{ECEF}, y_{ECEF}, and z_{ECEF} are the positions in the ECEF frame; φ, λ, and h are the initial altitude, longitude, and altitude, respectively; N is the radius curvature in prime vertical; a is the semi-major Earth axis; b is the semi-minor Earth axis; e^2 is the eccentricity of Earth; and f is the flattening.

Appendix A.6. Conversion from ECEF to NED

$$P_{NED} = R(P_{ECEF} - P_{REF}) \tag{A2}$$

$$R = \begin{bmatrix} -sin\varphi_{REF}cos\lambda_{REF} & -sin\varphi_{REF}sin\lambda_{REF} & cos\varphi_{REF} \\ -sin\lambda_{REF} & cos\lambda_{REF} & 0 \\ -cos\varphi_{REF}cos\lambda_{REF} & -cos\varphi_{REF}sin\lambda_{REF} & -sin\varphi_{REF} \end{bmatrix}$$
$$x_{REF} = (N+h_{REF})\cos\varphi_{REF}\cos\lambda_{REF}$$
$$y_{REF} = (N+h_{REF})\cos\varphi_{REF}\sin\lambda_{REF}$$
$$z_{REF} = \left[N\left(1-e^2\right)+h\right]\sin\varphi_{REF}$$

where P_{REF} represents the vector with the initial LLA coordinates in an ECEF coordinate system.

Appendix A.7. Coordinate Frame Transformations

To convert the body frame defined previously to the NED frame, a DCM (direction cosine matrix) is used, as follows:

$$C_b^n = \begin{bmatrix} cos\theta\ cos\psi & -\ cos\varphi\ sin\psi + sin\varphi\ sin\theta\ cos\psi & sin\varphi\ sin\psi + cos\varphi\ sin\theta\ cos\psi \\ cos\theta\ sin\psi & cos\varphi\ cos\psi + sin\varphi\ sin\theta\ sin\psi & -sin\varphi\ cos\psi + cos\varphi\ sin\theta\ sin\psi \\ -sin\theta & sin\varphi\ cos\theta & cos\varphi\ cos\theta \end{bmatrix} \quad (A3)$$

where C_b^n is the DCM matrix and θ , ψ, and φ are the pitch, yaw, and roll angle in radians, respectively.

Appendix B. Metrics

To assess the two fusion architectures, the following metrics were used:

$$\text{Vertical error} = \sqrt{\left(D_i - \hat{D}_i\right)^2} \quad (A4)$$

$$\text{Horizontal error} = \sqrt{\left(N_i - \hat{N}_i\right)^2 + \left(E_i - \hat{E}_i\right)^2} \quad (A5)$$

$$\text{3D error} = \sqrt{\left(N_i - \hat{N}_i\right)^2 + \left(E_i - \hat{E}_i\right)^2 + \left(D_i - \hat{D}_i\right)^2} \quad (A6)$$

$$\begin{aligned} RMSE_N &= \sqrt{\tfrac{1}{n}\sum_i^n \left(N_i - \hat{N}_i\right)^2} = \sqrt{\tfrac{1}{n}\sum_i^n \Delta N_i{}^2} \\ RMSE_E &= \sqrt{\tfrac{1}{n}\sum_i^n \left(E_i - \hat{E}_i\right)^2} = \sqrt{\tfrac{1}{n}\sum_i^n \Delta E_i{}^2} \\ RMSE_D &= \sqrt{\tfrac{1}{n}\sum_i^n \left(D_i - \hat{D}_i\right)^2} = \sqrt{\tfrac{1}{n}\sum_i^n \Delta D_i{}^2} \end{aligned} \quad (A7)$$

$$RMSE_{NED} = \sqrt{RMSE_N{}^2 + RMSE_E{}^2 + RMSE_D{}^2} = \sqrt{\frac{1}{n}\sum_i^n \left(\Delta N_i{}^2 + \Delta E_i{}^2 + \Delta D_i{}^2\right)} \quad (A8)$$

where N_i, E_i, and D_i are the ground truth positioning output from Unreal Engine in NED coordinates, while $\hat{N}_i$, $\hat{E}_i$, and $\hat{D}_i$ are the estimated NED position.

References

1. Janke, C.; de Haag, M.U. Implementation of European Drone Regulations—Status Quo and Assessment. *J. Intell. Robot. Syst. Theory Appl.* **2022**, *106*, 33. [CrossRef]
2. Cunliffe, A.M.; Anderson, K.; DeBell, L.; Duffy, J.P. A UK Civil Aviation Authority (CAA)-approved operations manual for safe deployment of lightweight drones in research. *Int. J. Remote Sens.* **2017**, *38*, 2737–2744. [CrossRef]
3. Lee, D.; Hess, D.J.; Heldeweg, M.A. Safety and privacy regulations for unmanned aerial vehicles: A multiple comparative analysis. *Technol. Soc.* **2022**, *71*, 102079. [CrossRef]
4. Yang, B.; Hawthorne, T.L.; Torres, H.; Feinman, M. Using object-oriented classification for coastal management in the east central coast of Florida: A quantitative comparison between UAV, satellite, and aerial data. *Drones* **2019**, *3*, 60. [CrossRef]
5. Guan, S.; Zhu, Z.; Wang, G. A Review on UAV-Based Remote Sensing Technologies for Construction and Civil Applications. *Drones* **2022**, *6*, 117. [CrossRef]
6. Lewicka, O.; Specht, M.; Specht, C. Assessment of the Steering Precision of a UAV along the Flight Profiles Using a GNSS RTK Receiver. *Remote Sens.* **2022**, *14*, 6127. [CrossRef]
7. Hemerly, M. MEMS IMU Stochastic Error Modelling. *Syst. Sci. Control Eng.* **2016**, *5*, 1–8. [CrossRef]
8. Balamurugan, G.; Valarmathi, J.; Naidu, V.P.S. Survey on UAV navigation in GPS denied environments. In Proceedings of the International Conference on Signal Processing, Communication, Power and Embedded System, SCOPES 2016—Proceedings, Paralakhemundi, India, 3–5 October 2016; Institute of Electrical and Electronics Engineers Inc.: New York, NY, USA, 2016; pp. 198–204. [CrossRef]
9. Scaramuzza, D.; Fraundorfer, F. Tutorial: Visual odometry. *IEEE Robot. Autom. Mag.* **2011**, *18*, 80–92. [CrossRef]
10. Aqel, M.O.A.; Marhaban, M.H.; Saripan, M.I.; Ismail, N.B. Review of visual odometry: Types, approaches, challenges, and applications. *SpringerPlus* **2016**, *5*, 1897. [CrossRef]
11. Das, S. Simultaneous Localization and Mapping (SLAM) using RTAB-MAP. *arXiv* **2018**, arXiv:1809.02989.

12. Servières, M.; Renaudin, V.; Dupuis, A.; Antigny, N. Visual and Visual-Inertial SLAM: State of the Art, Classification, and Experimental Benchmarking. *J. Sens.* **2021**, *2021*, 2054828. [CrossRef]
13. Hening, S.; Ippolito, C.; Krishnakumar, K.; Stepanyan, V.; Teodorescu, M. 3D LiDAR SLAM integration with GPS/INS for UAVs in urban GPS-degraded environments. In Proceedings of the AIAA Information Systems-AIAA Infotech at Aerospace, Grapevine, TX, USA, 9–13 January 2017. [CrossRef]
14. Couturier, A.; Akhloufi, M.A. A review on absolute visual localization for UAV. *Robot. Auton. Syst.* **2021**, *135*, 103666. [CrossRef]
15. Saranya, K.C.; Naidu, V.P.S.; Singhal, V.; Tanuja, B.M. Application of Vision Based techniques for UAV Position Estimation. In Proceedings of the 2016 International Conference on Research Advances in Integrated Navigation Systems (RAINS), Bangalore, India, 6–7 May 2016; pp. 1–5. [CrossRef]
16. Lawrence, P.J.; Berarducci, M.P. Comparison of Federated and Centralized Kalman Filters with Fault Detection Considerations. In Proceedings of the 1994 IEEE Position, Location and Navigation Symposium—PLANS'94, Las Vegas, NV, USA, 11–15 April 1994; pp. 703–710. [CrossRef]
17. Carlson, N.A. Federated Square Root Filter for Decentralized Parallel Processes. *IEEE Trans. Aerosp. Electron. Syst.* **1990**, *26*, 517–525. [CrossRef]
18. Quinchia, A.G.; Falco, G.; Falletti, E.; Dovis, F.; Ferrer, C. A comparison between different error modeling of MEMS applied to GPS/INS integrated systems. *Sensors* **2013**, *13*, 9549–9588. [CrossRef] [PubMed]
19. Zhuang, Y.; Sun, X.; Li, Y.; Huai, J.; Hua, L.; Yang, X.; Cao, X.; Zhang, P.; Cao, Y.; Qi, L.; et al. Multi-sensor integrated navigation/positioning systems using data fusion: From analytics-based to learning-based approaches. *Inf. Fusion* **2023**, *95*, 62–90. [CrossRef]
20. Dai, H.F.; Bian, H.W.; Wang, R.Y.; Ma, H. An INS/GNSS integrated navigation in GNSS denied environment using recurrent neural network. *Def. Technol.* **2020**, *16*, 334–340. [CrossRef]
21. Guang, X.; Gao, Y.; Liu, P.; Li, G. IMU data and GPS position information direct fusion based on LSTM. *Sensors* **2021**, *21*, 2500. [CrossRef] [PubMed]
22. Zhang, G.; Xu, P.; Xu, H.; Hsu, L.T. Prediction on the Urban GNSS Measurement Uncertainty Based on Deep Learning Networks with Long Short-Term Memory. *IEEE Sens. J.* **2021**, *21*, 20563–20577. [CrossRef]
23. Negru, S.A.; Geragersian, P.; Petrunin, I.; Zolotas, A.; Grech, R. GNSS/INS/VO fusion using Gated Recurrent Unit in GNSS denied environments. In Proceedings of the AIAA Science and Technology Forum and Exposition, AIAA SciTech Forum 2023, National Harbor, MD, USA, 23–27 January 2023. [CrossRef]
24. Geragersian, P.; Petrunin, I.; Guo, W.; Grech, R. An INS/GNSS fusion architecture in GNSS denied environments using gated recurrent units. In Proceedings of the AIAA Science and Technology Forum and Exposition, AIAA SciTech Forum 2022, San Diego, CA, USA, 3–7 January 2022. [CrossRef]
25. Al Bitar, N.; Gavrilov, A.; Khalaf, W. Artificial Intelligence Based Methods for Accuracy Improvement of Integrated Navigation Systems During GNSS Signal Outages: An Analytical Overview. *Gyroscopy Navig.* **2020**, *11*, 41–58. [CrossRef]
26. Jwo, D.J.; Biswal, A.; Mir, I.A. Artificial Neural Networks for Navigation Systems: A Review of Recent Research. *Appl. Sci.* **2023**, *13*, 4475. [CrossRef]
27. Jwo, D.J.; Chen, J.J. GPS/INS Navigation Filter Designs Using Neural Network with Optimization Techniques. In *Advances in Natural Computation, Proceedings of the Second International Conference, ICNC 2006, Xi'an, China, 24–28 September 2006*; Springer: Berlin/Heidelberg, Germany, 2006; Volume 4221, pp. 461–471. [CrossRef]
28. Jwo, D.-J.; Chang, C.-S.; Lin, C.-H. Neural Network Aided Adaptive Kalman Filtering for GPS Applications. In Proceedings of the 2004 IEEE International Conference on Systems, Man and Cybernetics, The Hague, The Netherlands, 10–13 October 2004; pp. 3686–3691. [CrossRef]
29. Zhang, Q.; Li, B. A Low-cost GPS/INS Integration Based on UKF and BP Neural Network. In Proceedings of the Fifth International Conference on Intelligent Control and Information Processing, Dalian, China, 18–20 August 2014; pp. 100–107. [CrossRef]
30. Tang, Y.; Jiang, J.; Liu, J.; Yan, P.; Tao, Y.; Liu, J. A GRU and AKF-Based Hybrid Algorithm for Improving INS/GNSS Navigation Accuracy during GNSS Outage. *Remote Sens.* **2022**, *14*, 752. [CrossRef]
31. Zhang, C.; Zhao, X.; Pang, C.; Li, T.; Zhang, L. Adaptive fault isolation and system reconfiguration method for GNSS/INS integration. *IEEE Access* **2020**, *8*, 17121–17133. [CrossRef]
32. Wang, Y.; Huynh, G.; Williamson, C. Integration of Google Maps/Earth with microscale meteorology models and data visualization. *Comput. Geosci.* **2013**, *61*, 23–31. [CrossRef]
33. Cai, G.; Chen, B.M.; Lee, T.H. *Unmanned Rotorcraft Systems*; Springer Science & Business Media: London, UK, 2011.
34. Alkendi, Y.; Seneviratne, L.; Zweiri, Y. State of the Art in Vision-Based Localization Techniques for Autonomous Navigation Systems. *IEEE Access* **2021**, *9*, 76847–76874. [CrossRef]
35. Sánchez, J.; Monzón, N.; Salgado, A. An analysis and implementation of the harris corner detector. *Image Process. Line* **2018**, *8*, 305–328. [CrossRef]
36. Harris, C.; Stephens, M. A Combined Corner and Edge Detector. In Proceedings of the Alvey Vision Conference, Manchester, UK, 31 August–2 September 1998; pp. 23.1–23.6. [CrossRef]
37. Features from Accelerated Segment Test (FAST) Deepak Geetha Viswanathan. Available online: https://api.semanticscholar.org/CorpusID:17031649 (accessed on 3 December 2023).

38. Bay, H.; Ess, A.; Tuytelaars, T.; van Gool, L. Speeded-Up Robust Features (SURF). *Comput. Vis. Image Underst.* **2008**, *110*, 346–359. [CrossRef]
39. Lowe, D.G. Distinctive Image Features from Scale-Invariant Keypoints. *Int. J. Comput. Vis.* **2004**, *60*, 91–110. [CrossRef]
40. Rublee, E.; Rabaud, V.; Konolige, K.; Bradski, G. ORB: An efficient alternative to SIFT or SURF. In Proceedings of the 2011 International Conference on Computer Vision, Barcelona, Spain, 6–13 November 2011; pp. 564–2571. [CrossRef]
41. Tareen, S.A.K.; Saleem, Z. A comparative analysis of SIFT, SURF, KAZE, AKAZE, ORB, and BRISK. In Proceedings of the 2018 International Conference on Computing, Mathematics and Engineering Technologies: Invent, Innovate and Integrate for Socioeconomic Development, iCoMET 2018—Proceedings, Sukkur, Pakistan, 3–4 March 2018; pp. 1–10. [CrossRef]
42. Rosin, P.L. Measuring Corner Properties. *Comput. Vis. Image Underst.* **1999**, *73*, 291–307. [CrossRef]
43. Muja, M.; Lowe, D. FLANN-Fast Library for Approximate Nearest Neighbors User Manual. Available online: https://www.fit.vutbr.cz/~ibarina/pub/VGE/reading/flann_manual-1.6.pdf (accessed on 3 December 2023).
44. Hartley, R.; Zisserman, A. *Multiple View Geometry in Computer Vision*; Cambridge University Press: Cambridge, UK, 2004. [CrossRef]
45. Foley, J.D.; Fischler, M.A.; Bolles, R.C. Graphics and Image Processing Random Sample Consensus: A Paradigm for Model Fitting with Apphcatlons to Image Analysis and Automated Cartography. *Commun. ACM* **1981**, *24*, 381395. [CrossRef]
46. Helgesen, H.H.; Leira, F.S.; Bryne, T.H.; Albrektsen, S.M.; Johansen, A. Real-time Georeferencing of Thermal Images using Small Fixed-Wing UAVs in Maritime Environments. *ISPRS J. Photogramm. Remote Sens.* **2019**, *154*, 84–97. [CrossRef]
47. Xiang, H.; Tian, L. Method for automatic georeferencing aerial remote sensing (RS) images from an unmanned aerial vehicle (UAV) platform. *Biosyst. Eng.* **2011**, *108*, 104–113. [CrossRef]
48. George, A.; Koivumäki, N.; Hakala, T.; Suomalainen, J.; Honkavaara, E. Visual-Inertial Odometry Using High Flying Altitude Drone Datasets. *Drones* **2023**, *7*, 36. [CrossRef]
49. Qin, T.; Li, P.; Shen, S. VINS-Mono: A Robust and Versatile Monocular Visual-Inertial State Estimator. *IEEE Trans. Robot.* **2017**, *34*, 1004–1020. [CrossRef]
50. Dai, J.; Hao, X.; Liu, S.; Ren, Z. Research on UAV Robust Adaptive Positioning Algorithm Based on IMU/GNSS/VO in Complex Scenes. *Sensors* **2022**, *22*, 2832. [CrossRef]
51. Cai, G.; Chen, B.M.; Lee, T.H. Coordinate systems and transformations. In *Unmanned Rotorcraft Systems*; Advances in Industrial Control; Springer International Publishing: London, UK, 2011; pp. 23–34, ISBN 9780857296344. [CrossRef]

sensors

MDPI

Article

Real-Time Object Detection from UAV Inspection Videos by Combining YOLOv5s and DeepStream

Shidun Xie [1], Guanghong Deng [1,*], Baihao Lin [1], Wenlong Jing [2,*], Yong Li [2,*] and Xiaodan Zhao [1]

[1] Guangdong Engineering Technology Research Center of UAV Remote Sensing Network, Guangzhou iMapCloud Intelligent Technology Co., Ltd., Guangzhou 510095, China; xieshidun@geoai.com (S.X.); linbaihao@geoai.com (B.L.); zhaoxiaodan@geoai.com (X.Z.)

[2] Guangdong Province Engineering Laboratory for Geographic Spatiotemporal Big Data, Key Laboratory of Guangdong for Utilization of Remote Sensing and Geographical Information System, Guangdong Open Laboratory of Geospatial Information Technology and Application, Guangzhou Institute of Geography, Guangdong Academy of Sciences, Guangzhou 510070, China

* Correspondence: dengguanghong@geoai.com (G.D.); jingwl@lreis.ac.cn (W.J.); liyong@gdas.ac.cn (Y.L.); Tel.: +86-18848956622 (G.D.); +86-18612649525 (W.J.); +86-13610257606 (Y.L.)

Abstract: The high-altitude real-time inspection of unmanned aerial vehicles (UAVs) has always been a very challenging task. Because high-altitude inspections are susceptible to interference from different weather conditions, interference from communication signals and a larger field of view result in a smaller object area to be identified. We adopted a method that combines a UAV system scheduling platform with artificial intelligence object detection to implement the UAV automatic inspection technology. We trained the YOLOv5s model on five different categories of vehicle data sets, in which mAP50 and mAP50-95 reached 93.2% and 71.7%, respectively. The YOLOv5s model size is only 13.76 MB, and the detection speed of a single inspection photo reaches 11.26 ms. It is a relatively lightweight model and is suitable for deployment on edge devices for real-time detection. In the original DeepStream framework, we set up the http communication protocol to start quickly to enable different users to call and use it at the same time. In addition, asynchronous sending of alarm frame interception function was added and the auxiliary services were set up to quickly resume video streaming after interruption. We deployed the trained YOLOv5s model on the improved DeepStream framework to implement automatic UAV inspection.

Keywords: UAVs; YOLOv5; object detection; DeepStream; route planning

1. Introduction

Real-time object detection is not only a challenging task in computer vision, but also a hot topic in industrial applications, for example, in object tracking [1], autonomous driving [2], medical image processing [3], agricultural machine vision applications [4], etc. Real-time detection requires lightweight convolutional neural networks and equipment that can process floating point operations faster. Equipment that implements real-time detection usually includes mobile GPU or CPU servers and various neural processing units (NPUs). Different edge processing machines focus on the acceleration of different modules. In this paper, we propose a real-time detection system mainly applied to cloud or mobile devices.

In recent years, different real-time detection models have been proposed, suitable for different edge devices. MCUNet [5] and NanoDet [6] are mainly designed to produce lower-power microcontrollers and improve edge CPU inference speed. The You Only Look Once (YOLO) series of algorithms have better accuracy and faster reasoning speed, and are widely used in the industrial field. YOLOv1 [7] is a typical one-stage object detector, and based on this, a series of improvements were made, resulting in YOLOv2 [8], and YOLOv3 [9], which have faster detection speeds and higher detection accuracy. YOLOv4 [10] redesigns the three independent architectures of the trunk, neck and head to make them better trained on a single GPU. At present, YOLOv5 [11], YOLOX [12], PPYOLOE [13], etc., have extremely competitive performance in real-time detection and deployment. More recently, real-time object detectors have mainly focused on simple and efficient structural design, so that real-time detection effects can be achieved when used on the CPU [14–16]. The simple framework design with good performance is mainly based on MobileNet [17–19], ShuffleNet [20,21], or GhostNet [22]. However, real-time detectors developed based on GPU [23,24] mostly use ResNet [25] or DLA [26], combined with the CSPNet [27] strategy to optimize the architecture and improve detector performance.

With the continuous development of lightweight real-time detection models, some types of industrial application equipment accompanied by algorithms are also constantly being introduced. The development and application of UAVs have made many fields more intelligent, such as industrial inspection [28], intelligent substation inspection [29], agricultural applications [30], etc. Running convolutional neural networks on embedded systems has become a reality, and when combined with the application innovation of UAVs, there will be new ways of application in various fields. Tijtgat et al. [31] designed a system based on NVIDIA Jetson TX2 edge computing device running YOLOv2 to achieve real-time object detection with UAVs. Abdulghafoor et al. [32] proposed a method of combining edge computing devices with a DeepStream software development kit (DS-SDK) 4.0.2 [33] to implement a convolutional network model that can process video streams with high performance. In order to improve the practicality of a real-time video stream detection system, Guo et al. [34] proposed a novel region of interest detection (ROIDet) algorithm and designed a bandwidth-efficient multi-camera video streaming system for deep learning video analysis. Hossain et al. [35] proposed the joint implementation of the application of deep learning technology with a computer system integrated with the UAV, which can track and detect objects in real time. Vandersteen et al. [36] proposed a multi-data set learning strategy to optimize the real-time performance of detection on embedded hardware devices and improve detection efficiency. Haq et al. [37] deployed the DeepStream framework on the NVIDIA Jetson single-board computer to run deep learning algorithms, especially the YOLO algorithm. The study also verified that the DeepStream framework can run well in virtual machines, especially using Docker, which can further improve the performance of the model and the portability during the deployment process. Huu et al. [38] proposed a method based on the NVIDIA DS-SDK architecture, using multiple surveillance camera detection methods to implement the application of deep learning-based algorithms for vehicle monitoring. Ghaziamin et al. [39] deployed the object detection model to Nvidia Jetson devices and designed a passenger counting system. And after edge deployment through Nvidia DeepStream, it improved efficiency while saving the use of hardware resources. Smink et al. [40] used edge devices combined with the detection and tracking system of the NVIDIA DeepStream framework to implement a set of real-time tag-reading applications. Qaraqe et al. [41] designed an end-to-end security intelligent monitoring system that used the DeepStream software development kit (SDK) for real-time inference, which can have a significant impact on public safety and crowd management.

In summary, their method cannot simultaneously ensure detection accuracy and must also maintain detection speed. Especially in the UAV inspection process, the requirements for hardware robustness are relatively high. Existing real-time inspection technology, especially the use of fixed camera equipment for monitoring, will result in a small field of view, poor mobility of the equipment, the inability to realize real-time switching between different areas, and the existence of large monitoring blind spots and other problems. UAV high-altitude inspection can only solve the problems of vision and flexibility. Therefore, this article proposes a method that combines UAVs and artificial intelligence real-time detection to implement an automatic UAV real-time inspection system. The main contributions of this article are as follows:

1. The integration of UAV inspection, YOLOv5s object detection, and DeepStream framework realizes a new real-time object detection method.

2. The DeepStream service can be quickly started using the http communication protocol, and can be called and used by different users at the same time.
3. The asynchronous sending of the alarm frame interception function is implemented, making the real-time video stream smoother.
4. Auxiliary services can be set up to quickly resume video streaming after interruption.

2. Methods

2.1. DeepStream Software Development Kit

DS-SDK [42] is an intelligent video analysis suite assembled based on NVIDIA technology. It introduces deep neural networks and other complex processing tasks into stream processing pipelines to achieve near real-time analysis of video and other sensor data. The application framework has hardware-accelerated processing building blocks, so developers do not need to design an end-to-end solution from scratch and need only to focus on building the core deep learning network and video stream processing modules. Moreover, it builds an end-to-end video stream detection pipeline based on gstream, which simplifies the development and application of video streams. Algorithm personnel can quickly convert other video stream processing capabilities while focusing on the development of all components. The original DeepStream flow chart is shown in Figure 1.

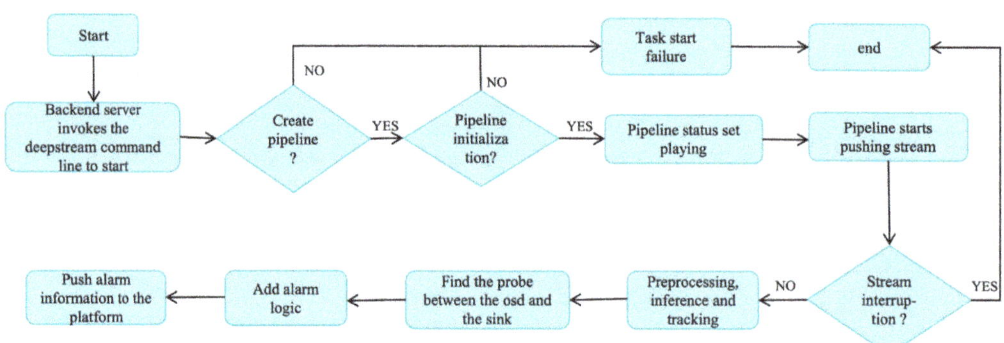

Figure 1. DeepStream architecture diagram. First, the back-end server calls the DeepStream command line to start, and then after a series of flow operations, it finally pushes alarm information to the platform.

In Figure 1, the basic workflow of DeepStream's original architecture is shown. First, we need to start a DeepStream video streaming service on the backend of the platform, immediately start creating the DeepStream pipeline, and determine the pipeline initialization status. If the pipeline initialization verification is successful, then the pipeline status is set to the playing state, which is the pipeline starting working status. Secondly, the pipeline starts to push the stream, and the algorithm performs video stream data pre-processing including normalization and scaling. Then, the video stream will be input into the YOLOv5 model for inference, and finally the DeepSORT [43,44] tracking algorithm is used to track the corresponding object. Add alarm logic to the object that needs to be identified, push the processed video stream to the platform service, and finally display the identification results in real time on the platform software 2.5.9.

The original DeepStream framework is connected to a UAV inspection video, and users cannot call different services and different recognition algorithms at any time during use. The communication protocol does not support UAV video streaming communication services. Therefore, we improved DeepStream, as shown in Figure 2.

As shown in Figure 2, this shows the improved workflow of the DeepStream architecture. First, the platform starts the DeepStream service command by sending an http request. Currently, once the algorithm service receives the platform request to start the command, it immediately starts the DeepStream service, starts parsing the request command parameters, determines whether the parameters meet the requirements, and generates a configuration file according to the corresponding task. Secondly, after the configuration is successful, start creating a sub-process to perform the task and create a DeepStream pipeline. Determine whether the pipeline is initialized. After the initialization is successful, set the pipeline status to the playing state. The pipeline starts pushing and the algorithm starts working, including pre-processing, reasoning, tracking, and other operations. After the object is detected, alarm logic is added. Finally, the video stream is pushed to the platform service, and the platform interface displays the stream inferred by the algorithm. In addition, we also added an auxiliary service, whose purpose is to monitor whether the DeepStream service is disconnected. If a disconnection occurs, the auxiliary service will immediately kill the task and restart the task immediately.

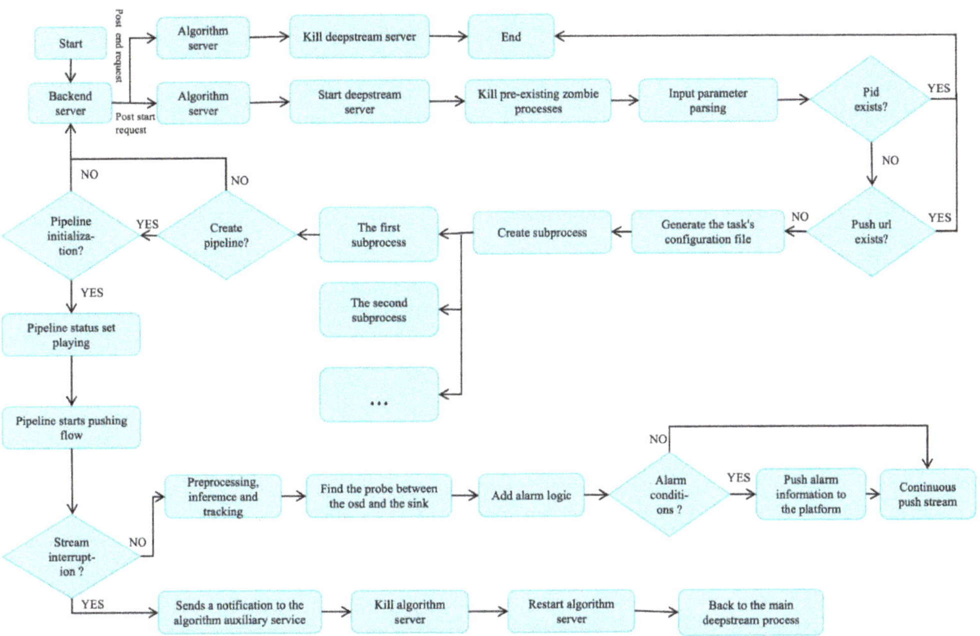

Figure 2. Improved DeepStream architecture diagram. First, start the backend service, then go through more judgment initialization and other operations, and finally go through the backend service feedback.

2.2. The YOLO Algorithm

The YOLOv5 algorithm is an outstanding object detection algorithm and has good robustness to UAV inspection images. YOLOv5 contains five network models of different sizes, namely YOLOv5n, YOLOv5s, YOLOv5m, YOLOv5l and YOLOv5x. This article verifies the performance of different YOLOv5 models and selects the most suitable detector for deploying the DeepStream architecture. The algorithm structure flow chart is shown in Figure 3.

As shown in Figure 3, there are two extremely important elements, data processing and data training. The first is the data processing stage. In the process of collecting data, we need to plan the UAV inspection route. In the areas that need to be inspected, let the UAVs automatically inspect and collect photos or videos according to the planned route. During the process of collecting images with a UAV, we flew in different time periods, different weather, and different lighting conditions to achieve a better performance of the model. To make the model more robust to the complex background of high-altitude inspections, we planned the flight area to include different areas such as villages, water bodies, and towns. The collected UAV inspection videos need to be manually edited and the required object information retained and extracted into images. Images were manually filtered and labeled using LabelImg software 1.8.6. Finally, the data were divided into a training set and a test set.

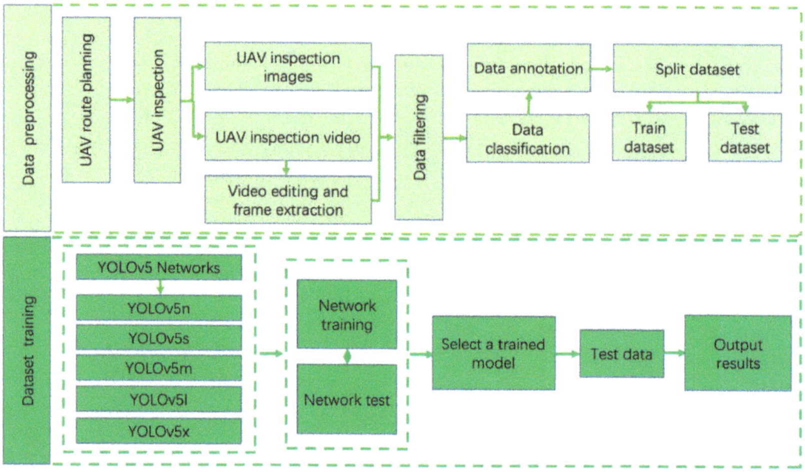

Figure 3. Algorithm training architecture diagram. It consists of two stages: data preprocessing and data training.

The second stage is that of model training. In the first stage of data processing, the processed data were divided into a training set and test set. We were in the data collection phase. Effectively planning routes for different scenarios and collecting data under different conditions fully ensures the richness of model training samples and can effectively prevent over-fitting during the training process. We evaluated the performance of YOLOv5n, YOLOv5s, YOLOv5m, YOLOv5l, and YOLOv5x models in the process of UAV high-altitude inspection. The performance was evaluated on the training set and test set divided from the same data set. Finally, we selected a model with strong robust performance for real-time detection based on its real-time detection effect.

Since UAVs fly at a high altitude, they will keep shaking, the field of view when flying at high altitude is large, and the shooting area is also large, resulting in a smaller target to be detected. There are difficulties in using object detection algorithms to detect smaller objects. This article mainly used YOLOv5s, a lightweight object detection model, combined with DeepStream architecture to achieve real-time video stream detection. To better implement a complete set of real-time target detection system architecture, we made a series of improvements to the original DreamStream and combined with the YOLOv5s model to achieve target detection. YOLOv5s is designed to detect the goals that require inspection. It is designed to achieve automatic detection capabilities. The main improvement part is the design of the overall logical architecture of DeepStream, and the improved experimental results are also very effective.

2.3. Evaluation Indicators

Due to the high-altitude inspection of UAVs, as the height of the UAV flight increases, the inspection field of view becomes wider. It is easily affected by the complex low-altitude environment, causing certain false and missed detections. Therefore, we used precision and recall measuring to assess whether a detector had good performance. The precision and recall formulas are shown in (1) and (2):

$$P = \frac{TP}{TP + FP} \tag{1}$$

$$R = \frac{TP}{TP + FN} \tag{2}$$

where P and R represent precision and recall, respectively. TP represents the true positives; FP represents the false positives; and FN represents the false negatives.

The F_1 value combines the two indicators of precision and recall to comprehensively evaluate the effect of the model. As shown in Formula (3).

$$F_1 = \frac{2PR}{P + R} \tag{3}$$

In order to further obtain a better real-time video stream detection model, two evaluation indicators, AP (average precision) and mAP (mean average precision), were used. The Formulas (4) and (5) of AP and mAP are shown as follows:

$$AP = \int_0^1 P(R)dR \tag{4}$$

$$mAP = \frac{\sum_{q=1}^{Q} AP(q)}{Q} \tag{5}$$

where Q represents the number of categories.

3. Experiment and Results

3.1. Experimental Conditions

3.1.1. Data Acquisition and Transmission

This article mainly collected the data we needed through UAV inspection. A UAV can achieve automatic flight collection through the automatic command and dispatch platform. The command and dispatch interface are shown in Figure 4.

Figure 4. UAV command and dispatch platform. This is a UAV ready to take off displayed on the UAV monitoring platform.

It can be seen from Figure 4 that the flight of a UAV realizes automatic inspection through the dispatch of the command and dispatch system. Through the command and dispatch system interface, the status of the UAV and its slot can be monitored in real time. We can see a white drone nest, the drone nest door has been opened, the platform has been raised, and a black UAV is waiting to fly on it. There are three buttons available on the left side of the command and dispatch system. The main functions include UAV control, landing control, system functions, power control, emergency control, etc. It can not only command the takeoff and landing of the drone, but also control the angle and focus of the gimbal. When the drone is flying, the battery power can be monitored in real time. When an emergency occurs, such as low battery or automatic hovering, it can also return to home with one click.

The data used in this article mainly use images or videos collected during UAV inspections to annotate them. The annotated original image is shown in Figure 5. This article used LabelImg to label five different categories of vehicle models. The labels are sjc, hc, wjj, dzj, and dc. Where sjc represents a car, hc represents a truck, wjj represents an excavator, dzj represents a pile driver, and dc represents a crane. To ensure the effectiveness of model training in this article, we increased the richness of samples as much as possible. Therefore, in the process of collecting images, we planned routes in different areas and different route altitudes. While ensuring flight safety, the drone's flight speed and different flight mileage were set.

Figure 5. Original image of the data set. Images of different scenes collected by UAV from a high-altitude perspective.

Figure 6 shows the data set used for training in this article, with a total of 3208 images. There were 2887 images in the training set and 321 images in the test set. We divided the training set and the test set at a ratio of 9:1. To ensure the effectiveness of the training model, the divided data sets were independent of each other.

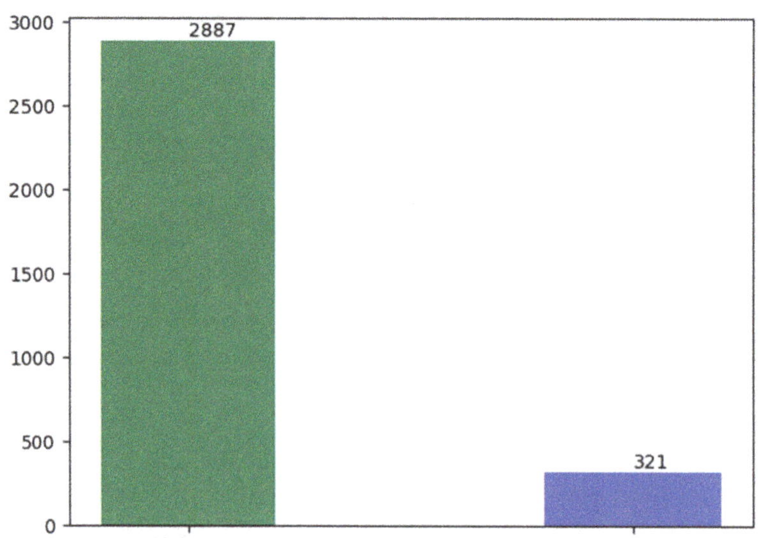

Figure 6. Dataset statistics. In the process of training, the data set, the data set is divided into a training set and a test set.

3.1.2. Inspection Routes

To realize the automatic take-off and landing flight of a UAV, you first need to plan an inspection route. We planned four different UAV inspection routes as shown in Figure 7, and the parameters of UAV inspections on different routes are shown in Table 1.

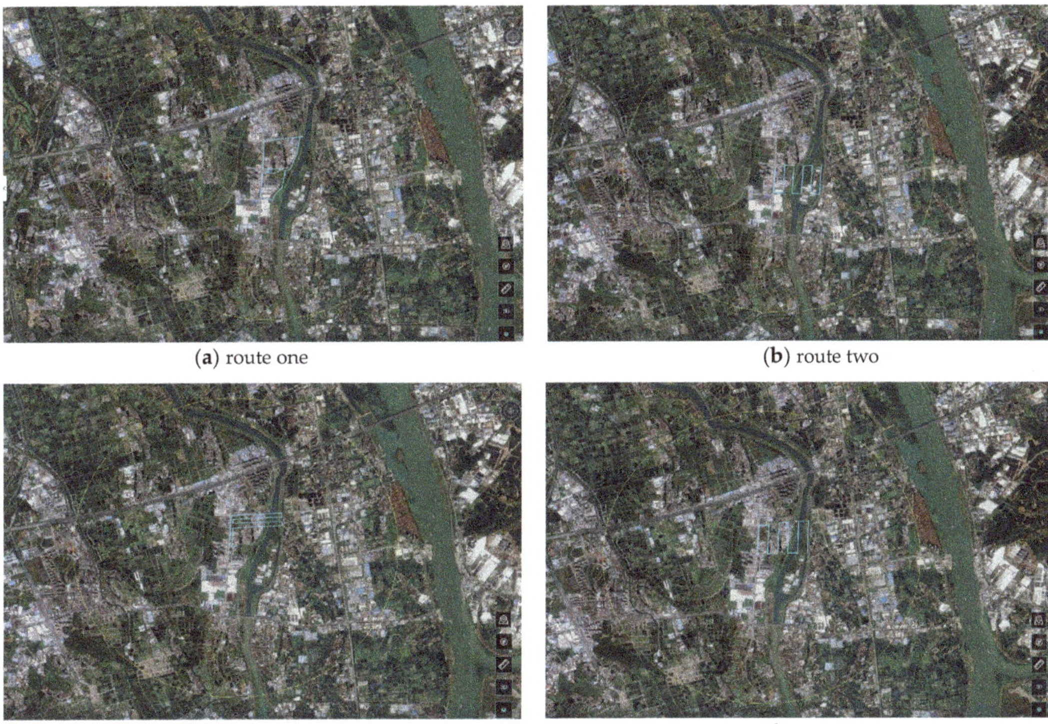

(**a**) route one (**b**) route two (**c**) route three (**d**) route four

Figure 7. Different route planning. The flight routes of UAV include different areas such as towns and water bodies. The blue line is the planned UAV route.

Table 1. Flight parameters of different routes.

Routes	Flight Length/m	Flight Minutes	Route Altitude/m	Taking off and Landing Height/m	Round Trip Speed/m/s	Route Speed /m/s
One	4150.8	11.07	127	129	10	8
Two	4314.8	17.95	127	131	10	5
Three	4912.8	18.93	150	155	10	5
Four	4728.6	19.42	127	131	10	5

It can be concluded from Figure 7 that in order to realize automatic UAV inspection, we planned four inspection routes on the UAV command and dispatch platform. Among them, the blue linear area is the area where the UAV needs to fly according to the route.

The routes are all included in a yellow circular area, which is displayed on a map interface. We can clearly see that there are residential areas and river areas next to the planned routes. To increase the richness of the sample, our routes span multiple regions to collect images. Table 1 shows the flight parameters of four different inspection routes, such as flight length, flight duration, route altitude, round-trip speed, route speed, take-off and landing altitude, etc.

3.1.3. Experiment Platform

Server-side: Ubuntu 18.04, Intel® Silver 4210 CPU@2.20 GHz, NVIDIA GeForce RTX A100(80 GB) GPU. The model framework is Pytorch 1.10.0, and the related software is CUDA11.1, CUDNN 8.0.5 and Python 3.8.

Ubuntu 20.04, Intel® Xeon® Gold 6278C CPU@2.60 GHZ, NVIDIA Tesla T4(16G) GPU. The related software is CUDA 11.6, CUDNN 8.4.0.

3.2. *Experimental Results*

YOLOv5 Detection Results

After the model training was completed, to verify the detection effects of different models, we used images containing five different categories to test the detection performance of five different models. The detection results of different models are displayed in Figure 8.

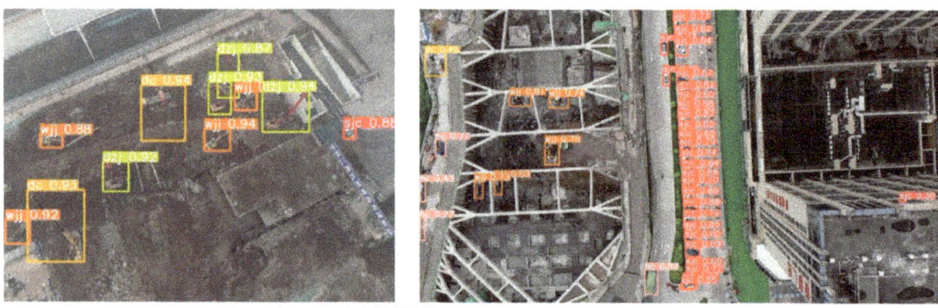

(**a**) YOLOv5n

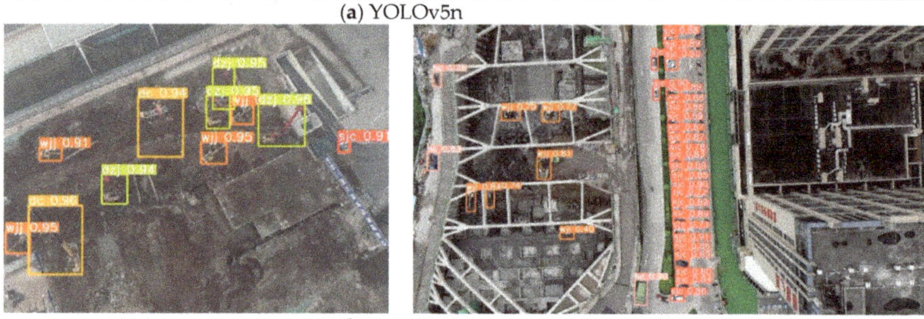

(**b**) YOLOv5s

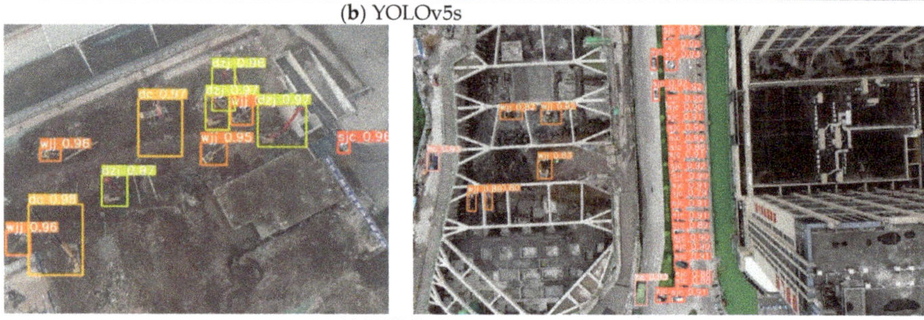

(**c**) YOLOv5m

Figure 8. *Cont.*

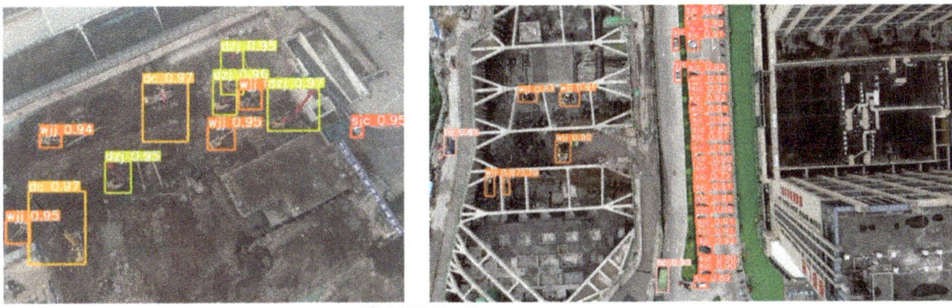

(**d**) YOLOv5l

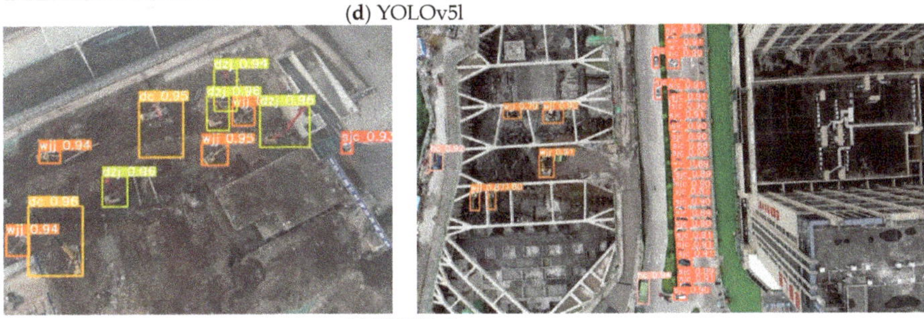

(**e**) YOLOv5x

Figure 8. YOLOv5 detection results. The prediction of (**a**) YOLOv5n; (**b**) YOLOv5s; (**c**) YOLOv5m; (**d**) YOLOv5l; and (**e**) YOLOv5x. It shows the detection results of the YOLOv5 model on this data set.

From Figure 8, we can obtain the detection results of five different YOLOv5 models. By comparing the results, it can be concluded that YOLOv5n can, to a certain extent, identify five different categories of vehicles. However, the roof of the building was misidentified. YOLOv5s, YOLOv5m, YOLOv5l, and YOLOv5x can all produce better detection results. We can clearly see that different YOLOv5 detection models can detect five different types of objects, namely, car, truck, excavator, pile driver, and crane. This experiment shows the effectiveness of the training method in this paper and its ability to accurately detect different types of vehicles.

3.3. Comparative Requirements

In order to further verify the robust performance of this model, we compared two different models, mAP50 and mAP50-95, as shown in Tables 2 and 3, respectively.

Table 2. Comparison of mAP50 detection results of different models.

	sjc/%	hc/%	wjj/%	dc/%	dzj/%	mAP50/%
Yolov5n	85.8	71.6	96.7	96.8	99.5	90.1
Yolov5s	91.2	80.9	97.7	96.6	99.5	93.2
Yolov5m	91.8	83.4	96.7	97.3	99.5	93.7
Yolov5l	93.1	83.7	97.2	97.6	99.5	94.2
Yolov5x	93.2	86.1	96.9	97.8	99.5	94.7

Table 3. Comparison of mAP50-95 detection results of different models.

	sjc/%	hc/%	wjj/%	dc/%	dzj/%	mAP50–95/%
Yolov5n	51.7	42.0	67.6	75.1	86.0	64.5
Yolov5s	62.2	54.7	74.7	78.5	88.6	71.7
Yolov5m	66.3	61.1	77.4	79.6	89.6	74.8
Yolov5l	68.9	63.1	79.2	81.3	90.3	76.6
Yolov5x	69.7	65.3	79.3	82.4	91.1	77.6

It can be seen from Table 2 that the mAP50 of YOLO5n, YOLOv5s, YOLOv5m, YOLOv5l, and YOLOv5x reached 90.1%, 93.2%, 93.7%, 94.2%, and 94.7%, respectively. Among them, the mAP50 detection accuracy of the YOLOv5s model is 3.1% higher than that of YOLOv5n, but the difference in mAP50 accuracy with YOLOv5m, YOLOv5l, and YOLOv5x is minimal.

It can be seen from Table 3 that the mAP50-95 of YOLO5n, YOLOv5s, YOLOv5m, YOLOv5l, and YOLOv5x reached 64.5%, 71.7%, 74.8%, 76.6%, and 77.6%, respectively. Among them, the detection accuracy of mAP50-95 of the YOLOv5n model is extremely low. This experiment shows that YOLOv5n is less robust than the other four trained models.

To train a better real-time object detector, we must not only compare its detection accuracy but also its detection speed. Therefore, the detection speeds of different models are shown in Table 4.

Table 4. Comparison of the inference speed of different models.

	Yolov5n	Yolov5s	Yolov5m	Yolov5l	Yolov5x
Model Size/MB	3.7	13.76	40.27	88.57	165.13
Speed/ms	11.1	11.26	15.77	17.44	20.29

Table 4 shows the detection speed of YOLOv5n, YOLOv5s, YOLOv5m, YOLOv5l, and YOLOv5x on NVIDIA A100 GPU, which are 11.1 ms, 11.26 ms, 15.77 ms, 17.44 ms, and 20.29 ms, respectively. The model sizes of YOLOv5n, YOLOv5s, YOLOv5m, YOLOv5l, and YOLOv5x are 3.7 MB, 13.76 MB, 40.27 MB, 88.57 MB, and 165.13 MB, respectively. Compared with YOLOv5n, the inference speed of YOLOv5s is 0.16ms slower, but the inference speed is faster than YOLOv5s, YOLOv5m, YOLOv5l, and YOLOv5x. To verify the performance of a model during the training process, you can observe whether the model's loss gradually converges. The training loss and the verification loss of this model are shown in Figure 9.

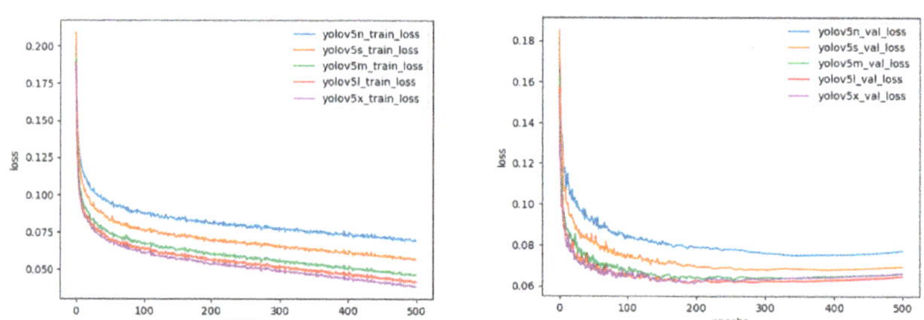

(**a**) training loss (**b**) validation loss

Figure 9. YOLOv5 model loss curve.

It can be concluded from Figure 7 that during the training period of 500 epochs, the training loss and verification loss of YOLOv5n, YOLOv5s, YOLOv5m, YOLOv5l, and YOLOv5x gradually decreased and reached a certain degree of convergence. At the same time, the effectiveness of the training method in this article is verified.

The accuracy of a model in detecting objects is measured during the training process, and the mAP50 and mAP50-95 of the model can be tested during the iteration process. The performance of mAP50 and mAP50-95 of the YOLOv5 model during the training iteration process is shown in Figure 10.

It can be concluded from Figure 10 that the mAP50 detection accuracy of the YOLOv5s model trained in this article is significantly higher than the YOLOv5n model in 500 epochs of training but is not much different from the mAP50 performance of the YOLOv5m,

YOLOv5l, and YOLOv5x models. However, the mAP50-95 performance of the YOLOv5s model is significantly better than the YOLOv5n model, but lower than the YOLOv5m, YOLOv5l, and YOLOv5x models.

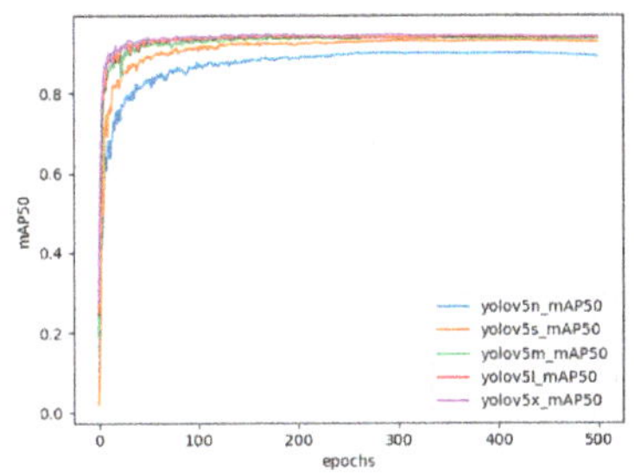

(**a**) mAP50

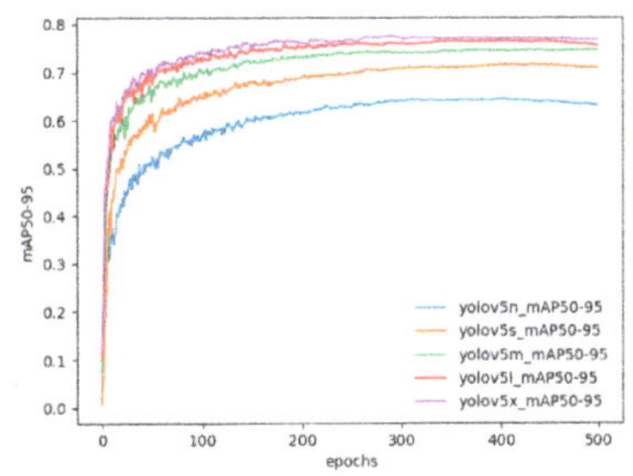

(**b**) mAP50-95

Figure 10. YOLOv5 model accuracy curve.

The confusion matrix is a common visualization tool for measuring the performance of object detectors. You can clearly see the performance of the various indicators in Figure 11. It shows the confusion matrix diagrams of the five different models, YOLOv5n, YOLOv5s, YOLOv5m, YOLOv5l, and YOLOv5x, respectively.

It can be seen from Figure 11 that the data in each grid of the confusion matrix are the result of normalization. We can measure the performance capability of a model only by comparing the size of the diagonal value. The five category values of YOLOv5s are 0.92, 0.80, 0.97, 0.99, and 1.00, respectively. The results of the YOLOv5n model tested in five different categories were 0.87, 0.71, 0.96, 0.98, and 0.99, respectively. Therefore, the YOLOv5s model performance is significantly better than YOLOv5n. At the same time, it can also be concluded that compared with YOLOv5m, YOLO v5l, and YOLOv5x, the performance gap is not obvious.

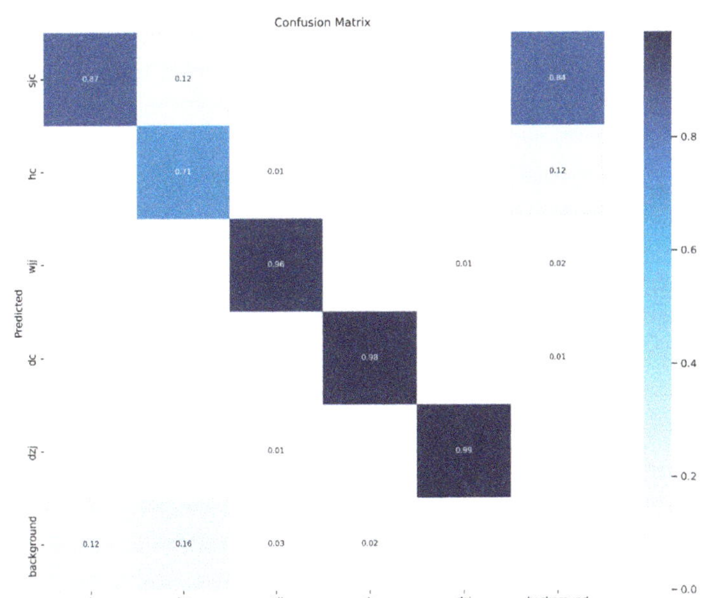

(**a**) YOLOv5n

Figure 11. *Cont.*

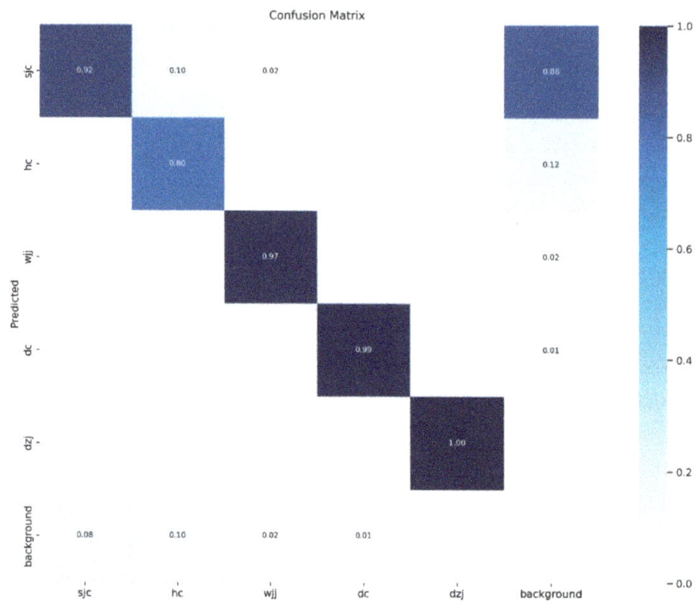

(**b**) YOLOv5s

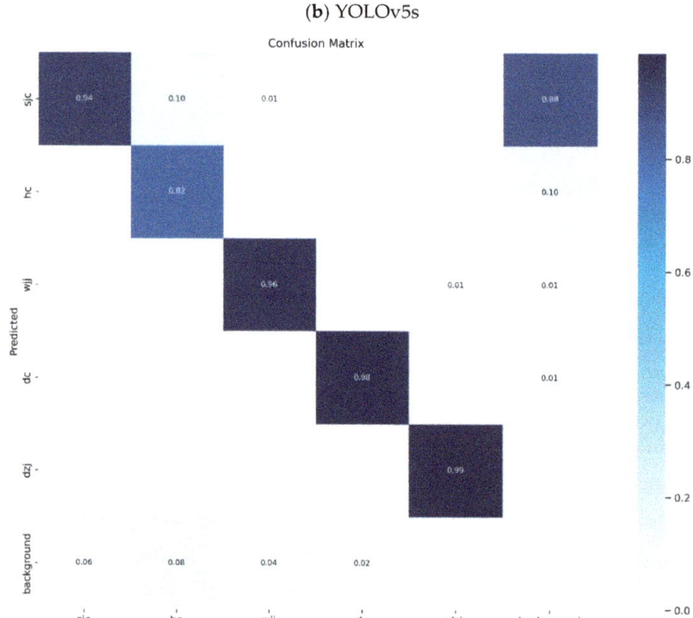

(**c**) YOLOv5m

Figure 11. *Cont.*

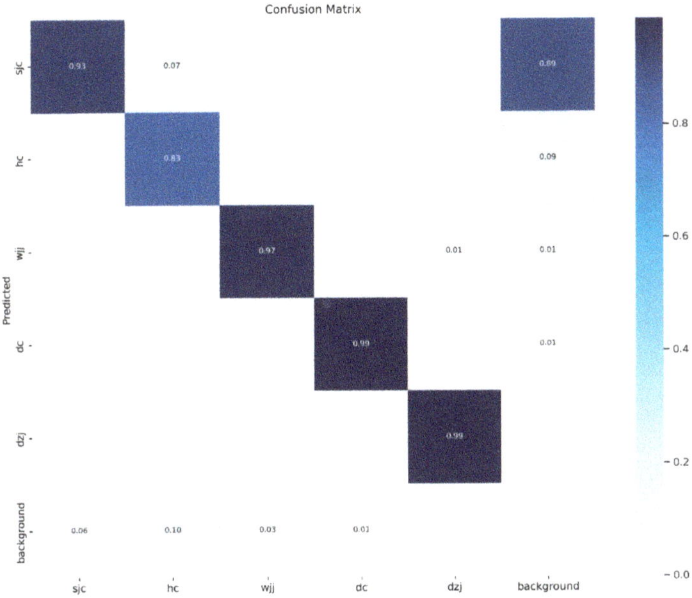

(**d**) YOLOv5l

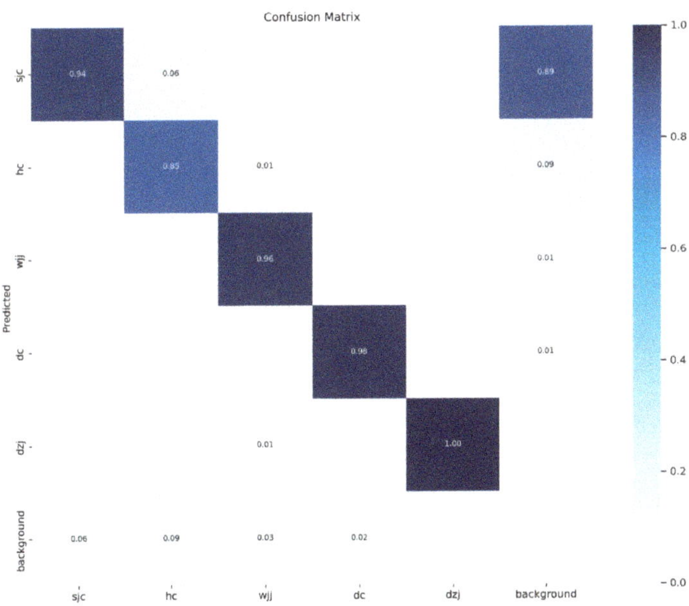

(**e**) YOLOv5x

Figure 11. Confusion matrixes.

The precision and recall curves are a basic indicator of the robustness of a model. During the process of training and testing the model, when the precision is close to zero, the recall is close to 1. When the recall is close to 0, the precision is close to 1, as shown in Figure 12.

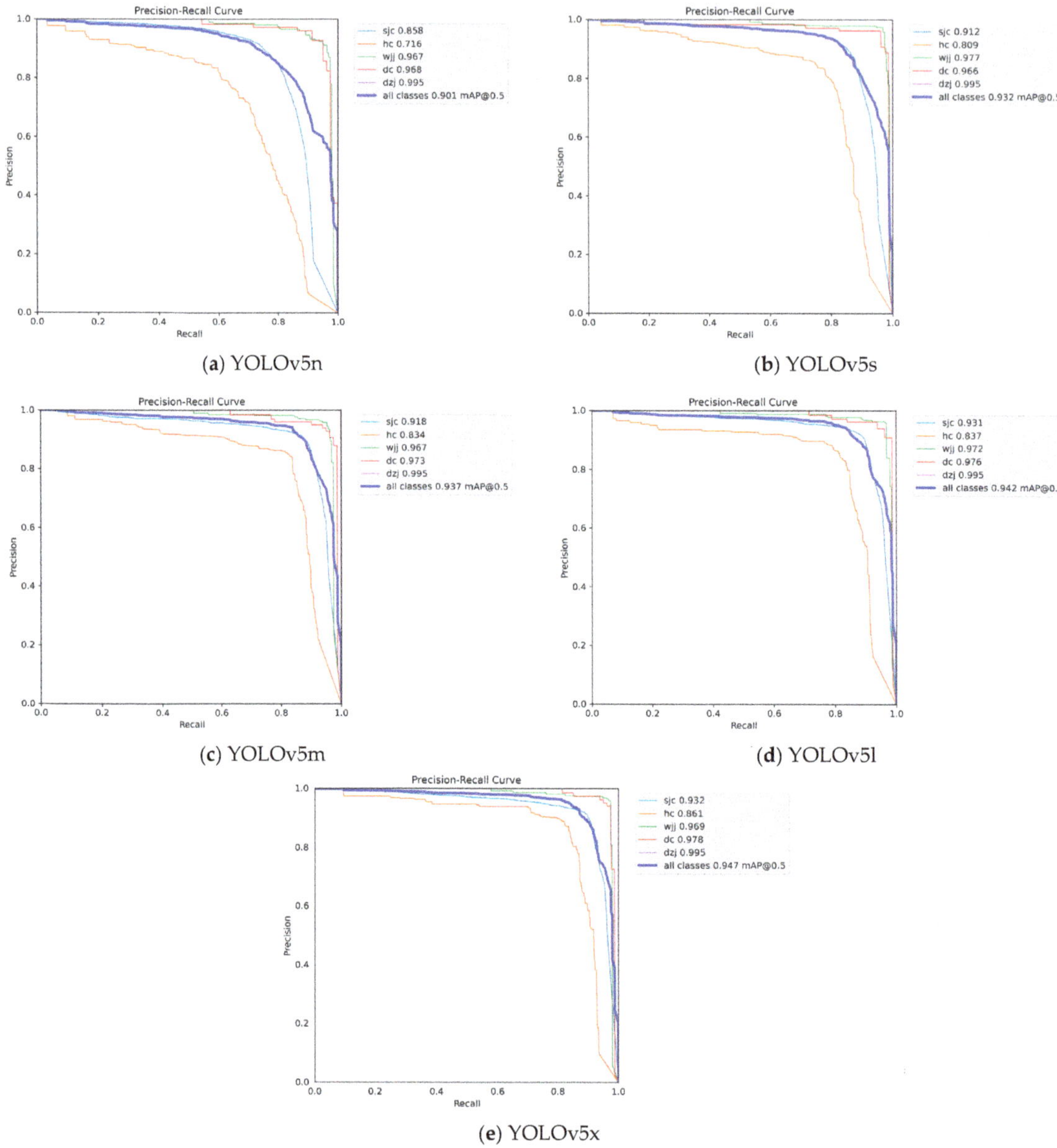

(**a**) YOLOv5n (**b**) YOLOv5s (**c**) YOLOv5m (**d**) YOLOv5l (**e**) YOLOv5x

Figure 12. The precision and recall curves.

Figure 12 shows the precision and recall curves of five different models. It can be clearly seen from Figure 12 that the fitting curves of five different models of YOLOv5 all show good results. This experiment shows the effectiveness of the training method in this article.

In order to combine precision and recall performance metrics, this paper also shows the F_1 value to measure the performance of the detector, and the confidence level to measure the probability of a object prediction. The F_1 value and the confidence curve are shown in Figure 13.

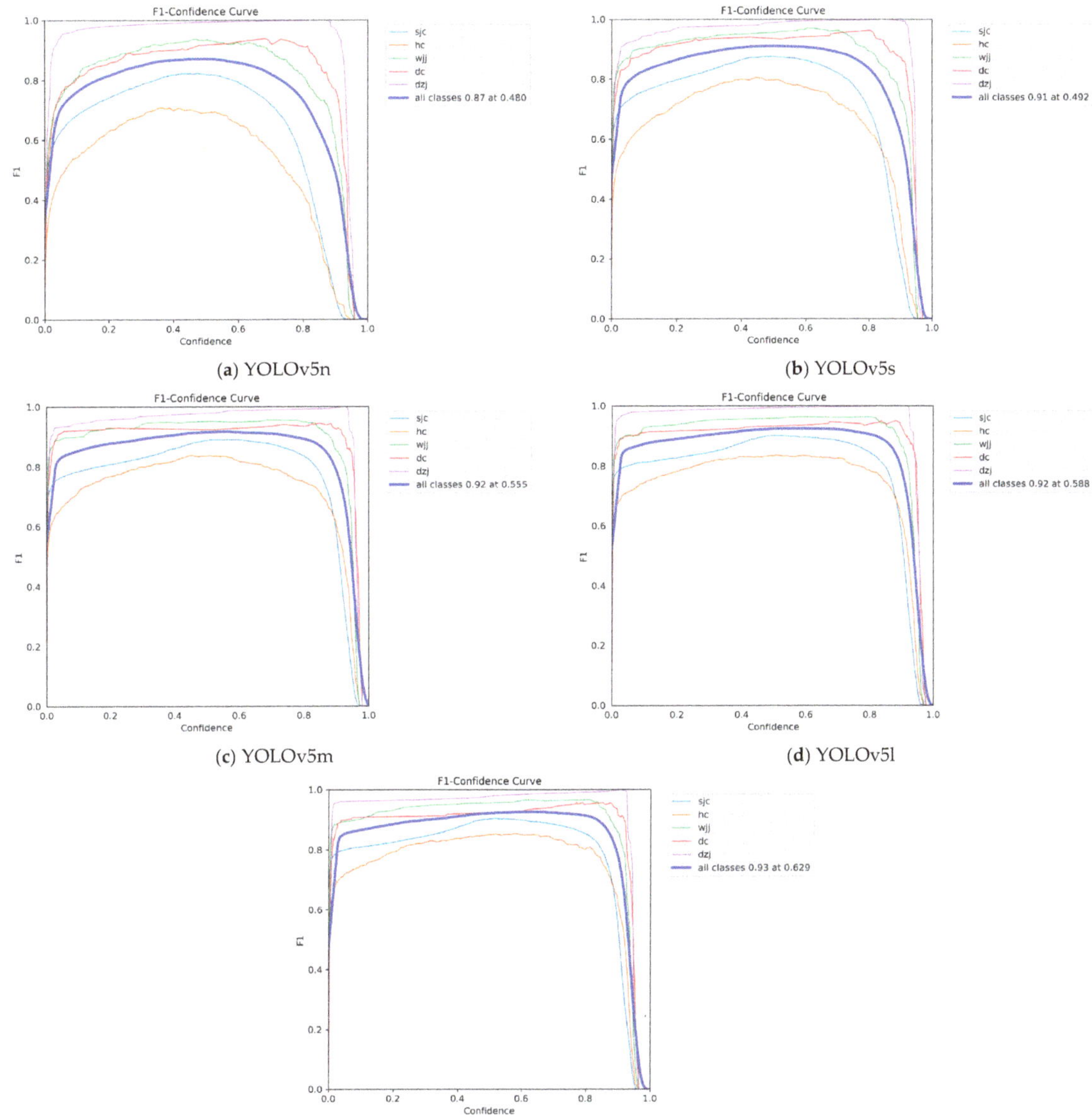

(**a**) YOLOv5n (**b**) YOLOv5s (**c**) YOLOv5m (**d**) YOLOv5l (**e**) YOLOv5x

Figure 13. F_1 value and confidence curve.

It can be seen from Figure 13 that the F_1 values of five different models change as the confidence level changes. It shows the F_1 score performance under different confidence classification thresholds. It can be seen from the figure that the model in this article also has a higher F_1 value at a higher confidence level, indicating that the model trained in this article has higher robustness.

We trained and tested five different YOLOv5 models on the same data set, and finally selected the YOLOv5s model, which is relatively lightweight. to the correct selection of a data set is extremely important for model training. Therefore, in order to further verify the effectiveness of the design of this article, Figure 14 shows the correlation of labels.

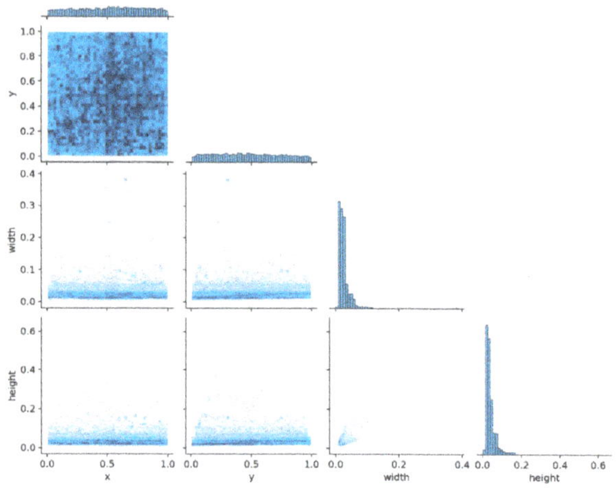

Figure 14. The correlation of labels.

Figure 14 shows the correlation distribution diagram of the data set labels we used. From the figure, we can clearly see the distribution of the center point, width, and height of the image annotation target. It shows that the labeled data samples used by this algorithm are diverse and the samples are very rich. It is effective for model training.

3.4. UAV Inspection Interface Display

During the flight inspection process of the UAV, the object detection algorithm service is called to deploy the trained model on an NVIDIA Tesla T4 GPU cloud server. We deployed a set of vehicle detection methods on this server using a set of algorithms called by the DeepStream framework. When the UAV conducts inspections along the designated route, we can clearly see the detection results processed by the UAV in real time on the dispatch platform. Figure 15 shows the real-time detection results of the UAV during automatic inspection according to the planned route.

Figure 15. UAV inspection interface. It shows a scene diagram of real-time detection and tracking of high-altitude inspections, and the UAV flight route can be seen in the lower right corner.

In Figure 15, a display of real-time inspection pictures is shown. During this type of high-altitude inspection by drones, the drones are looking down at the ground angle. Information such as buildings, ponds, and vehicles on the ground can be clearly seen. It can also be clearly seen that the YOLOv5s model trained in this article can automatically detect the required targets during the automatic drone inspection process, and correctly distinguish the target area and the background area. In the lower right corner of the image, you can obtain the inspection route of the drone and the specific location information of the drone flying on this route.

3.5. Mathematical Analysis of DeepStream Service Resources and Startup Time

To verify the quality of a framework, we not only consider its algorithm detection capabilities, but also consider its reasonable utilization of effective resources. Therefore, it is necessary to compare the real-time video streams before and after the improvement to check the resource overhead of the system architecture. The main comparison before and after improvement is as follows:

1. Before the improvement, it was necessary to bind drone equipment and specify execution tasks. In terms of resource consumption, a Telsa T4 GPU, 16G graphics card server can only bind up to six devices and tasks to be executed. That is, the binding of tasks limits the reasonable utilization of GPU and memory resources and cannot be dynamically adjusted according to task requirements. In short, the performance of the hardware before the improvement was very poor.
2. After improvement, our real-time video stream detection system does not need to bind devices and tasks; tasks are no longer bound, and resource allocation can be dynamically adjusted according to real-time needs. Not only can multiple devices be controlled at the same time, but different tasks can be switched freely. Therefore, after the improvement, the flexibility and the scalability of the system have been greatly improved.

DeepStream startup time mainly consists of two time elements. One is the time from the request to the pipeline initialization, and the other is the time from the pipeline initialization to the pipeline state switching to the playing state. The sum of the two is the startup time. The initial application of the DeepStream service has a startup time of approximately 7 s. Therefore, there is huge room for improvement. Based on this, we tested the startup of video streams in different definitions. The definition is divided into five levels, namely ultra-high definition (Ultra HD), ultra definition (UD), high definition (HD), standard definition (SD), and smooth, as shown in Table 5.

Table 5. Startup time test.

	The First Stage/s	The Second Stage/s	Total/s	Clarity
1	2.21	0.60	2.81	Ultra HD
2	2.24	0.28	2.52	Ultra HD
3	2.27	1.24	3.51	UD
4	2.27	0.50	2.77	UD
5	2.19	3.52	5.71	HD
6	2.21	3.42	5.63	HD
7	2.21	3.03	5.24	SD
8	2.29	3.24	5.53	SD
9	2.23	3.40	5.63	Smooth
10	2.22	3.51	5.73	Smooth

As shown in Table 5, during the first test, the first-stage startup time of UHD was 2.21 s, and the second-stage startup time was 0.60 s. The total time consumed is 2.81 s. The total time for the second test of UHD video is 2.52 s. The overall consumption time is significantly faster than other low-definition videos. Moreover, the startup time of the optimized model architecture is significantly faster than the original DeepStream processing time.

4. Conclusions

In this paper, we trained the YOLOv5s model to detect five different types of vehicles, and combined UAV automatic inspection technology to achieve the automatic real-time detection of UAVs. The main contributions of this article are as follows:

1. The integration of UAV inspection, YOLOv5s object detection, and the DeepStream framework realizes a new real-time object detection method.
2. The DeepStream service can be quickly started using the http communication protocol, and can be called and used by different users at the same time.
3. The asynchronous sending of the alarm frame interception function is implemented, making the real-time video stream smoother.
4. Auxiliary services are set up to quickly resume video streaming after interruption.

5. Future Work

The main contribution of this article is to combine UAV automatic inspection technology and artificial intelligence technology to achieve real-time video stream detection. However, how to design a more complete and efficient UAV video stream detection system still requires more research and also faces huge challenges. For example, more complex backgrounds, how to effectively remove background interference, high-altitude UAVs with smaller viewing angle objects, how to improve the performance of small object detectors, how to make models more lightweight and more suitable for deployment to edge devices, etc.

Author Contributions: Conceptualization, S.X. and G.D.; methodology, B.L.; software, S.X.; validation, G.D., B.L. and X.Z.; formal analysis, X.Z.; investigation, Y.L.; resources, W.J.; data curation, S.X.; writing—original draft preparation, G.D.; writing—review and editing, B.L.; visualization, S.X.; supervision, Y.L.; project administration, G.D.; funding acquisition, W.J. All authors have read and agreed to the published version of the manuscript.

Funding: This research was funded by the GDAS' Project of Science and Technology Development (2022GDASZH-2022010202, 2022GDASZH-2022010111, 2022GDASZH-2022020402-01), National Key R&D Program of China (2022YFF0711602), National Natural Science Foundation of China (42271479), the Science and Technology Program of Guangdong (2021B1212100006), the Science and Technology Program of Guangzhou (202206010098), and the National Key Research and Development Program of China (2023YFB3905702).

Institutional Review Board Statement: Not applicable.

Informed Consent Statement: Not applicable.

Data Availability Statement: Data are contained within the article.

Acknowledgments: We thank the Guangzhou Institute of Geography, Guangdong Academy of Sciences.

Conflicts of Interest: Shidun Xie, Guanghong Deng, Baihao Lin, and Xiaodan Zhao are employees of Guangzhou iMapCloud Intelligent Technology Co., Ltd. Wenlong Jing, Yong Li have no conflicts of interest to declare.

References

1. Zhang, Y.; Wang, C.; Wang, X.; Zeng, W.; Liu, W. FAIRMOT: On the fairness of detection and reidentification in multiple object tracking. *Int. J. Comput. Vis.* **2021**, *129*, 3069–3087. [CrossRef]
2. Feng, D.; Haase-Schütz, C.; Rosenbaum, L.; Hertlein, H.; Glaeser, C.; Timm, F.; Wiesbeck, W.; Dietmayer, K. Deep multi-modal object detection and semantic segmentation for autonomous driving: Datasets, methods, and challenges. *IEEE Trans. Intell. Transp. Syst.* **2020**, *22*, 1341–1360. [CrossRef]
3. Jaeger, P.F.; Kohl, S.A.A.; Bickelhaupt, S.; Isensee, F.; Kuder, T.A.; Schlemmer, H.-P.; Maier-Hein, K.H. Retina U-Net: Embarrassingly simple exploitation of segmentation supervision for medical object detection. *arXiv* **2020**, arXiv:1811.08661.
4. Li, Z.; Zhou, Y.; Lyu, S.; Huang, Y.; Yi, Y.; Zhao, C. Design of Fruit-Carrying Monitoring System for Monorail Transporter in Mountain Orchard. *J. Circuits Syst. Comput.* **2023**, *32*, 2350264. [CrossRef]
5. Lin, J.; Chen, W.-M.; Lin, Y.; Gan, C.; Han, S. MCUNet: Tiny deep learning on IoT devices. *Adv. Neural Inf. Process. Syst. (NeurIPS)* **2020**, *33*, 11711–11722.

6. Lyu, R. NanoDet-Plus. 2021. Available online: https://github.com/RangiLyu/nanodet/releases/tag/v1.0.0-alpha-1 (accessed on 26 December 2021).
7. Redmon, J.; Divvala, S.; Girshick, R.; Farhadi, A. You only look once: Unified, real-time object detection. In Proceedings of the IEEE Conference on Computer Vision and Pattern Recognition, Las Vegas, NV, USA, 27–30 June 2016; pp. 779–788.
8. Redmon, J.; Farhadi, A. Yolo9000: Better, faster, stronger. In Proceedings of the IEEE Conference on Computer Vision and Pattern Recognition, Honolulu, HI, USA, 21–26 July 2017; pp. 7263–7271.
9. Redmon, J.; Farhadi, A. Yolov3: An incremental improvement. *arXiv* **2018**, arXiv:1804.02767.
10. Bochkovskiy, A.; Wang, C.-Y.; Liao, H.-Y.M. Yolov4: Optimal speed and accuracy ofobject detection. *arXiv* **2020**, arXiv:2004.10934.
11. YOLOv5. Available online: https://github.com/ultralytics/yolov5 (accessed on 10 June 2020).
12. Ge, Z.; Liu, S.; Wang, F.; Li, Z.; Sun, J. Yolox: Exceeding yolo series in 2021. *arXiv* **2021**, arXiv:2107.08430.
13. Xu, S.; Wang, X.; Lv, W.; Chang, Q.; Cui, C.; Deng, K.; Wang, G.; Dang, Q.; Wei, S.; Du, Y.; et al. Pp-yoloe: An evolved version of yolo. *arXiv* **2022**, arXiv:2203.16250.
14. Wu, B.; Li, C.; Zhang, H.; Dai, X.; Zhang, P.; Yu, M.; Wang, J.; Lin, Y.; Vajda, P. FBNetv5: Neural architecture search for multiple tasks in one run. *arXiv* **2021**, arXiv:2111.10007.
15. Xiong, Y.; Liu, H.; Gupta, S.; Akin, B.; Bender, G.; Wang, Y.; Kindermans, P.-J.; Tan, M.; Singh, V.; Chen, B. MobileDets: Searching for object detection architectures for mobile accelerators. In Proceedings of the IEEE/CVF Conference on Computer Vision and Pattern Recognition (CVPR), Nashville, TN, USA, 20–25 June 2021; pp. 3825–3834.
16. Yu, G.; Chang, Q.; Lv, W.; Xu, C.; Cui, C.; Ji, W.; Dang, Q.; Deng, K.; Wang, G.; Du, Y.; et al. PP-PicoDet: A better real- time object detector on mobile devices. *arXiv* **2021**, arXiv:2111.00902.
17. Howard, A.; Sandler, M.; Chu, G.; Chen, L.-C.; Chen, B.; Tan, M.; Wang, W.; Zhu, Y.; Pang, R.; Vasudevan, V.; et al. Searching for MobileNetV3. In Proceedings of the IEEE/CVF Conference on Computer Vision and Pattern Recognition (CVPR), Long Beach, CA, USA, 15–20 June 2019; pp. 1314–1324.
18. Howard, A.G.; Zhu, M.; Chen, B.; Kalenichenko, D.; Wang, W.; Weyand, T.; Andreetto, M.; Adam, H. MobileNets: Efficient convolutional neural networks for mobile vision applications. *arXiv* **2017**, arXiv:1704.04861.
19. Sandler, M.; Howard, A.; Zhu, M.; Zhmoginov, A.; Chen, L.-C. MobileNetV2: Inverted residuals and linear bottlenecks. In Proceedings of the IEEE/CVF Conference on Computer Vision and Pattern Recognition (CVPR), Salt Lake City, UT, USA, 18–23 June 2018; pp. 4510–4520.
20. Ma, N.; Zhang, X.; Zheng, H.-T.; Sun, J. ShuffleNet V2: Practical guidelines for efficient CNN architecture design. In Proceedings of the European Conference on Computer Vision (ECCV), Munich, Germany, 8–14 September 2018; pp. 116–131.
21. Zhang, X.; Zhou, X.; Lin, M.; Sun, J. ShuffleNet: An extremely efficient convolutional neural network for mobile devices. In Proceedings of the IEEE/CVF Conference on Computer Vision and Pattern Recognition (CVPR), Salt Lake City, UT, USA, 18–23 June 2018; pp. 6848–6856.
22. Han, K.; Wang, Y.; Tian, Q.; Guo, J.; Xu, C.; Xu, C. GhostNet: More features from cheap operations. In Proceedings of the IEEE/CVF Conference on Computer Vision and Pattern Recognition (CVPR), Seattle, WA, USA, 13–19 June 2020; pp. 1580–1589.
23. Wang, C.-Y.; Yeh, I.-H.; Liao, H.-Y.M. You only learn one representation: Unified network for multiple tasks. *arXiv* **2021**, arXiv:2105.04206.
24. Zhou, X.; Wang, D.; Krähenbühl, P. Objects as points. *arXiv* **2019**, arXiv:1904.07850.
25. He, K.; Zhang, X.; Ren, S.; Sun, J. Deep residual learning for image recognition. In Proceedings of the IEEE/CVF Conference on Computer Vision and Pattern Recognition (CVPR), Las Vegas, NV, USA, 27–30 June 2016; pp. 770–778.
26. Yu, F.; Wang, D.; Shelhamer, E.; Darrell, T. Deep layer aggregation. In Proceedings of the IEEE/CVF Conference on Computer Vision and Pattern Recognition (CVPR), Salt Lake City, UT, USA, 18–23 June 2018; pp. 2403–2412.
27. Wang, C.-Y.; Liao, H.-Y.M.; Wu, Y.-H.; Chen, P.-Y.; Hsieh, J.-W.; Yeh, I.-H. CSP- Net: A new backbone that can enhance learning capability of CNN. In Proceedings of the IEEE/CVF Conference on Computer Vision and Pattern Recognition Workshops (CVPRW), Salt Lake City, UT, USA, 18–23 June 2018; pp. 390–391.
28. Luque-Vega, L.F.; Castillo-Toledo, B.; Loukianov, A.; Gonzalez-Jimenez, L.E. Power line inspection via an unmanned aerial system based on the quadrotor helicopter. In Proceedings of the MELECON 2014—2014 17th IEEE Mediterranean Electrotechnical Conference, Beirut, Lebanon, 13–16 April 2014; IEEE: Piscataway, NJ, USA, 2014; pp. 393–397.
29. Deng, G.; Huang, T.; Lin, B.; Liu, H.; Yang, R.; Jing, W. Automatic Meter Reading from UAV Inspection Photos in the Substation by Combining YOLOv5s and DeeplabV3+. *Sensors* **2022**, *22*, 7090. [CrossRef]
30. Candiago, S.; Remondino, F.; De Giglio, M.; Dubbini, M.; Gattelli, M. Evaluating multispectral images and vegetation indices for precision farming applications from uav images. *Remote Sens.* **2015**, *7*, 4026–4047. [CrossRef]
31. Tijtgat, N.; Van Ranst, W.; Volckaert, B.; Goedemé, T.; De Turck, F. Embedded real-time object detection for a UAV warning system. In Proceedings of the IEEE International Conference on Computer Vision Workshops, Venice, Italy, 22–29 October 2017; pp. 2110–2118.
32. Abdulghafoor, N.H.; Abdullah, H.N. Real-time moving objects detection and tracking using deep-stream technology. *J. Eng. Sci. Technol.* **2021**, *16*, 194–208.
33. NVIDIA DeepStream SDK 4.0.2 Release. Available online: https://docs.nvidia.com/metropolis/deepstream/dev-guide (accessed on 20 December 2019).

34. Guo, H.; Tian, B.; Yang, Z.; Chen, B.; Zhou, Q.; Liu, S.; Nahrstedt, K.; Danilov, C. DeepStream: Bandwidth efficient multi-camera video streaming for deep learning analytics. *arXiv* **2023**, arXiv:2306.15129.
35. Hossain, S.; Lee, D.-J. Deep Learning-Based Real-Time Multiple-Object Detection and Tracking from Aerial Imagery via a Flying Robot with GPU-Based Embedded Devices. *Sensors* **2019**, *19*, 3371. [CrossRef]
36. Vandersteen, M.; Van Beeck, K.; Goedemé, T. Super accurate low latency object detection on a surveillance UAV. In Proceedings of the 16th International Conference on Machine Vision Applications (MVA), Tokyo, Japan, 27–31 May 2019.
37. Haq, M.A.; Fahriani, N. Improving YOLO Object Detection Performance on Single-Board Computer using Virtual Machine. *Emerg. Inf. Sci. Technol.* **2024**, *5*. [CrossRef]
38. Huu, P.N.; Anh, B.N.; Minh, Q.T. Proposing Smart System for Detecting and Monitoring Vehicle Using Multiobject Multicamera Tracking. *Int. J. Digit. Multimed. Broadcast.* **2024**, *2024*, 6667738. [CrossRef]
39. Ghaziamin, P.; Bajaj, K.; Bouguila, N.; Patterson, Z. A Privacy-Preserving Edge Computing Solution for Real-Time Passenger Counting at Bus Stops using Overhead Fisheye Camera. In Proceedings of the 2024 IEEE 18th International Conference on Semantic Computing (ICSC), Laguna Hills, CA, USA, 5–7 February 2024; IEEE: Piscataway, NJ, USA, 2024; pp. 25–32.
40. Smink, M.; Liu, H.; Döpfer, D.; Lee, Y.J. Computer Vision on the Edge: Individual Cattle Identification in Real-Time With ReadMyCow System. In Proceedings of the IEEE/CVF Winter Conference on Applications of Computer Vision, Waikoloa, HI, USA, 3–8 January 2024; pp. 7056–7065.
41. Qaraqe, M.; Elzein, A.; Basaran, E.; Yang, Y.; Varghese, E.B.; Costandi, W.; Rizk, J.; Alam, N. Public Vision: A Secure Smart Surveillance System for Crowd Behavior Recognition. *IEEE Access* **2024**, *12*, 26474–26491. [CrossRef]
42. NVIDIA DeepsTream SDK 6.1.0 Release. Available online: https://docs.nvidia.com/metropolis/deepstream/6.1/dev-guide/ (accessed on 19 May 2022).
43. Wojke, N.; Bewley, A.; Paulus, D. Simple online and realtime tracking with a deep association metric. In Proceedings of the 2017 IEEE international conference on image processing (ICIP), Beijing, China, 17–20 September 2017; IEEE: Piscataway, NJ, USA, 2017; pp. 3645–3649.
44. Kapania, S.; Saini, D.; Goyal, S.; Thakur, N.; Jain, R.; Nagrath, P. Multi object tracking with UAVs using deep SORT and YOLOv3 RetinaNet detection framework. In Proceedings of the 1st ACM Workshop on Autonomous and Intelligent Mobile Systems, Bangalore, India, 11 January 2020; pp. 1–6.

sensors

MDPI

Article

UAV-UGV Collaborative Localisation with Minimum Sensing

A. H. T. Eranga De Silva *,† and Jayantha Katupitiya

School of Mechanical and Manufacturing Engineering, University of New South Wales, Sydney, NSW 2052, Australia; j.katupitiya@unsw.edu.au
* Correspondence: ahteranga@gmail.com
† Current address: Department of Mechanical Engineering, University of Moratuwa, Moratuwa 10400, Sri Lanka.

Abstract: This paper presents a novel methodology to localise Unmanned Ground Vehicles (UGVs) using Unmanned Aerial Vehicles (UAVs). The UGVs are assumed to be operating in a Global Navigation Satellite System (GNSS)-denied environment. The localisation of the ground vehicles is achieved using UAVs that have full access to the GNSS. The UAVs use range sensors to localise the UGV. One of the major requirements is to use the minimum number of UAVs, which is two UAVs in this paper. Using only two UAVs leads to a significant complication that results an estimation unobservability under certain circumstances. As a solution to the unobservability problem, the main contribution of this paper is to present a methodology to treat the unobservability problem. A Constrained Extended Kalman Filter (CEKF)-based solution, which uses novel kinematics and heuristics-based constraints, is presented. The proposed methodology has been assessed based on the stochastic observability using the Posterior Cramér–Rao Bound (PCRB), and the results demonstrate the successful operation of the proposed localisation method.

Keywords: constrained Kalman Filter; posterior Cramér–Rao bound; UAV–UGV collaboration; GNSS denied localisation

1. Introduction

Quite often, ground vehicles operate in GNSS-denied environments. In such cases, methodologies are required to enable the localisation of ground vehicles. This paper presents a method to accurately localise the ground vehicles using UAVs. It is assumed that sufficiently accurate GNSS locations of the UAVs are available, and that the UAVs always operate above the tree canopies; for example, localising a UGV that travels under a forest tree canopy can be localised using some UAVs which are flying above the forest tree canopy. In such situations, UAVs can be deployed to collaboratively estimate the location of the UGVs in real time.

The motivations behind developing the research outputs presented in this paper are (1) to develop a UGV localisation method that uses a minimum number of UAVs for UGV localisation; (2) to avoid unobservabilities, which may arise when UAVs are used for UGV localisation; (3) to use the proposed UAV–UGV collaborative system in adverse environment/field conditions. Such UAV–UGV collaborative systems can be used to localise UGVs in battlefields and disaster zones. For example, a UGV can be sent to a high-priority rescue mission where the GNSS reception is weak, and the environment has thick smoke and flames. Moreover, in a rescue scenario during bushfires, such a system can be used to rescue people who are surrounded by bushfires. In such an unfortunate situation, the firetrucks must be operating autonomously, since the firefighters cannot be sent to rescue and the field conditions are adverse. Similarly, UGVs that are used for farming may need the assistance of UAVs for their localisation when their GNSS reception is poor. For example, if a farm is covered by a tall tree canopy, the farming UGVs operating under such a tall tree canopy will not have sufficient GNSS reception. Thereupon, UAVs with sufficient GNSS reception, which hover/fly over the tree canopy, can be used to collaboratively localise farming UGVs operating under the tall tree canopy. Henceforth, a robust localisation method, which has been presented in this paper, is a necessity to navigate such UGVs.

Since there is a motivation to use the proposed UAV–UGV collaborative system in adverse environment/field conditions, sensing methods must be robust against dust, smoke, darkness, high heat, glare, etc. RADAR, LIDAR, Vision Camera, IR camera, and ultrasonic ranging are prospective remote sensing methods. The vision cameras are unable to perform the localisation properly in smoke and also during the night time. Due to the high heat fluxes present, IR cameras will jam the localisation if the localisation is supposed to be performed on a hostile battlefront with frequent glares. Ultrasonic ranging can have accuracy problems due to the impracticality of distinguishing the UGVs clearly from the other sound disturbances. LIDAR seems to show a promising ranging solution, even in vegetation clutter [1], but the inability to penetrate through the dense smoke makes it unsuitable for sensing during a situation like a forest fire [2].

However, RADAR technology shows promising results, as it is not affected by adverse environmental conditions such as bad weather [3] or smoke [4]. RADAR image processing is very cumbersome, and requires expert human intervention to interpret the RADAR scan images. Ultra Wide Band (UWB) sensing is also a RADAR range-finding technique. Unlike RADAR, it does not have to choose a location where surroundings induce minimal clutter, since UWB signals can sense through clutter [5]. In UWB sensing, the large bandwidth enhances reliability as the sensing signal contains different frequencies, which increases the possibility that at least a handful of the emitted signal can go through/around obstacles, and the high bandwidth offers improved ranging accuracy [6]. Moreover, UWB sensors are capable of delivering range measurements at Non-Line-of-Sight (NLOS) situations (e.g., ranging through a forest canopy) without significant degradation of the range measurements [5,7]. Often, UWB sensors can be operated in RADAR mode or range sensing mode. Again, if the UWB sensors are used in the RADAR mode, sensing data processing is very complex, e.g., identifying a UGV travelling under a tree canopy using RADAR mode UWB images, which are acquired from a UAV that is flying above the tree canopy. On the contrary, when UWB sensors are used in the range sensing mode, such data processing complexities do not arise, while the sensing robustness is also safeguarded. Therefore, the proposed UAV–UGV collaborative localisation method is a UWB range-only localisation method.

In UWB ranging, Time of Arrival (TOA) techniques are providing less complex, reliable and cost-effective solutions. There are three commonly used TOA techniques, namely: (i) basic two-way ranging TOA, (ii) synchronous two-way TOA, and (iii) asynchronous two-way TOA. The basic two-way TOA expects ideal instrumentation, which results in low accuracy [6]. In the synchronous two-way TOA method, the time delay in returning back the response to the initial signal sender has been compensated [8]. In addition to the advantages of the synchronous two-way TOA method, the asynchronous two-way TOA method has compensated for frequency and/or phase mismatches between the UWB transceivers [9]. Concerning the aforementioned advantages, UWB range sensing is assumed to be performed by an asynchronous two-way TOA ranging algorithm. Most modern UWB sensors that use the asynchronous two-way ranging method have achieved ranging accuracy up to ± 2 cm [7,10]. Henceforth, the observer model in the localisation algorithm does not have to account for either the time delay in range sensing or the frequency and/or the phase mismatches between the UWB transceivers [9].

In a practical application of collaboratively localising a UGV using UAVs, reducing the number of UAVs that have to be utilised for the task is equally important. By reducing the number of UAVs, the capital cost that has to be spent can be reduced. Furthermore, the operational costs can also be reduced since the electricity power cost is low when a lesser number of UAVs are to be airborne. Due to the limited flight time of UAVs, additional UAVs are kept by users to run UAV operations without interruption. The additional UAVs are utilised for the operation while the battery swapping is performed. In that regard, if the number of UAVs required for a UAV operation is minimised, the additional number of UAVs that have to be purchased for the application can also be reduced.

However, as the number of drones is reduced to a minimum of two, processing of the range data for UGV localisation faces substantial challenges. The main problem to be addressed is the ambiguity of the localisation due to the loss of system observability. This paper addresses this problem, and shows the successful localisation of ground vehicles using the proposed method. In the literature, the aforementioned problem is known as the "flip ambiguity phenomenon", and it has been researched in Wireless Sensor Networks (WSN) and in tracking/localisation. In [11], flip ambiguity has been overcome by using a high number of location anchors/nodes in the WSN so that the ambiguously localised nodes can be identified, and their localisation is supposed to be refined to avoid the flip ambiguity. Since two UAVs are used as anchors in this research, identification of the ambiguous localisation using many UAVs is not possible. In [12], flip ambiguity in intra-localisation of UAVs in a UAV swarm has been addressed, along with the measurement errors. The solution is based on geometric constraints in a 2D plane like in a WSN, which are based on the range measurement constraints, communication range constraints and kinematic information constraints. In [13], an Extended Kalman Filter (EKF) has been designed to localise a GNSS unavailable UAV in a UAV swarm, and the flip ambiguity in localisation has been overcome by estimating the angular velocity of the UAV. However, UAVs have low process noise in their motion. Nevertheless, in a noisy process situation, such as in a UGV motion on a farm/forest ground, angular velocity estimations will have significant deviations, so that granting the angular velocity estimation as crisp information to address the flip ambiguity in localisation will not be a reliable solution to a UGV localisation. Therefore, in this research, a constrained state estimation-based method is developed to address the localisation issues arising from the flip ambiguity.

The system observability was analysed in a deterministic approach to identify the unobservable situations in the proposed localisation method. Based on that, a methodology was developed to successfully avoid the localisation errors caused by the unobservability. In this research, constrained stochastic estimation has been used for localisation. The constraints mitigate the challenges that arise when only two UAVs are utilised for localisation. In order to check the ambiguity aversion performance while using the constrained stochastic location estimation, stochastic observability has been analysed during temporary unobservable scenarios using simulations and experiments.

Due to the strictly/narrowly focused operational scenario considered in this research, the authors did not find comparable past research works/methods that possess similar system implementations. Therefore, the authors believe that the system design of this UAV–UGV collaborative localisation is novel. In the constrained state estimation-based localisation method, all of the kinematics-based constraints are newly formulated for localisation ambiguity aversion. Moreover, when the UGV localisation is performed by UAVs in real time, a method has to be developed to validate the efficacy of the CEKF-based localisation method presented. Thereupon, a novel analytical method is formulated to show the efficacy of the CEKF-based localisation method using the Constrained Posterior Cramér–Rao Bound (CPCRB).

The rest of this paper is organised as follows: Section 2 describes the preliminaries of the motion model, observation model and the EKF-based localisation. Section 3 describes the unobservability identification method of the EKF-based localisation. Sections 4 and 5 explain the proposed method of overcoming localisation unobservabilities and how the presented unobservability aversion techniques assure the unobservability aversion, respectively. Finally, Sections 6 and 7 present simulation results and experiment results, respectively.

2. Problem Statement

Localisation of a UGV that is travelling on a horizontal planar terrain is supposed to be carried out using two multi-rotor UAVs using range measurements. The overall vehicle and sensor arrangement is as depicted in Figure 1. Since the heading of the UGV is also an important aspect in real-time navigation, two UWB sensors are attached to the ground vehicle at two different locations along the centreline of the ground vehicle at a constant height from the ground. In this paper, the UWB sensors that are attached to the UGV will be named as UWB tags, and are located at (x_f, y_f) and (x_r, y_r). Furthermore, UWB range sensors are mounted on each of the drones to obtain the range measurements to two range sensors mounted on the back and on the front of the UGV. UWB sensors that are mounted on each UAV will be named as UWB anchors, and are located at (x_1, y_1, z_1) and (x_2, y_2, z_2). The UWB anchors that are fixed in the drones fetch the range readings R_{1f}, R_{1r}, R_{2f} and R_{2r} between both of the UWB tags on the UGV. For this UGV localisation task, the global positions of the drones are to be known with sufficiently high accuracy. In the next section, the system models that are associated with the localisation are presented.

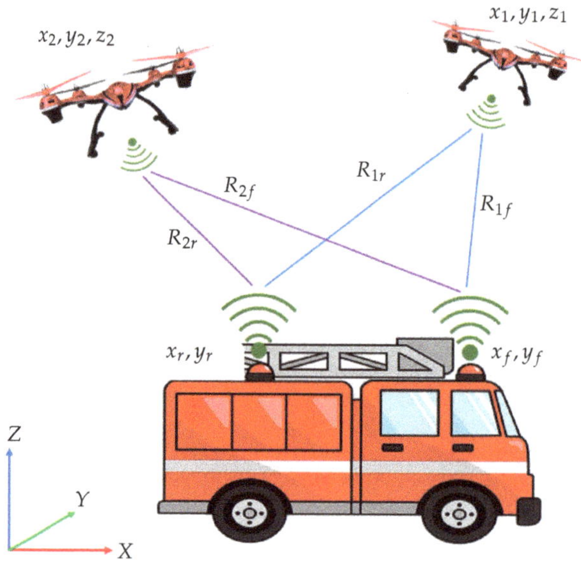

Figure 1. 2D localisation of a UGV using drones.

2.1. System Models

2.1.1. Motion Model for UGV Motion Simulation

Taking $\mathbf{x}_a = [x_a \; y_a \; \theta_a]^T$ as the state of the UGV in the real world (i.e., simulated actual state), where x_a and y_a are the geometric centroid's coordinates of the UGV and θ_a is the heading angle of the UGV in radians, the reduced model of the UGV, by assuming the steering angle (v_2), speed (v_1), step time (T), and vehicle length (l) are known, can be formulated as [14]:

$$\begin{bmatrix} x_a \\ y_a \\ \theta_a \end{bmatrix}_{k+1} = \begin{bmatrix} T & 0 & 0 \\ 0 & T & 0 \\ 0 & 0 & T \end{bmatrix} \begin{bmatrix} v_1 \cos(\theta_a) \\ v_1 \sin(\theta_a) \\ (v_1/l)\tan(v_2) \end{bmatrix} + \begin{bmatrix} x_a \\ y_a \\ \theta_a \end{bmatrix}_k \tag{1}$$

This motion model will be used to simulate the motion of the UGV when validating the presented localisation method using numerical simulations. The forward UWB tag and the rear UWB tag are fixed right on the centreline of the UGV, in the front and in the rear of the UGV. Therefore, the middle position of the forward and the rear UWB tags coincides with the exact midpoint of the UGV on the horizontal plane. Hence, the real-world forward UWB tag's position (x_f, y_f) and the real-world rear UWB tag's position (x_r, y_r) can be found, if x_a, y_a and θ_a are known. Using the solution of (1), x_a, y_a and θ_a can be calculated when simulation parameters v_1, v_2, l and T are known. Therefore, simulating the motion of the UWB tags based on the simulated motion of the UGV can be performed.

2.1.2. Motion Model for Localisation and Heading Estimation

For the location and the heading estimation problem, a Continuous Velocity (CV) model [15] can be assumed for a 2D localisation scenario depicted in Figure 1. The CV model has been used by assuming that the localised vehicles are not highly manoeuvred in both linear and angular movements. Let the tag positions' state vector $\mathbf{x}$ be defined as $\begin{bmatrix} x_f \; \dot{x}_f \; y_f \; \dot{y}_f \; x_r \; \dot{x}_r \; y_r \; \dot{y}_r \end{bmatrix}^T$. Hence, when considering the UGV state with respect to the earth-fixed inertial frame, the kinematic equation of the estimation model is given by:

$$\begin{aligned} \widehat{\mathbf{x}}_{k+1}^{-} &= \begin{bmatrix} 1 & T & 0 & 0 & 0 & 0 & 0 & 0 \\ 0 & 1 & 0 & 0 & 0 & 0 & 0 & 0 \\ 0 & 0 & 1 & T & 0 & 0 & 0 & 0 \\ 0 & 0 & 0 & 1 & 0 & 0 & 0 & 0 \\ 0 & 0 & 0 & 0 & 1 & T & 0 & 0 \\ 0 & 0 & 0 & 0 & 0 & 1 & 0 & 0 \\ 0 & 0 & 0 & 0 & 0 & 0 & 1 & T \\ 0 & 0 & 0 & 0 & 0 & 0 & 0 & 1 \end{bmatrix} \widehat{\mathbf{x}}_k^{+} + \nu \\ \widehat{\mathbf{x}}_{k+1}^{-} &= \mathbf{A}\,\widehat{\mathbf{x}}_k^{+} + \nu \end{aligned} \tag{2}$$

where ν is the process noise. The hat notation depicts that the respective variable is estimated from the localisation algorithm. The subscript k denotes the time step, and superscripts "+" and "−" are used to denote the *a posteriori* estimation and the *a priori* estimation, respectively.

Since the observations are taken as the ranges between UWB tags on the UGV and UWB anchors on the drones, the equation of the state dynamics should have to be written relative to the UAVs. Nevertheless, the state dynamics in (2) have not been written relative to the UAVs. By projecting the state of UAVs to the ground plane (as the altitude of the UGV is not used), the following derivation justifies why it is not necessary to use a system dynamics equation written related to UAV frames.

The state of a given UAV, projected on to the ground plane, is defined as ${}_g\mathbf{u}_{n,k}$; $n \in \{1,2\}$, representing either of the UAVs by the subscript n. Therefore,

$${}_g\mathbf{u}_{n,k} = \begin{bmatrix} x_{n,k} \; \dot{x}_{n,k} \; y_{n,k} \; \dot{y}_{n,k} \; x_{n,k} \; \dot{x}_{n,k} \; y_{n,k} \; \dot{y}_{n,k} \end{bmatrix}^T$$

If the UGV states are written relative to the ground projected body fixed frames of the UAVs by assuming that the evolution of UAV states can be modelled using a discrete nonlinear/linear transition $b_n(\cdot)$ of the current state, based on the exact UAV motion model, then Equation (2) can be written as:

$$\begin{aligned} \hat{\mathbf{x}}_{k+1}^{-} &= \mathbf{A}\,\hat{\mathbf{x}}_k^{+} + {}_g\mathbf{u}_{1,k+1} - b_1({}_g\mathbf{u}_{1,k}) \\ \hat{\mathbf{x}}_{k+1}^{-} &= \mathbf{A}\,\hat{\mathbf{x}}_k^{+} + {}_g\mathbf{u}_{2,k+1} - b_2({}_g\mathbf{u}_{2,k}) \end{aligned} \tag{3}$$

since the UAV coordinates are known with high accuracy. In ${}_g\mathbf{u}_{n,k}$, it can be seen that the first four elements are identical to the last four elements. This is because, when the UGV states are written relative to the ground projected body-fixed frames of a specific UAV, both front tag elements and rear tag elements in $\mathbf{x}$ have to be written relative to that specific UAV.

Based on the assumption that the UAV states' evolution can be modelled using a discrete nonlinear transition $b_n(\cdot)$ of the current state based on the exact UAV motion model, the relative state transition of the UGV in (3) reduces to its initial form as in (2). In other words, it is justifiable to use (2) to represent the discrete state transition of the UGV for this range-based localisation, without writing (2) relative to each UAV.

2.1.3. Measurement Model

The measurement model for the EKF is formulated by getting expressions for the squared range measurements using the coordinates of both UAVs and UGVs.

The state of either of the UAVs will be taken as $\mathbf{u}_{n,k} = [x_{n,k}\ y_{n,k}\ z_{n,k}]^T;\ n \in \{1,2\}$, elements of which will be used in later sections. For simplicity, the time step subscript will be omitted whenever it is insignificant. The observation vector is $h(\mathbf{x}) = [h_1\ h_2\ h_3\ h_4]^T$, where h_1, h_2, h_3 and h_4 are the squared range measurements.

If $R_{nm} = \sqrt{z_n^2 + (x_n - x_m)^2 + (y_n - y_m)^2}$ and $m \in \{f, r\}$, which is the range between the UGV tag m and the UAV anchor n, then the overall observation equation can be written as:

$$\mathbf{y} = h(\mathbf{x}) = \begin{bmatrix} R_{1f}^2 \\ R_{1r}^2 \\ R_{2f}^2 \\ R_{2r}^2 \end{bmatrix} + \eta \tag{4}$$

where η is the measurement noise of the range sensors. Since this measurement equation is nonlinear, the Jacobian with respect to the estimation state vector has to be calculated when implementing the EKF using the backward numerical differentiation. Thereupon, the Jacobian of the measurement function $h(x)$, evaluated at *a priori* is given by Equation (5).

$$\mathbf{H} = \left.\frac{\partial h(\mathbf{x})}{\partial \mathbf{x}}\right|_{\hat{\mathbf{x}}_k^-} \tag{5}$$

2.2. Localisation Using the Kalman Filter

Due to the nonlinearity in the observation/measurement model, and also because the localisation is performed sequentially in real-time, the discrete-time version of the EKF [16] is used. In order to use an EKF, the initial $\hat{\mathbf{x}}_{k-1}^+$ and the initial $\mathbf{P}_{k-1}^+$, which is the error covariance associated with the *a posteriori* estimates $\hat{\mathbf{x}}_{k-1}^+$, has to be provided to the EKF algorithm. A random point near the vicinity of the UGV can be given as the initial $\hat{\mathbf{x}}_{k-1}^+$ for the UGV. Initially, a diagonal matrix, which has fairly high values in the diagonal elements, can be given for $\mathbf{P}_{k-1}^+$ for all UGVs, indicating the initial value of $\hat{\mathbf{x}}_{k-1}^+$ is substantially uncertain. In the following sections, EKF-related symbols have the usual notation.

Following the estimation from the EKF, the forward UWB tag position estimate $(\hat{x}_f, \hat{y}_f)$ and the rear UWB tag position estimate $(\hat{x}_r, \hat{y}_r)$ can be used to find the location (i.e., geometrical centroid) of the UGV and its heading using:

$$(\hat{x}_a, \hat{y}_a) = \left[\frac{\hat{x}_f + \hat{x}_r}{2}, \frac{\hat{y}_f + \hat{y}_r}{2}\right] \tag{6}$$

$$\hat{\theta}_a = \tan^{-1}\left(\frac{\hat{y}_f - \hat{y}_r}{\hat{x}_f - \hat{x}_r}\right) \tag{7}$$

as previously mentioned in Section 2.1. In (6), it is assumed that the UWB range sensors are mounted on the UGV centreline from an equal distance from the centre of the UGV.

3. Unobservability in Localisation

When observability lapses in a system, the optimal estimators fail in state estimation. If such a situation occurs in localisation, it is defined as an unobservability in localisation. Therefore, a thorough observability analysis is essential in order to guarantee a fail-safe state estimation during the localisation. The next two sections will explain how to identify the localisation singularities/unobservabilities in the UAV–UGV collaborative localisation scenario based on deterministic observability.

3.1. Deterministic Observability

The deterministic observability analyses and checks whether the state of a system can be determined without any ambiguity, based on the system's outputs [17], where a state estimator cannot give an accurate estimate about the system state at a deterministic observability lapsed situation. Nevertheless, the observability of a nonlinear system is not a global attribute in the entire state space as the relationship between the measurement space, and the state space is not one-to-one [18]. Thereupon, deterministic local observability has been defined as: "A system is locally observable at a state x_0, if there exists a neighbourhood $\mathcal{N}$ of x_0 such that every state, which belongs to $\mathcal{N}$, other than x_0 is distinguishable from x_0. Finally, the system is locally observable if it is locally observable at each state" [19].

For a generic affine continuous-time nonlinear system:

$$\begin{aligned} \dot{\mathbf{x}} &= f(\mathbf{x}) + \sum_{\varrho=1}^{\vartheta} u_\varrho g_\varrho(\mathbf{x}) \\ \mathbf{y} &= h(\mathbf{x}) \end{aligned} \tag{8}$$

where $\mathbf{x} \in \mathbb{R}^\zeta$ is the system state vector, $\mathbf{y} \in \mathbb{R}^\vartheta$ is the output (observation) vector, $g_1(\mathbf{x}), g_2(\mathbf{x}), \ldots, g_\vartheta(\mathbf{x})$ are known vector fields, and the control input is $\mathbf{u} = [u_1 \ \ldots \ u_\vartheta]^T$; if the current state $(\mathbf{x}_0)$ is given and the expression

$$\begin{aligned} &(\nabla L_{z_s} L_{z_{s-1}} \ldots L_{z_1} h_j)(\mathbf{x}_0) \\ &s \geq 0 \quad z_i \in \{ f, g_1, g_2, \ldots, g_\vartheta \} \end{aligned} \tag{9}$$

is calculated at x_0, where $L(\cdot)$ is the Lie derivative, then System (8) is locally observable in the neighbourhood of x_0 if there are n linearly independent row vectors in this set (i.e., full in rank).

Since the proposed localisation method is used for real-time sequential state estimation, achieving deterministic local observability all the time is a requirement to maintain an accurate state estimation without any singularities/ambiguities.

3.2. Identification of Singularities

Since our aim is to localise a non-manoeuvring UGV, $\mathbf{u}$ in (8) is not known to the estimation algorithm. Henceforth, the expression in (9) reduces to an observability matrix,

$$\mathbf{O}(\mathbf{x}) = \begin{bmatrix} \dfrac{\partial L_f^0 h(\mathbf{x})}{\partial \mathbf{x}} & \cdots \cdots \cdots & \dfrac{\partial L_f^{\zeta-1} h(\mathbf{x})}{\partial \mathbf{x}} \end{bmatrix}^T \tag{10}$$

where $L_f h(\mathbf{x})$ is the Lie derivative of the function $h(\mathbf{x})$ by the function $f(\mathbf{x})$ and $L_f^\zeta h(\mathbf{x})$ is the $(\zeta+1)^{th}$ Lie derivative of the function $h(\mathbf{x})$ by the function $f(\mathbf{x})$.

If the system is locally observable at every time, the rank of $\mathbf{O}(\mathbf{x})$ should be equal to n (i.e., full rank) [18,19]. In this localisation scenario, $\mathbf{O}(\mathbf{x})$ for our system, when written as in (10), is a 32×8 matrix that uses (4) as the observation equation $h(\mathbf{x})$ and the linear continuous form of the difference equation (2) as the system equation $f(\mathbf{x})$ for UGV dynamics. To find the singularities, $\mathbf{O}(\mathbf{x})^T$ is used for mathematical convenience instead of $\mathbf{O}(\mathbf{x})$ using the identity $rank(\mathbf{O}(\mathbf{x})) = rank(\mathbf{O}(\mathbf{x})^T)$. By analysing at which UGV states result in loss of full rank of $\mathbf{O}(\mathbf{x})^T$, the singularities can be identified. The upper triangular matrix of the LU-decomposed $\mathbf{O}(\mathbf{x})$ is in the row echelon form [20], therefore if any of its diagonal elements is zero at any system state, $\mathbf{O}(\mathbf{x})$ cannot be a full rank matrix, and hence the system becomes locally unobservable.

By analysing the diagonal elements of the upper triangular matrix of the LU-decomposed $\mathbf{O}(\mathbf{x})^T$, the singularity is identified to be taking place when UGV states satisfy either

$$\begin{bmatrix} (y_2 - y_1) \\ 0 \\ (x_1 - x_2) \\ 0 \\ 0 \\ 0 \\ 0 \\ 0 \end{bmatrix}^T \cdot \hat{\mathbf{x}}_k = x_1 y_2 - x_2 y_1 \tag{11}$$

or

$$\begin{bmatrix} 0 \\ 0 \\ 0 \\ 0 \\ (y_2 - y_1) \\ 0 \\ (x_1 - x_2) \\ 0 \end{bmatrix}^T \cdot \hat{\mathbf{x}}_k = x_1 y_2 - x_2 y_1 \tag{12}$$

Hence, the location where the localisation ambiguity/singularity of the observer occurs according to (11) or (12) is the vertical shaded blue surface in Figure 2, which intercepts both UAV positions.

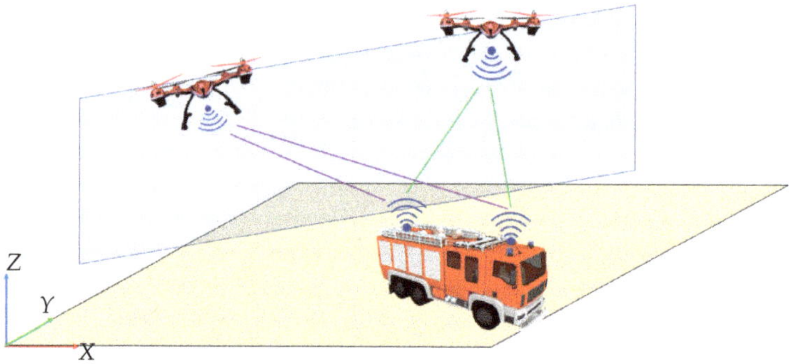

Figure 2. Ambiguity region of the location observer.

What happens due to this localisation singularity during a real-time localisation is that when a UGV range sensing tag intercepts the shaded blue surface in Figure 2, the optimal estimator fails to estimate the UGV location, i.e., the optimal estimator cannot determine in which side of the shaded blue surface, the intercepted UGV range sensing tag is at.

4. Aversion of Ambiguities

Based on the identified singularities in Section 3, avoiding ambiguity will be of interest in accurately localising the UGVs in real time. Nevertheless, the motion of the UGV and UAVs is subjective to the mission objectives that cannot be altered due to observability issues. Therefore, developing a method which enables a UGV to successfully pass through the ambiguity region in Figure 2 is the key to eradicating the ambiguity errors in localisation. Thereupon, CEKF has been utilised by imposing constraints based on heuristics, covariance-based accuracy margins and inter-dependency of state variables.

Even if the system model depicts the real system to a greater extent, KFs with constraints [21–24] have been used in estimation problems to improve the estimation accuracy by avoiding the unrealistic state estimates irrespective of the system nonlinearities [21].

In this CEKF-based UAV–UGV collaborative localisation method, the estimation projection method has been used for the state estimation, where the unconstrained *a posteriori* estimate of the EKF $\hat{\mathbf{x}}_k^+$ is projected into the constrained space and obtaining the constrained estimate $\tilde{\mathbf{x}}_k^+$ by:

$$\tilde{\mathbf{x}}_k^+ = argmin_{\mathbf{x}}(\mathbf{x} - \hat{\mathbf{x}}_k^+)^T \mathbf{W}(\mathbf{x} - \hat{\mathbf{x}}_k^+) \tag{13}$$

such that

$$\mathbf{D}\,\mathbf{x} = \mathbf{d} \quad and/or \quad \mathbf{D}\,\mathbf{x} \leq \mathbf{d}$$

are the linear equality constraints and linear inequality constraints, respectively.

The nonlinear constraints can also be handled in the same way by linearising the nonlinear constraints using the first-order Taylor expansion of the constraints [21,25]. In (13), the matrix **W** is the weighting matrix. The value of **W** is equal to the inverse of the *a posteriori* covariance matrix of estimation if the projection is based on the maximum probability approach. If the projection is based on the least-squares approach, the matrix **W** is equal to the identity matrix. In the proposed localisation method, the maximum probability state projection method is followed ($\therefore\ \mathbf{W} = (\mathbf{P}_k^+)^{-1}$). Due to the maximum probability constraining approach, and also because the EKF is unbiased, the overall CEKF state estimation process is still a minimum variance estimation approach. Figure 3 illustrates the difference between the two projection approaches intuitively.

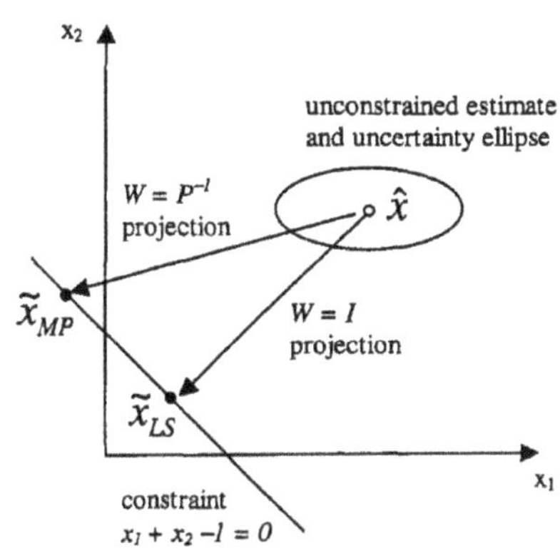

Figure 3. An example which shows the difference between the Maximum Probability (MP) approach and the Least Squares (LS) approach in projecting the estimate into the constraint space [25].

In Figure 3, the least squares method has been followed to calculate $\tilde{x}_{LS}$ by projecting the estimate $\hat{x}$ onto the constraint space. In that method, there is no concern about safeguarding the minimum variance objective of the estimator when imposing constraints.

Imposing constraints to an EKF (i.e., CEKF) is a quadratic optimisation problem as formulated in (13). Active set-based quadratic programming can be utilised to find the constrained estimate $\tilde{\mathbf{x}}_k^+$, by identifying the active constraints in each step of the optimisation. Hildreth's quadratic programming procedure, which is simple and reliable in real-time implementation can be utilised to solve this quadratic optimisation problem [26–28].

In the following three sections, the novel constraints, which are the main contributions of this paper, have been formulated with derivations. The symbol σ_χ depicts the standard deviation of the a posteriori estimation of a variable χ.

Note: The constraints of the CEKF, explained in this section, are to be imposed only after the EKF's covariance matrix of estimation is sufficiently converged. All of the constraints are formulated based on $\pm 3\sigma$ uncertainty (i.e., 99.73% confidence level).

4.1. The Position Constraint

Assuming that the UGV motion can be successfully modelled by the CV motion model, a position constraint can be imposed on the UGV state estimation based on the $\pm 3\sigma$ uncertainty of the position caused by the uncertainty of the velocity. For instance, if the estimated position along x-direction $x_{f,k-1}$ and the estimated velocity along x-direction $\dot{x}_{f,k-1}$ of the UGV's front UWB tag at the previous time step is known, then the position of that tag at this time step can be kinematically anticipated as $x_{f,k-1} + \dot{x}_{f,k-1}\,T$. However, $x_{f,k-1}$ and $\dot{x}_{f,k-1}$ are stochastic variables. Therefore, deterministic anticipation is inappropriate, and have to take $\sigma_{\dot{x}_{f,k-1}}$ to stochastically anticipate the position of that tag. Hence, a compound inequality can be written for that tag as:

$$\dot{x}_{f,k-1}\,T \;-\; 3\sigma_{\dot{x}_{f,k-1}}\,T \;\le\; x_{f,k} \;-\; x_{f,k-1} \le \dot{x}_{f,k-1}\,T \;+\; 3\sigma_{\dot{x}_{f,k-1}}\,T \tag{14}$$

and a kinematic constraint can be formulated at the $x_{f,k}$ estimation step based on this compound inequality. In such a kinematically constrained EKF-based estimator, $x_{f,k-1}$ and $\dot{x}_{f,k-1}$ are constrained estimates, and have to be written as $\tilde{x}_{f,k-1}$ and $\dot{\tilde{x}}_{f,k-1}$, respectively. Then, $x_{f,k}$ will be the free variable for the optimisation in the CEKF, which was explained in Section 4. Hence, (14) can be modified as:

$$\dot{\tilde{x}}_{f,k-1}\,T \;-\; 3\sigma_{\dot{x}_{f,k-1}}\,T \;\le\; x_{f,k} \;-\; \tilde{x}_{f,k-1} \le \dot{\tilde{x}}_{f,k-1}\,T \;+\; 3\sigma_{\dot{x}_{f,k-1}}\,T \tag{15}$$

The constraints in (15) can be extended to all the other position variables ($x_{r,k}$, $y_{f,k}$ and $y_{r,k}$) of the UGV tag positions' state vector ($\mathbf{x}$) by writing another three compound inequalities analogous to (15). Altogether, all four compound inequalities should be imposed simultaneously in the CEKF, and also, the compound inequalities have to be re-arranged as inequalities such that the consolidated inequality becomes:

$$\mathbf{L}\,\mathbf{x} \le \mathbf{M}\,\tilde{\mathbf{x}}_{k-1} + \mathbf{N}\,\mathrm{diag}\left(\mathbf{P}^{+}_{k-1}\right)^{\circ 1/2} \tag{16}$$

where the 8×8 diagonal matrices $\mathbf{L}, \mathbf{M}$ and $\mathbf{N}$ independently have the matrices $\mathbf{L}_0, \mathbf{M}_0$ and $\mathbf{N}_0$ as their block diagonal matrices, respectively, where

$$\mathbf{L}_0 = \begin{bmatrix} 1 & 0 \\ -1 & 0 \end{bmatrix}, \mathbf{M}_0 = \begin{bmatrix} 1 & T \\ -1 & -T \end{bmatrix}, \mathbf{N}_0 = \begin{bmatrix} 0 & T \\ 0 & T \end{bmatrix},$$

and $(\,\cdot\,)^{\circ 1/2}$ is the element-wise square root of a matrix.

Further derivations have to be performed to (16) in order to implement its constraints in the form of $\mathbf{D}_1\mathbf{x} \le \mathbf{d}_1$. First, (16), must be re-arranged such that (16) can be obtained as a nonlinear inequality in the form of $g_i(\mathbf{x}) \le \mathbf{b}_i$, where the $\mathbf{b}_i$ vector consists of all the variables in (16), which are independent of $\mathbf{x}$. In order to linearise (16), a first-order Taylor expansion at the *a priori* estimate can be used [25]. Hence, the Jacobian of (16) can be written as

$$g_i{}'(\mathbf{x}) = \frac{\partial g_i(\mathbf{x})}{\partial \mathbf{x}} = \begin{bmatrix} 1 & 0 & 0 & 0 & 0 & 0 & 0 & 0 \\ -1 & 0 & 0 & 0 & 0 & 0 & 0 & 0 \\ 0 & 0 & 1 & 0 & 0 & 0 & 0 & 0 \\ 0 & 0 & -1 & 0 & 0 & 0 & 0 & 0 \\ 0 & 0 & 0 & 0 & 1 & 0 & 0 & 0 \\ 0 & 0 & 0 & 0 & -1 & 0 & 0 & 0 \\ 0 & 0 & 0 & 0 & 0 & 0 & 1 & 0 \\ 0 & 0 & 0 & 0 & 0 & 0 & -1 & 0 \end{bmatrix} \tag{17}$$

Using the techniques presented in [21,24,25], (16) can be written as a linear inequality as

$$g_i'(\hat{\mathbf{x}}_k^-)\,\mathbf{x} \leq \mathbf{b}_i - g_i(\hat{\mathbf{x}}_k^-) + g_i'(\hat{\mathbf{x}}_k^-)\hat{\mathbf{x}}_k^- \tag{18}$$

in the form of $\mathbf{D}_1\mathbf{x} \leq \mathbf{d}_1$.

4.2. The Heading Constraint

Assuming that the UGV motion can be modelled by the CV motion model, a heading constraint can be imposed based on the 3σ uncertainty of the heading of a UGV range sensing tag, based on a UGV tag's velocity triangle. This is a constraint which is based on the kinematics of the UGV on a planar terrain. For simplicity, first, the subsequent derivation was performed by considering a given UWB tag on the UGV as a particle moving on a Cartesian plane. In Figure 4, the position plane is the Cartesian coordinate frame, relative to which a UGV tag's positions are denoted. Moreover, the velocity plane is the Cartesian coordinate frame, relative to which a UGV tag's velocity vectors are denoted. The velocity plane is an instantaneous coordinate frame, the origin of which is placed on the respective UGV tag's estimated position $(\widetilde{x}_{k-1}, \widetilde{y}_{k-1})$ at the previous time step. The velocity of a UGV tag is drawn in a solid black arrow, as shown in Figure 4 on the velocity plane.

$$\tan^{-1}\left(\frac{\widetilde{\dot{y}}_{k-1} - 3\sigma_{\dot{y}_{k-1}}}{\widetilde{\dot{x}}_{k-1} + 3\sigma_{\dot{x}_{k-1}}}\right) \leq \tan^{-1}\left(\frac{y_k - \widetilde{y}_{k-1}}{x_k - \widetilde{x}_{k-1}}\right) \leq \tan^{-1}\left(\frac{\widetilde{\dot{y}}_{k-1} + 3\sigma_{\dot{y}_{k-1}}}{\widetilde{\dot{x}}_{k-1} - 3\sigma_{\dot{x}_{k-1}}}\right) \tag{19}$$

$$\begin{gathered} \frac{\widetilde{\dot{y}}_{k-1} - 3\sigma_{\dot{y}_{k-1}}}{\widetilde{\dot{x}}_{k-1} + 3\sigma_{\dot{x}_{k-1}}} \leq \frac{(y_k - \widetilde{y}_{k-1})/T}{(x_k - \widetilde{x}_{k-1})/T} \leq \frac{\widetilde{\dot{y}}_{k-1} + 3\sigma_{\dot{y}_{k-1}}}{\widetilde{\dot{x}}_{k-1} - 3\sigma_{\dot{x}_{k-1}}} \\ \frac{\widetilde{\dot{y}}_{k-1} - 3\,\sigma_{\dot{y}_{k-1}}}{\widetilde{\dot{x}}_{k-1} + 3\,\sigma_{\dot{x}_{k-1}}} \leq \frac{\dot{y}_k}{\dot{x}_k} \leq \frac{\widetilde{\dot{y}}_{k-1} + 3\,\sigma_{\dot{y}_{k-1}}}{\widetilde{\dot{x}}_{k-1} - 3\,\sigma_{\dot{x}_{k-1}}} \end{gathered} \tag{20}$$

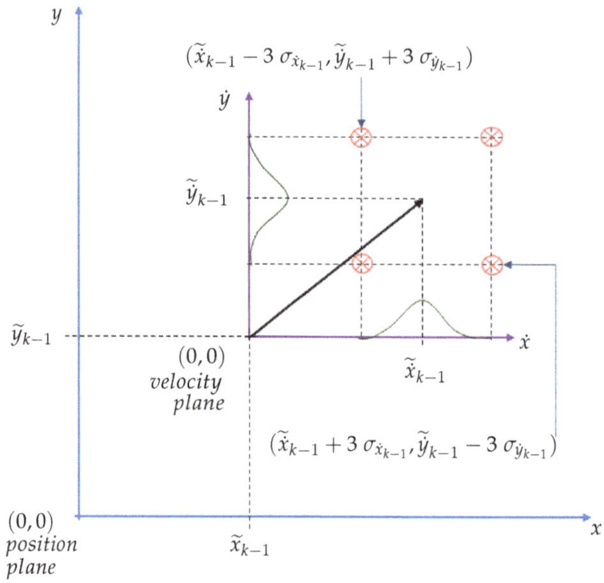

Figure 4. Illustration of the heading constraint.

When x, y, $\dot{x}$ and $\dot{y}$ denote the x-position of a tag, y-position of a tag, the velocity of a tag along the x-axis and the velocity of a tag along the y-axis, respectively; if an assumption is made (for simplicity) that the direction of a UGV tag's velocity is in the first quadrant of the velocity plane as shown in Figure 4, then the compound inequality (19) can be obtained. In the compound inequality (19), the expression

$$\tan^{-1}\left(\frac{y_k - \tilde{y}_{k-1}}{x_k - \tilde{x}_{k-1}}\right)$$

denotes the direction angle of the relative position vector of the current tag position estimate with respect to the previous tag position estimate. Nevertheless, this direction angle cannot be a deterministic variable due to the uncertainty of the velocity estimates. Referring to the velocity plane in Figure 4, the maximum and the minimum angle of the estimated velocity vector at the previous time step can be easily identified based on the velocity estimation uncertainties' 3σ edge limits. Hence, the left-hand side expression and the right-hand side expression of the compound inequality (19) can be formulated. Furthermore, (19) can be simplified to (20), where a compound inequality can be obtained, which can be used to impose as a constraint on the UGV state's velocity variables in the EKF-based localisation.

If the compound inequality constraint in (20) is generalised such that the constraint can be imposed while the UGV tag's velocity is in any quadrant of the velocity plane, then a common pattern can be observed. Hence, a consolidated inequality expression for the front UGV tag can be written as

$$\mathbf{W_f}\begin{bmatrix} x_{f_k} \\ \dot{x}_{f_k} \\ y_{f_k} \\ \dot{y}_{f_k} \end{bmatrix} < \begin{bmatrix} 0 \\ 0 \end{bmatrix} \tag{21}$$

where

$$\mathbf{W_f} = \begin{bmatrix} 0 & w_{f_{12}} & 0 & w_{f_{14}} \\ 0 & w_{f_{22}} & 0 & w_{f_{24}} \end{bmatrix}$$

and the elements of $\mathbf{W_f}$ are:

$$\begin{aligned} w_{f_{12}} &= -\left(\tilde{\dot{y}}_{f,k-1} + 3\sigma_{\dot{y}_{f,k-1}}\,\mathrm{sgn}(\tilde{\dot{x}}_{f,k-1})\right) \\ w_{f_{22}} &= \left(\tilde{\dot{y}}_{f,k-1} - 3\sigma_{\dot{y}_{f,k-1}}\,\mathrm{sgn}(\tilde{\dot{x}}_{f,k-1})\right) \\ w_{f_{14}} &= \left(\tilde{\dot{x}}_{f,k-1} - 3\sigma_{\dot{x}_{f,k-1}}\,\mathrm{sgn}(\tilde{\dot{y}}_{f,k-1})\right) \\ w_{f_{24}} &= -\left(\tilde{\dot{x}}_{f,k-1} + 3\sigma_{\dot{x}_{f,k-1}}\,\mathrm{sgn}(\tilde{\dot{y}}_{f,k-1})\right) \end{aligned}$$

If (21) is extended to both tags of the UGV, a linear inequality in the form of $\mathbf{D}_2\,\mathbf{x} \leq \mathbf{d}_2$ can be obtained, which can be written as

$$\begin{bmatrix} \mathbf{W_f} & 0 \\ 0 & \mathbf{W_r} \end{bmatrix} \mathbf{x} < \begin{bmatrix} 0 \\ 0 \\ 0 \\ 0 \end{bmatrix} \tag{22}$$

where $\mathbf{W}_r$ is the equivalent weight matrix in (21) written for the rear tag of the UGV.

4.3. Node Separation Constraint

Since the UGV is a rigid body and the localisation sensor nodes are rigidly attached to the UGV in a specific separation distance, a heuristic equality constraint can be imposed based on this aspect. If the localisation tag positions' coordinates are taken from the UGV state, the equality constraint can be formulated as

$$(\,x_f - x_r)^2 + (\,y_f - y_r\,)^2 = l^2 \tag{23}$$

in the form of $g_e(\,\mathbf{x}\,) = \mathbf{b}_e$, where l is the sensor node separation distance. In this research, it is assumed that the vehicle length is equal to the sensor node separation. In order to linearise (23), first-order Taylor expansion at the *a priori* estimate can be used, such that

$$g'_e(\,\mathbf{x}\,) = \frac{\partial g_e}{\partial \mathbf{x}} = \begin{bmatrix} 2\left(x_f - x_r\right) \\ 0 \\ 2\left(y_f - y_r\right) \\ 0 \\ -2\left(x_f - x_r\right) \\ 0 \\ -2\left(y_f - y_r\right) \\ 0 \end{bmatrix}^T$$

Hence, the linearised equality constraint can be obtained in the form of $\mathbf{D_e}\,\mathbf{x} = \mathbf{d}_e$ as follows:

$$g'_e(\,\hat{\mathbf{x}}_k^-\,)\,\mathbf{x} = \mathbf{b}_e - g_e(\,\hat{\mathbf{x}}_k^-\,) + g'_e(\,\hat{\mathbf{x}}_k^-\,)\,\hat{\mathbf{x}}_k^-$$

In order to successfully localise the UGV without undergoing any localisation singularities/ambiguities, the three constraints which are discussed have to be imposed concurrently in the CEKF-based UAV–UGV collaborative localisation, where the final expression encompassing all the constraints will be:

$$\begin{gathered} \begin{bmatrix} \mathbf{D}_1 \\ \mathbf{D}_2 \end{bmatrix} \mathbf{x} \leq \begin{bmatrix} \mathbf{d}_1 \\ \mathbf{d}_2 \end{bmatrix} \\ \mathbf{D}_e\mathbf{x} = \mathbf{d}_e \end{gathered} \tag{24}$$

5. Observability Enhancement of the CEKF

Tolerance to the singularities should be an essential feature in the proposed CEKF, in order to maintain the accuracy of localisation. Therefore, observability improvement has to be analysed while the proposed constraints in the previous subsection are imposed in an EKF-based localisation. Apart from the deterministic observability, which is used to analyse fully deterministic systems (or by assuming a system is fully deterministic), stochastic observability is used to analyse the reliability in CEKF for localisation in this particular system as the system is deterministically unobservable.

Stochastic observability: different to a deterministic system, a system can be observable, theoretically at least, if an appropriate random process is driving the system [29]. The random process can also be influenced by a reverse effect [29].

Stochastic observability implies that there exists a state estimator/filter of which the state estimation variance is bounded [30]. Hence, we can accept the state estimation of the CEKF if the error covariance bound of the proposed CEKF is sufficient in eradicating the state ambiguity in estimation.

Stochastic observability of the CEKF is analysed in real-time using the posterior Cramér–Rao bound. The most common version of the posterior Cramér–Rao bound in the context of Kalman filters is the bound, which is computed using the Fisher information matrix. However, calculating the estimation error covariance bound using the Fisher information matrix is incorrect as the state estimation is constrained in the CEKF. Therefore, the posterior Cramér–Rao bound must be computed in real-time, and then compensation must be made to the result to reflect the effects of the CEKF constraints on the estimation error covariance bound.

5.1. Posterior Cramér–Rao Bound

The information of a given estimator is defined as the inverse of the covariance matrix of estimation [31]. Hence, the information matrix $\mathbf{J}_k$ of a given estimator is defined as the inverse of the estimation error covariance matrix [32]:

$$\mathbf{J}_k = (\mathbf{P}_k)^{-1} \tag{25}$$

On the other hand, PCRB has been defined as

$$PCRB = E\Big\{[g(\overline{\mathbf{y}}) - \overline{\mathbf{x}}][g(\overline{\mathbf{y}}) - \overline{\mathbf{x}}]^T\Big\} \geq \mathbf{J}_k^{-1} \tag{26}$$

where $g(\overline{\mathbf{y}})$ is a function of the observation vector $\overline{\mathbf{y}}$, which delivers the output of a state estimator. The output vector of the function $g(\overline{\mathbf{y}})$ is an estimate of the state vector $\overline{\mathbf{x}}$. From (25) and (26), it can be seen that the covariance matrix of estimation that can be obtained using an estimation algorithm is bounded by the PCRB. Moreover, for an unbiased estimator such as the Kalman filter, Equation (26) becomes an equality [31].

In a real-time estimation process, calculating the PCRB directly using the expected values in (26) cannot be performed, because $\overline{\mathbf{x}}$ is not available. Therefore, PCRB should be calculated using the probability distribution functions of the state and the estimation of the state instead. If $p_{\overline{\mathbf{y}},\overline{\mathbf{x}}}(\overline{\mathbf{y}},\overline{\mathbf{x}})$ is the joint probability density of $\overline{\mathbf{y}}$ and $\overline{\mathbf{x}}$, then the elements of the information matrix at a given time step are such that [32]:

$$\mathbf{J}_{ij} = E\left[-\frac{\partial^2 \log p_{\overline{\mathbf{y}},\overline{\mathbf{x}}}(\overline{\mathbf{y}},\overline{\mathbf{x}})}{\partial \overline{x}_i \partial \overline{x}_j}\right] \quad i,j = 1, \cdots, n \tag{27}$$

provided that the expectations and derivatives exist.

To calculate PCRB at time step k, if we define state vector $\overline{\mathbf{x}}_k = [\mathbf{x}_1, \ldots, \mathbf{x}_k]$ and the observation vector $\overline{\mathbf{y}}_k = [\mathbf{y}_1, \ldots, \mathbf{y}_k]$, where $\mathbf{x}_1, \ldots, \mathbf{x}_k$ and $\mathbf{y}_1, \ldots, \mathbf{y}_k$ are a sequence of states and a sequence of measurements/observations of a non-linear state estimation process, respectively; then, the joint probability distribution $p(\overline{\mathbf{x}},\overline{\mathbf{y}})$ in (27) can be expressed as:

$$p_{\overline{\mathbf{x}}_k,\overline{\mathbf{y}}_k}(\overline{\mathbf{x}}_k,\overline{\mathbf{y}}_k) = p(\mathbf{x}_0)\prod_{j=1}^{k} p(\mathbf{y}_j|\mathbf{x}_j)\prod_{i=1}^{k} p(\mathbf{x}_i|\mathbf{x}_{i-1}) \tag{28}$$

From (28), it is apparent that the $p_{\bar{\mathbf{x}}_k,\bar{\mathbf{y}}_k}(\bar{\mathbf{x}}_k, \bar{\mathbf{y}}_k)$ expression is expanding as the time increases. Hence, if $p_{\bar{\mathbf{x}}_k,\bar{\mathbf{y}}_k}(\bar{\mathbf{x}}_k, \bar{\mathbf{y}}_k)$ is used directly to calculate $\mathbf{J}$, the computational cost will increase as the time passes [31,32]. If there is a recursive calculation method to calculate the PCRB at the kth time step, using the PCRB of the time step $k-1$, it will be a more computationally efficient way than the aforementioned way of calculating the PCRB.

In [31], a Riccati-like information matrix calculation method has been proposed, which is a great achievement since the method helps the sequential calculation of the PCRB efficiently with lower computational power in real-time estimation processes. Based on the recursive PCRB calculation methods presented in [31], the information matrix $\mathbf{J}$ can be calculated for a nonlinear time-invariant system with additive Gaussian noises (v and η), such as:

$$\begin{aligned} \mathbf{x}_{k+1} &= f_k(\mathbf{x}_k) + v \\ \mathbf{y}_k &= h_k(\mathbf{x}_k) + \eta \end{aligned} \tag{29}$$

using

$$\begin{aligned} \mathbf{J}_{k+1} &= \left(\mathbf{Q}_k + \mathbf{A}_k \mathbf{J}_k^{-1} \mathbf{A}_k^T\right)^{-1} + \mathbf{H}_{k+1}^T \mathbf{R}_{k+1}^{-1} \mathbf{H}_{k+1} \\ &\text{where} \\ \mathbf{A} &= \frac{\partial f(\cdot)}{\partial \mathbf{x}} \\ \mathbf{H} &= \frac{\partial h(\cdot)}{\partial \mathbf{x}} \end{aligned} \tag{30}$$

when **A** and **B** are evaluated at the *a posteriori* estimate.

5.2. Constrained Posterior Cramér–Rao Bound

The constrained EKF method, which is used in this research (estimation projection), which has been explained in Section 4, is a minimum variance and unbiased method [22,33]. In the CEKF, the *a posteriori* state estimate is projected into the state space, which is feasible with respect to the constraints. Hence, the information addition during both the a posteriori estimation and constrained estimation have to be evaluated at the CEKF state estimation update. In [34], it has been shown that the classical PCRB is invalid for the constrained state estimation and, therefore, a modified PCRB has been derived for constrained state estimations. Moreover, it has been shown in [34,35] that only the active constraints are contributing to the decrement of the constrained PCRB. The constraints are functions of state variables and system parameters. If an estimator has c number of active equality constraints at a given time step, such that

$$\mathbf{G}_c = \begin{bmatrix} g_1(\cdot) \\ g_2(\cdot) \\ \vdots \\ g_c(\cdot) \end{bmatrix} = 0 \tag{31}$$

where only the active constraints are denoted by $g_1(\cdot), g_2(\cdot), \dots, g_c(\cdot)$, then the CPCRB for any unbiased constrained state estimator is defined as

$$CPCRB \ = \mathbf{C}_c \mathbf{J}^{-1} \tag{32}$$

where

$$\mathbf{C}_c = \mathbf{I} - \mathbf{J}^{-1} [\nabla \mathbf{G}_c]^T \left\{ [\nabla \mathbf{G}_c] \mathbf{J}^{-1} [\nabla \mathbf{G}_c]^T \right\}^+ [\nabla \mathbf{G}_c] \tag{33}$$

Note: In (33), $\{\cdot\}^+$ is the pseudo-inverse of a matrix and the gradient ∇ has to be calculated with respect to the state vector. In this research, the constraints $\mathbf{G}_1, \mathbf{G}_2$ and $\mathbf{G}_3$

are taken as the sensor node separation-based constraint, position constraint and heading constraint, respectively, whenever they become active constraints.

5.3. Observability Analysis of the Proposed CEKF-Based Localisation

Using the CPCRB, the variance of a given UWB tag's estimated travel direction θ can be calculated so that reliable navigation can be ensured if

$$3\sigma_\theta < (\alpha - \theta) \tag{34}$$

at every time step since the CEKF itself restricts the state estimation to $\pm 3\sigma_\theta$ as explained in Section 4 (see Figure 5). A safer operation can be ensured by incorporating a safety factor $k_{sf} > 1$ into this condition, such that Equation (34) will become:

$$3k_{sf}\sigma_\theta < (\alpha - \theta)$$

by providing the CEKF more robustness against uncertainties.

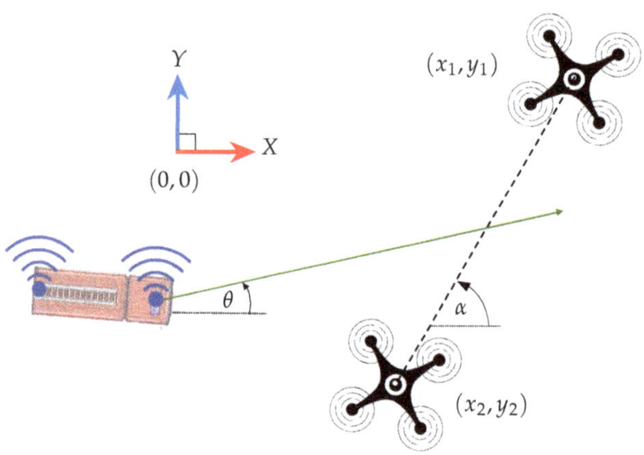

Figure 5. Plan view of the UAV–UGV system.

From (34), it is apparent that calculating σ_θ is the first step to assess the ambiguity aversion capability of the CEKF. For this purpose, we adopt the following relationship:

$$\sigma^2(\tan(\theta)) = \sigma^2\left(\frac{\tilde{y}^+}{\tilde{x}^+}\right) \tag{35}$$

where $\sigma^2(\cdot)$ is the variance.

Using the series expansion for $\tan(\theta)$ when $|\theta| \leq 45°$, (35) can be approximated by

$$\sigma^2\left(\theta + \frac{\theta^3}{3}\right) \approx \sigma^2\left(\frac{\tilde{y}^+}{\tilde{x}^+}\right) \tag{36}$$

where the $\tan(\cdot)$ approximation function behaves as shown in Figure 6.

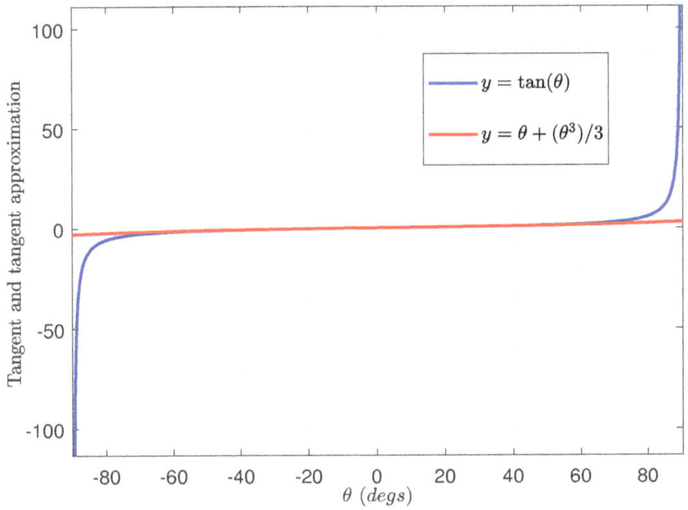

Figure 6. $\tan(\theta)$ approximation function.

Assuming the heading angle estimation's probability distribution is a Gaussian distribution, the LHS of Equation (36) can be estimated using the moment generating functions (non-central). The mean value of θ (i.e., μ_θ), which has to be known to be able to follow this method, can be calculated by

$$\mu_\theta = \tan^{-1}\left(\frac{\tilde{y}^+}{\tilde{x}^+}\right)$$

using the constrained *a posteriori* estimates. The RHS of (36) can be calculated using the Taylor series approximated variance of ratios [36]:

$$\sigma^2\left(\frac{\tilde{y}^+}{\tilde{x}^+}\right) = \left(\frac{\tilde{y}^+}{\tilde{x}^+}\right)^2 \left[\frac{\sigma^2\left(\tilde{y}^+\right)}{\left(\tilde{y}^+\right)^2} + \frac{\sigma^2\left(\tilde{x}^+\right)}{\left(\tilde{x}^+\right)^2} - 2\frac{\mathrm{cov}\left(\tilde{y}^+, \tilde{x}^+\right)}{\tilde{x}^+ \cdot \tilde{y}^+}\right] \tag{37}$$

The value of $\sigma^2\left(\tilde{y}^+ / \tilde{x}^+\right)$ in (37) can be calculated using the constrained estimations of the CEKF.

After calculating the LHS of (36) using moment-generating functions, (36) can be written as:

$$\frac{5}{3}\sigma^6(\theta) + \left(2 + 4\mu_\theta^2\right)\sigma^4(\theta) + \left(1 + \mu_\theta^2\right)^2 \sigma^2(\theta) \approx \sigma^2\left(\frac{\tilde{y}^+}{\tilde{x}^+}\right) \tag{38}$$

Hence, the solution to Equation (36) is a cubic equation, roots of which are σ_θ^2.

Solving Equation (38) as a cubic polynomial equation, σ_θ^2 can be calculated by only taking the positive real roots for granted. The value of σ_θ can be used to prove the efficacy of this CEKF-based localisation methodology, as mentioned earlier in this section. Moreover, σ_θ can be used as a safety indicator to alarm the UAVs to adjust the α angle by moving appropriately if $\alpha - \theta \leq 3\sigma_\theta$, which can lead to an erroneous localisation.

Caveat: Since the approximation in Equation (36) is not valid when $45^\circ < |\theta| \leq 90^\circ$, an intermediate coordinate transformation must be used in order to solve Equations (35)–(38), which can be reversed after obtaining the equations' solutions without affecting the CEKF/ EKF algorithms.

6. Simulation Results

The numerical simulations were performed to localise a UGV from UAVs using the CEKF method. Furthermore, the ambiguity aversion efficacy is tested for the simulated scenarios. In the simulations and experiments, the algorithmic workflow explained in Algorithm 1 was executed:

Algorithm 1 CEKF-based UAV–UGV collaborative localisation algorithm

initialise the EKF;
do
    obtain range measurements;
    estimate the UGV location using the EKF;
while *EKF estimation covariance matrix is not converged*
loop
    obtain range measurements
    estimate the UGV location using the EKF;
    calculate the constrained location estimation;
    calculate σ_θ;
    if $(\alpha - \theta) \leq 3k_{sf}\sigma_\theta$ **then**
        move the UAVs to adjust α;
    end
end loop

The simulation results are presented in the next two subsections.

6.1. Ambiguity Aversion by CEKF Method

Numerical simulations were carried out to assess a UGV localisation scenario on a two-dimensional plane using two stationary UAVs. The UGV is supposed to drive through the unobservable boundary in Figure 2.

Unconstrained EKF shows an error in localisation at the unobservable boundary (Figure 7), which ultimately results in a substantial localisation error, i.e., the mirror image of the actually traversed trajectory is given as the EKF estimated location. This happens because the range readings are identical for the real trajectory and its mirror image trajectory.

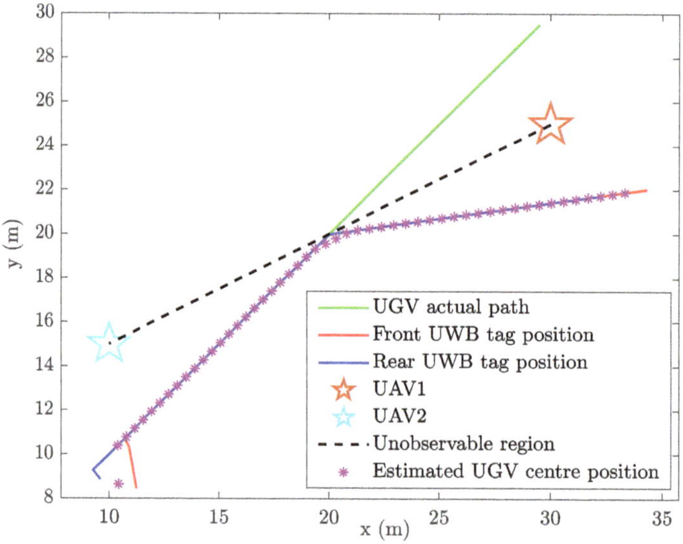

Figure 7. Unconstrained localisation of the UGV.

In an identical situation, the state-constrained localisation from CEKF does not show an error in localisation at the line of unobservability (Figure 8). Figure 9 shows the heading angle (θ) estimation result, which affirms that there have not been any singularities occurred during the CEKF-based localisation.

Another numerical simulation was performed where the UGV travels along a lengthy two-dimensional random path requiring the UAVs to move, i.e., the UAVs are also moving while the UGV localisation is carried out. CEKF is used to estimate the location of the UGV. In this simulation, the UGV successfully traverses across the line of unobservability twice without any erroneous localisation due to singularities.

Location estimation plot of the UGV, ground truth locations of the UGV and the respective UAV positions are shown in Figure 10. Root Mean Squared Error (RMSE) has been calculated while the UGV is travelling along the random path and the RMSE vs. time for both x and y coordinates is shown in Figure 11.

The sampling time of the simulation is 50 ms, which is typically an attainable step time in field vehicles' (e.g., unmanned fire truck) onboard computers while other peripheral devices are also operated/controlled by the same onboard computer. In order to measure the real-time performance of the CEKF localisation algorithm, the code was run in MATLAB2018b software and the physical time was measured for each time-step of the simulation using the inbuilt stopwatch timer facility (MATLAB commands: *tic, toc*). At the beginning of each iteration of the simulation, the *tic* command is called to record the physical time, and at the end of each iteration, the *toc* command is called to record the physical time. Using the time difference of the recorded physical timings, the iteration execution time is calculated. Based on performance results, which are plotted in Figure 12, it can be seen that the physical real-time calculations can be performed without any time lags to successfully localise a UGV using the presented CEKF-based localisation algorithm.

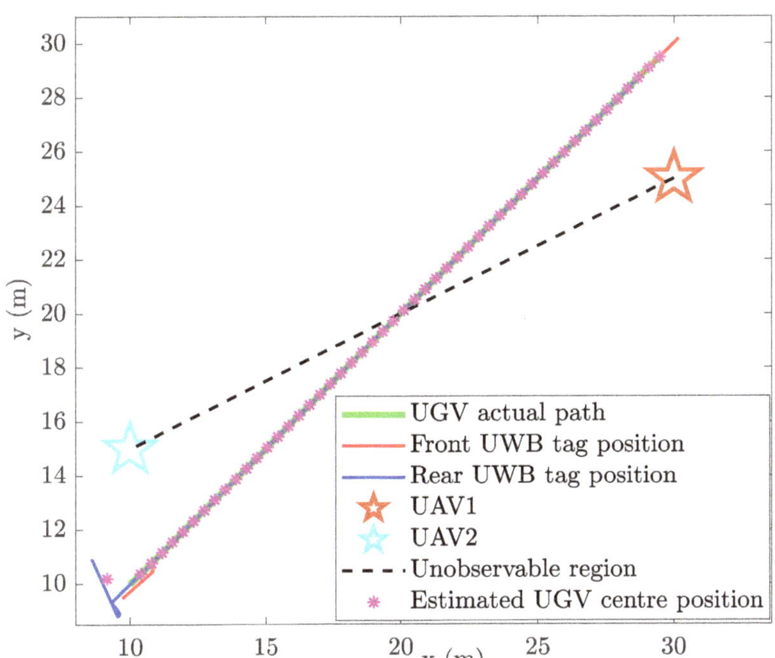

Figure 8. Constrained localisation of the UGV.

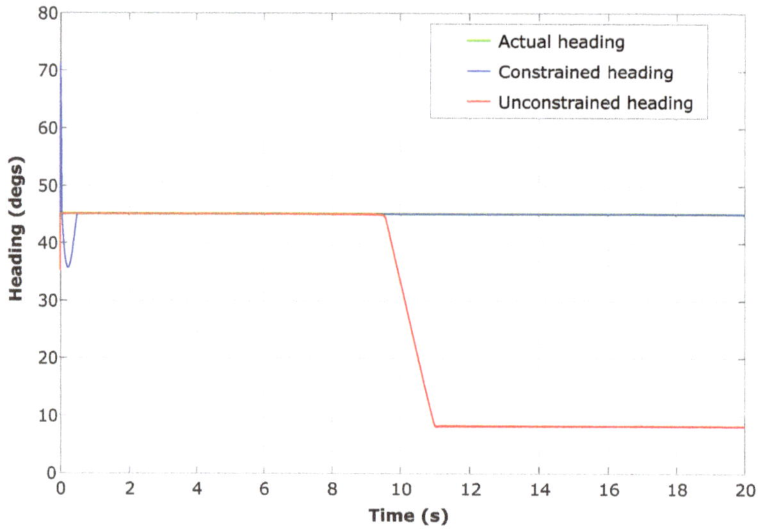

Figure 9. Estimated heading of a tag, which is fixed on the UGV.

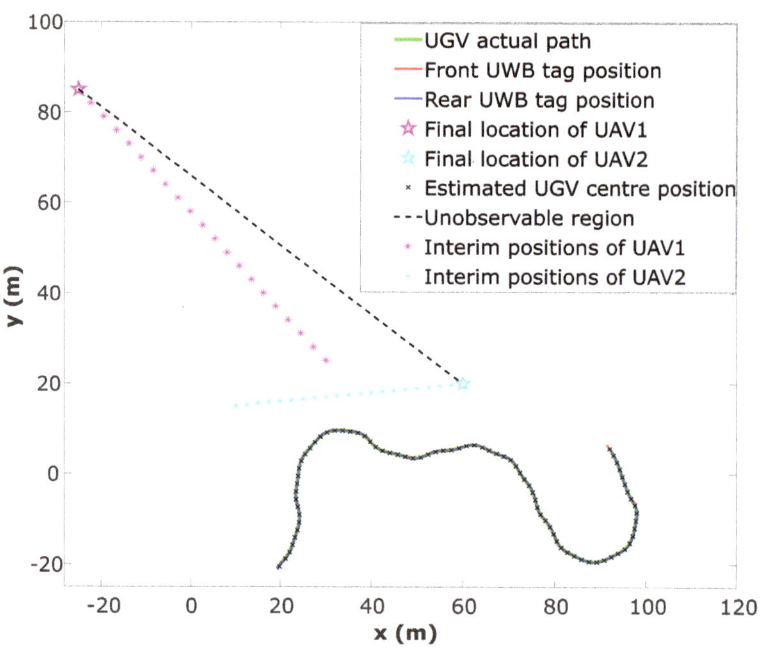

Figure 10. Constrained localisation of the UGV along a random path using the CEKF-based localisation method.

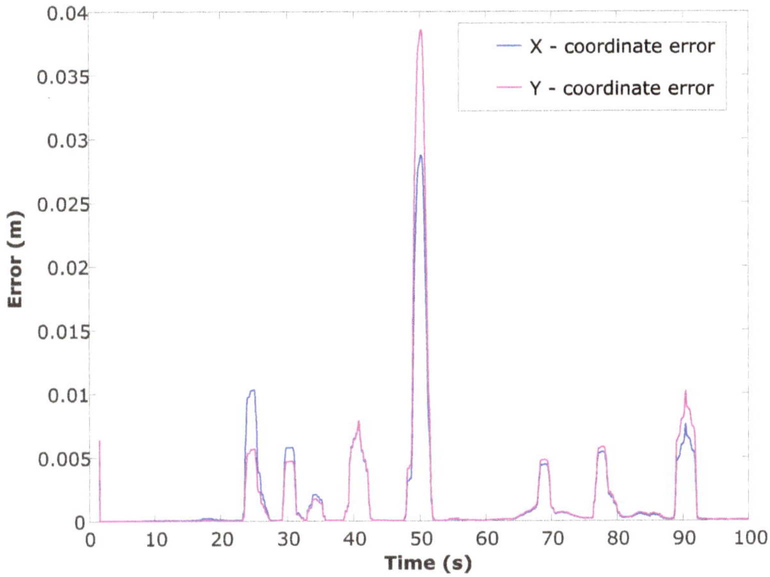

Figure 11. RMSE along the random travel path when the UGV is localised using the CEKF-based localisation method.

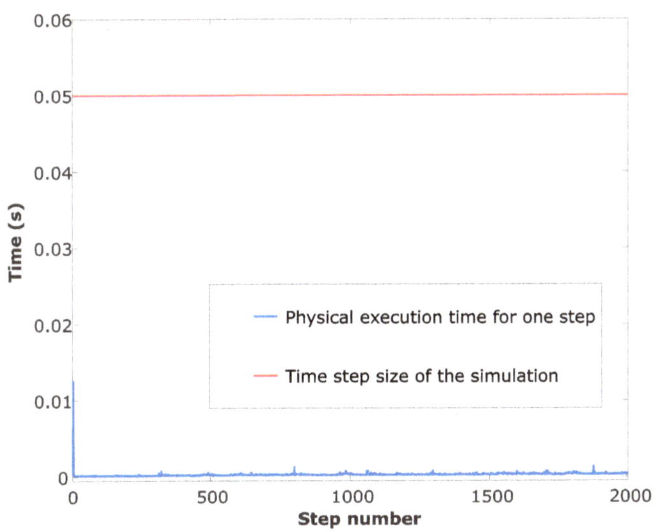

Figure 12. Real-time performance of the CEKF localisation algorithm.

6.2. Observability of CEKF

The diagonal of the CPCRB consists of the lower bounds of state estimation error variances for each and every element in the estimated state vector. Hence, the trace of the CPCRB matrix is an indicator of stochastic observability, i.e., smaller the trace value, the higher the stochastic observability and vice versa. The trace of the CPCRB calculated for the EKF localisation and for the CEKF localisation shows a difference in the trace of the covariance matrix of estimation error lower bound (i.e., CPCRB), as shown in Figure 13. Two peaks shown in blue show the increment of stochastic unobservability when the two UWB tags intercept the unobservable blue-coloured vertical plane shown in Figure 2. In contrast, the CEKF method has successfully minimised the adverse effects of unobservability.

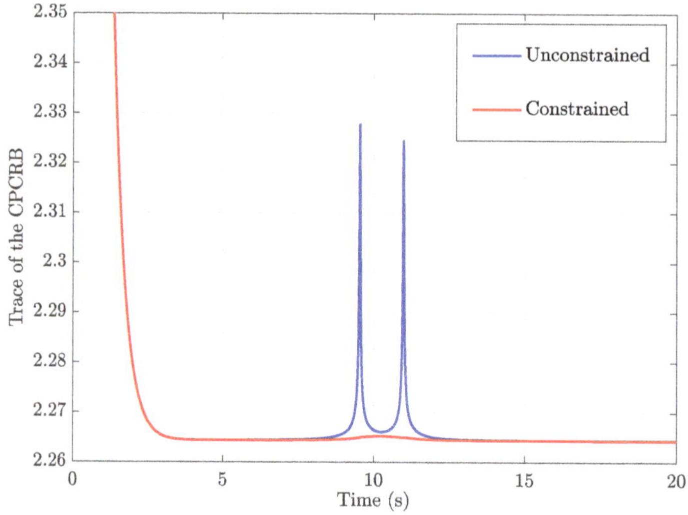

Figure 13. Trace of the CPCRB in unconstrained EKF and constrained EKF.

In this simulation, α was kept constantly at 26.6° by keeping the UAVs still, and θ was also kept constant at 45° by driving the UGV with a constant heading for simplicity. Figure 9 and 14 show a more steady θ estimated value despite the unobservability when the CEKF method is used for the localisation. However, when the unconstrained EKF is used for the localisation, a drastic deviation of the heading estimation is notable at 9.5 s when we observe the red curve of Figure 9. It happens at the same time when the UGV undergoes unobservability, as we can see that the first unobservability peak in Figure 13 has also taken place at 9.5 s. Moreover, when CEKF is used, σ_θ is significantly low without any sudden changes, even at/after the instance when the unobservability occurs. According to Figure 14, σ_θ is 31.1° before encountering the unobservability, and 23.9° after encountering the unobservability when using the unconstrained EKF. On the other hand, σ_θ is steadily maintained around 0.05°; when using the CEKF. Since σ_θ is around 0.05°, the condition for ambiguity aversion: $\alpha - \theta > 3\sigma_\theta$ (in (34)) is also strongly satisfied in this simulation. Hence, we can conclude that the CEKF has mitigated the ambiguities in localisation and contributed to increasing the localisation accuracy.

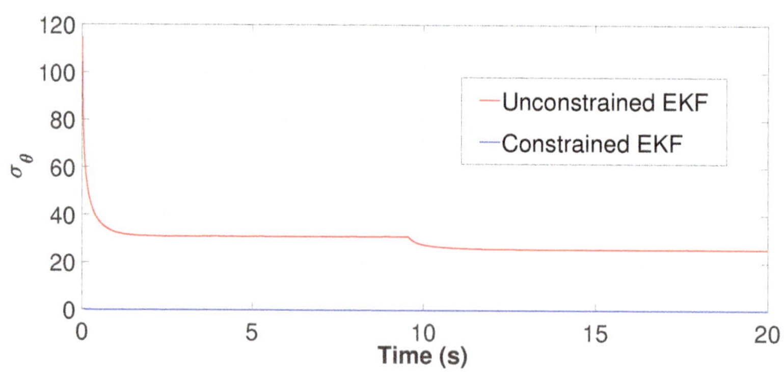

Figure 14. Estimated σ_θ, which is derived from the CPCRB.

7. Experiment Results

An experiment was performed to validate the CEKF-based UAV–UGV collaborative localisation method. Two DJI-M600Pro hexacopters (UAVs) were used for range measurement, position estimation and heading estimation of a stationary Antonio Carraro farm tractor (UGV). In Figure 15, an image of the overall experiment is shown. The experiment was performed on a farmland in Menangle, NSW, Australia.

Figure 15. UAV–UGV range-based collaborative localisation experiment.

The TREK-1000 range sensors manufactured by Decawave, Ireland were mounted on the front side of the vehicle and on the rear side of the UGV, as shown in Figure 16. The range sensors were mounted underneath the UAVs, as shown in Figure 17. Data logging and processing in each UAV were performed in Intel NUC onboard computers, which were fixed on each UAV as shown in Figure 17.

Figure 16. Range sensors on the UGV.

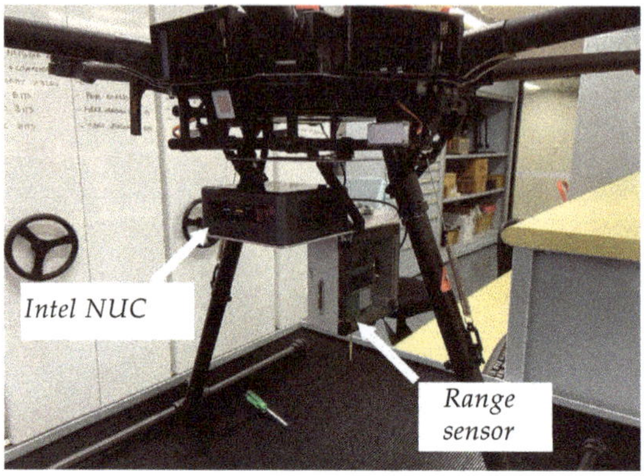

Figure 17. Range sensors on the UAV.

The next two sub-sections will discuss the results of a static UGV localisation experiment and a dynamic UGV localisation experiment.

7.1. Stationary UGV Localisation

While the UGV was stationary on the ground, the two UAVs were airborne, and the localisation began while the vehicles were maintaining their positions. While the localisation is in progress, the ground truth location and the heading of the UGV were recorded by the onboard computer of the UGV using an RTK GPS unit. The accuracy of the RTK GPS unit was ±2 cm. Figure 18 shows the estimated positions of the UGV during the experiment.

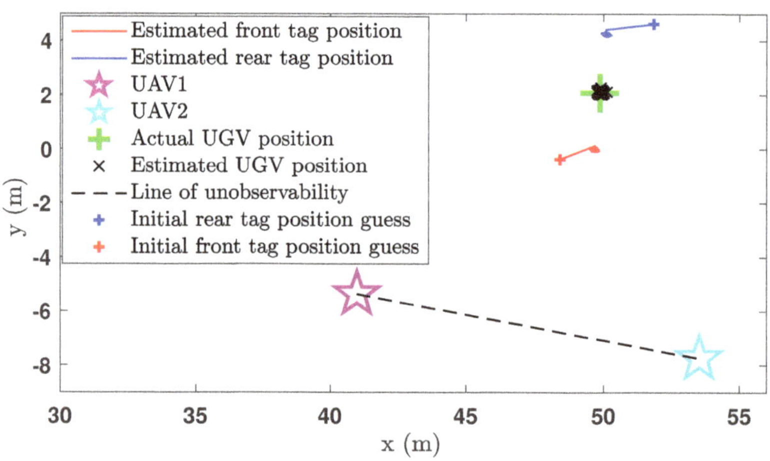

Figure 18. Position estimation plot of the CEKF-based UAV–UGV collaborative localisation when the UGV is not moving.

The green cross in Figure 18 is the ground truth middle position of the UGV, where the estimated middle position of the UGV is shown by the black cross marks. Actually, the middle position of the UGV was calculated at each iteration after estimating the positions of the front UWB tag and the rear UWB tag. The heading of the UGV was also estimated in the same way as shown in Figure 19, from which the heading estimation accuracy can be deduced as $\pm 4^{\circ}$. The UGV positioning error during the experiment is shown in Figure 20, which shows the positioning error of the localisation is less than 14 cm.

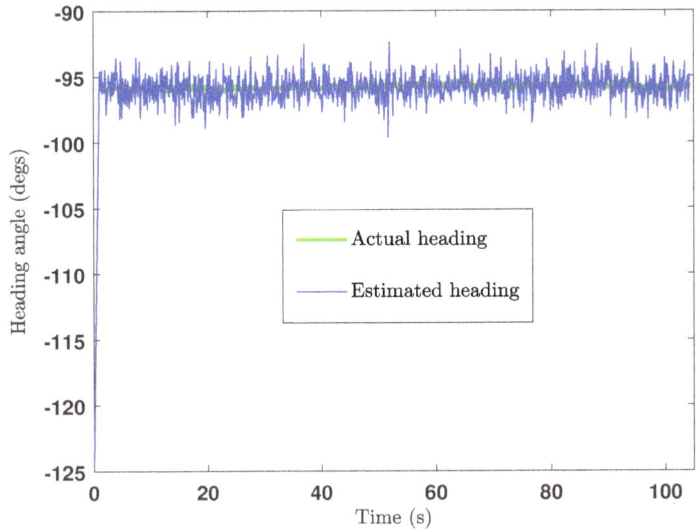

Figure 19. Heading estimation plot of the CEKF-based UAV–UGV collaborative localisation.

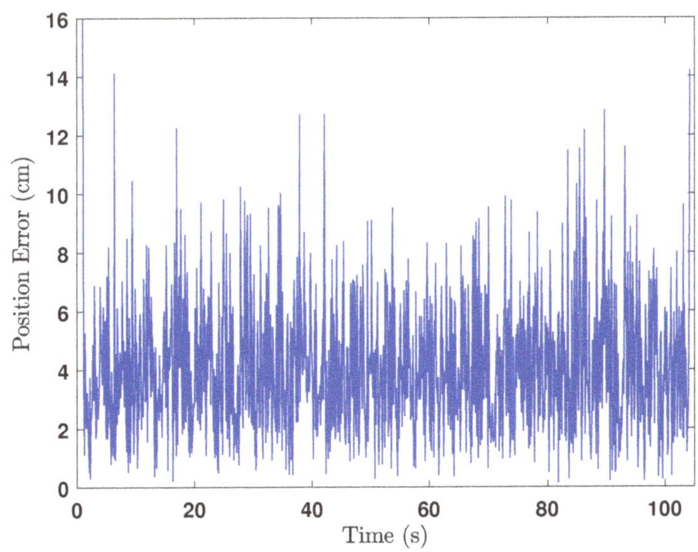

Figure 20. Error of the position estimation.

7.2. Non-Stationary UGV Localisation

This is a continuation of the experiment discussed in the previous section; thus same experimental instruments have been used. Initially, the UGV was stationary for 105 s, while the UAVs are airborne. Thereafter, the UGV started moving along the path marked by green dots in Figure 21. During the UGV movements, the UGV intercepted the line of unobservability several times, during which the CEKF-based localisation algorithm was able to successfully deal with the unobservability problem that was described earlier in Section 3. The CEKF-based localisation algorithm has localised the UGV from its starting position (49.88, 2.11) until its final location, which is marked by the green cross in Figure 21.

At the interceptions of the unobservable boundary, the CEKF constraints that are activated prevent the localisation algorithm from producing unrealistic location estimations which are against the kinematics of the UGV. For example, if the location estimation algorithm delivers a false location estimation due to the unobservability (as shown in Figure 7), where the false location estimation is on the mirror image path of the actual trajectory, then the constrains will identify it as an erroneous result and deliver the kinematically feasible correct location estimation. Hence, there have not been any erroneous localisation taken place during field experiments. Nevertheless, if the unconstrained EKF is used, erroneous location estimations will not be rectified as such.

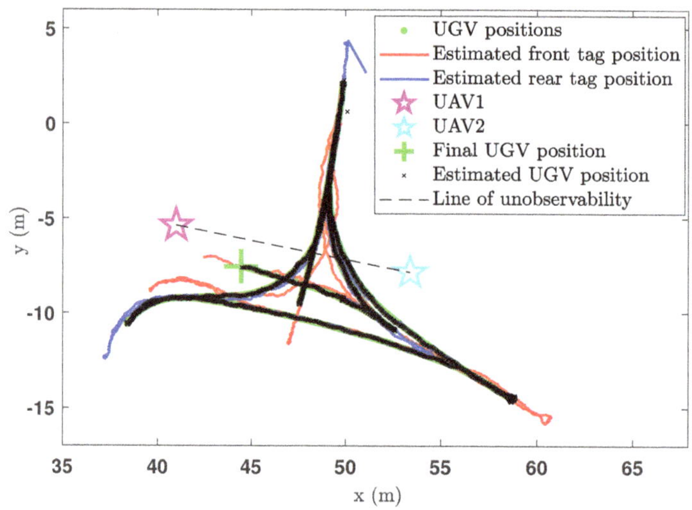

Figure 21. Position estimation plot of the CEKF-based UAV–UGV collaborative localisation when the UGV is moving.

The estimated UGV headings and the positions are as depicted in Figures 22 and 23, respectively. By observing Figures 22 and 23, we can deduce that the heading estimation accuracy is $\pm 4^{\circ}$ and the position estimation accuracy is ± 23 cm.

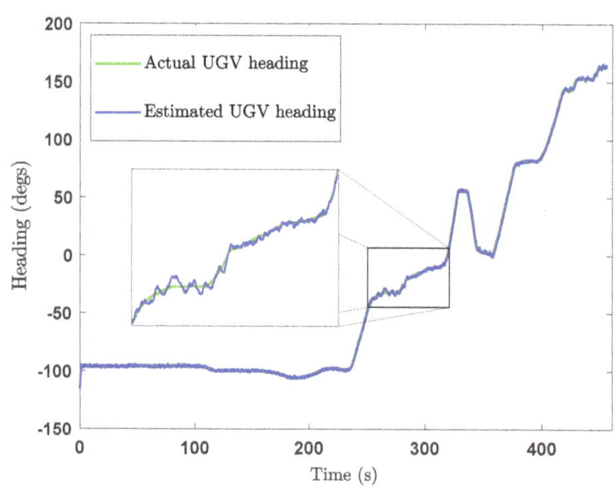

Figure 22. Heading estimation plot of the CEKF-based UAV–UGV collaborative localisation when the UGV is moving.

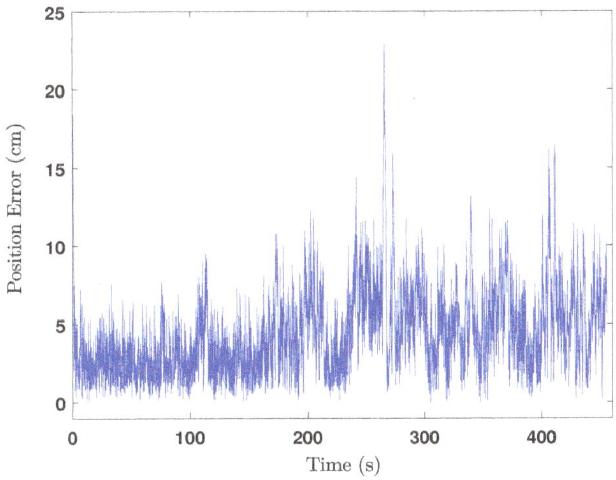

Figure 23. Error of the position estimation when the UGV is moving.

8. Conclusions and Future Works

In a UAV–UGV collaborative localisation with range-only observations, severe localisation errors can occur due to the unavailability of UAVs to be deployed to make sufficient range observations. It is well known that a unique global localisation to track a vehicle on a 2D terrain is possible when three or more drones are deployed to make range-only observations. However, when only two UAVs are deployed, there exists an unobservable region within which it is difficult to localise the UGVs using range-only observations when EKF is used.

The presented CEKF-based localisation method successfully avoids the ambiguity in localisation due to the unobservability. Successful numerical simulations and field experiments show the efficacy in localisation despite the localisation unobservability/ambiguity arising when only two UAVs are deployed for the collaborative localisation task.

The presented CEKF-based localisation method successfully localise a UGV travelling on a 2D terrain, using only two UAVs if an initial position guess (fairly close enough to the UGV) is provided. Nevertheless, the initial position guess must be in the same side where the actual UGV is at; with respect to the unobservability boundary (in Figure 2). After this initial position guess is provided, the subsequent location estimations will converge to the UGV location. Localising a UGV travelling on a 2D terrain, using less than three UAVs cannot be done due to the estimation singularities/ambiguities which may occur during localisation. While the location estimation is ongoing, the ambiguity aversion features of the presented localisation algorithm will rectify any localisation errors whenever the UGV encounters a localisation ambiguity. Because of that, localising a UGV travelling on a 2D terrain, using less than three UAVs is possible with the CEKF-based localisation method presented.

The stochastic observability analysis, which has been presented in this paper, gives solid evidence of how the CEKF-based UAV–UGV collaborative localisation method can overcome estimation singularities. CEKF-based localisation increases the stochastic observability of the localisation more than the EKF-based localisation method. Furthermore, the kinematic constraints, which take the state variables' variances, give an assurance of successful localisation despite the localisation singularity almost all the time.

However, there can be quite a few instances where UAVs must be slightly re-positioned to ensure a successful localisation. The CPCRB-based stochastic observability analysis framework, discussed in Section 5.3 can be used to identify the instances when the UAVs must be slightly re-positioned.

In conclusion, a UGV which is travelling on a horizontal plane can be localised (and the heading can also be estimated) by only two UAVs using the presented CEKF-based range-only localisation method.

In future works, the proposed system can be further researched to analyse localisation efficacy (1) for different range measurement sensitivities, (2) when occasional range sensing occlusions occur, and (3) in localising UGVs travelling in hilly terrains.

Author Contributions: Conceptualization, A.H.T.E.D.S. and J.K.; Methodology, A.H.T.E.D.S.; Software, A.H.T.E.D.S.; Validation, A.H.T.E.D.S.; Formal analysis, A.H.T.E.D.S.; Investigation, A.H.T.E.D.S.; Resources, J.K.; Data curation, A.H.T.E.D.S. and J.K.; Writing—original draft, A.H.T.E.D.S.; Writing—review & editing, J.K.; Supervision, J.K.; Project administration, J.K.; Funding acquisition, J.K. All authors have read and agreed to the published version of the manuscript

Funding: This research received funding from the University of New South Wales (UNSW), Sydney, Australia. No external funding was received.

Data Availability Statement: Data are contained within the article.

Acknowledgments: The authors acknowledge the facilitation provided by the University of New South Wales, especially the experimentation facilities provided. Further acknowledgements are for the Scientia research scholarship programme conducted by the University of New South Wales.

Conflicts of Interest: The authors declare no conflicts of interest.

References

1. Vandapel, N. Unmanned Ground Vehicle Navigation Using Aerial Ladar Data. *Int. J. Robot. Res.* **2002**, *25*, 31–51. [CrossRef] [CrossRef]
2. Starr, J.W. Rangefinding in Fire Smoke Environments. Ph.D. Thesis, Virginia Polytechnic Institute and State University, Blacksburg, VA, USA, 2015.
3. Yamauchi, B. All-weather perception for small autonomous UGVs. In Proceedings of the SPIE Defense and Security Symposium, Orlando, FL, USA, 16–20 March 2008. [CrossRef]
4. Sakai, M.; Aoki, Y.; Takagi, M. The support system of firefighters by detecting objects in smoke space. In Proceedings of the Optomechatronic Technologies, Sapporo, Japan, 5–7 December 2005. [CrossRef]
5. Steven T. Ostrowski. Ultra Wideband (UWB) Sensor Integration and Application in GPS-Compromised Environments. Master's Thesis, The Ohio State University, Columbus, OH, USA, 2015.
6. Kobayashi, H.; Molisch, A.F.; Poor, H.V.; Sahinoglu, Z. Localization via ultra-wideband radios: A look at positioning aspects for future sensor networks. *IEEE Signal Process. Mag.* **2005**, *22*, 70–84. [CrossRef]
7. Zhao, Y.; Zou, J.; Guo, J.; Huang, G.; Cai, L. A novel ultra-wideband double difference indoor positioning method with additional baseline constraint. *ISPRS Int. J. Geo-Inf.* **2021**, *10*, 634. [CrossRef] [CrossRef]
8. Kwak, M.; Chong, J. A New Double Two-way Ranging Algorithm for Ranging System. In Proceedings of the 2010 2nd IEEE International Conference on Network Infrastructure and Digital Content, Beijing, China, 24–26 September 2010; pp. 470–473. [CrossRef]
9. Wen, C.Y.; Chen, J.K.; Sethares, W. Asynchronous Two-Way Ranging Using Tomlinson-Harashima Precoding and UWB Signaling. *EURASIP J. Wirel. Commun. Netw.* **2008**, *2008*, 436865. [CrossRef] [CrossRef]
10. Al-hadhrami, S.; Al-ammar, M.A.; Al-khalifa, H.S. Ultra Wideband Indoor Positioning Technologies: Analysis and Recent Advances. *Sensors* **2016**, *76*, 707. [CrossRef] [CrossRef]
11. Kannan, A.A.; Mao, G.; Vucetic, B. Simulated annealing based wireless sensor network localization. *J. Comput.* **2006**, *1*, 15–22. [CrossRef] [CrossRef]
12. Guo, Q.; Zhang, Y.; Lloret, J.; Kantarci, B.; Seah, W.K. A Localization Method Avoiding Flip Ambiguities for Micro-UAVs with Bounded Distance Measurement Errors. *IEEE Trans. Mob. Comput.* **2019**, *18*, 1718–1730. [CrossRef] [CrossRef]
13. Mao, G.; Drake, S.; Anderson, B.D. Design of an extended kalman filter for UAV localization. In Proceedings of the 2007 Information, Decision and Control, Adelaide, SA, Australia, 12–14 February 2007; pp. 224–229. [CrossRef]
14. Kiss, D.; Tevesz, G. Autonomous Path Planning for Road Vehicles in Narrow Environments: An Efficient Continuous Curvature Approach. *J. Adv. Transp.* **2017**, *2017*, 2521638. [CrossRef] [CrossRef]
15. Bar-Shalom, Y.; Li, X.R.; Kirubarajan, T. *Estimation with Applications to Tracking and Navigation*; John Wiley & Sons, Inc.: Hoboken, NJ, USA, 2001. [CrossRef]
16. Sarkka, S. *Bayesian Filtering and Smoothing*; Cambridge University Press: Cambridge, UK, 2013; Volume 3.
17. Hwang, C.S. Observability and Information Structure of Nonlinear Systems. Ph.D. Thesis, Oregon State University, Corvallis, OR, USA, 1985.

18. Quan, Q. Observability and Kalman Filter. In *Introduction to Multicopter Design and Control;* Springer: Singapore , 2017; pp. 1–384. [CrossRef]
19. Vidyasagar, M. *Nonlinear Systems Analysis*, 2nd ed.; Prentice Hall: Upper Saddle River, NJ, USA, 2002.
20. Szabo, F. *The Linear Algebra Survival Guide;* Elsevier: Montreal, QC, Canada, 2015. [CrossRef]
21. Simon, D. Kalman filtering with state constraints: A survey of linear and nonlinear algorithms. *IET Control Theory Appl.* **2010**, *4*, 1303–1318. [CrossRef] [CrossRef]
22. Simon, D.; Chia, T.L. Kalman filtering with state equality constraints. *IEEE Trans. Aerosp. Electron. Syst.* **2002**, *38*, 128–136. [CrossRef] [CrossRef]
23. Teixeira, B.O.; Chandrasekar, J.; Tôrres, L.A.; Aguirre, L.A.; Bernstein, D.S. State estimation for linear and non-linear equality-constrained systems. *Int. J. Control* **2009**, *82*, 918–936. [CrossRef] [CrossRef]
24. Gupta, N.; Hauser, R. *Kalman Filtering with Equality and Inequality State Constraints;* Technical Report 07; Oxford University Computing Laboratory: Oxford, UK, 2008.
25. Simon, D. *Optimal State Estimation;* John Wiley & Sons: Hoboken, NJ, USA, 2006. [CrossRef]
26. Liuping Wang. *Model Predictive Control System Design and Implementation Using MATLAB;* Springer: Berlin/Heidelberg, Germany, 2009; pp. 25–26. [CrossRef]
27. Carignan, C.R. Trajectory optimization for kinematically redundant arms. *J. Robot. Syst.* **1991**, *8*, 221–248. [CrossRef] [CrossRef]
28. Carignan, C.R.; Tarrant, J.M. Constrained Trajectory Optimization for Kinematically Redundant Arms. In Proceedings of the Third International Symposium on Robotics and Manufacturing, Vancouver, BC, Canada, 18–20 July 1990.
29. Mohler, R.R.; Hwang, C.S. Nonlinear data observability and information. *J. Franklin Inst.* **1988**, *325*, 443–464. [CrossRef] [CrossRef]
30. Bageshwar, V.L.; Gebre-Egziabher, D.; Garrard, W.L.; Georgiou, T.T. Stochastic observability test for discrete-time kalman filters. *J. Guid. Control. Dyn.* **2009**, *32*, 1356–1370. [CrossRef] [CrossRef]
31. Tichavsky, P.; Muravchik, C.H.; Nehorai Arye. Posterior Cramer–Rao Bounds for Discrete-Time Nonlinear Filtering. *IEEE Trans. Signal Process.* **1998**, *46*, 1386–1396. [CrossRef] [CrossRef]
32. Šimandl, M.; Královec, J.; Tichavský, P. Filtering, predictive, and smoothing Cramér–Rao bounds for discrete-time nonlinear dynamic systems. *Automatica* **2001**, *37*, 1703–1716. [CrossRef] [CrossRef]
33. Musil, K. Methods for Constrained State Estimation: Comparison and Application to Zero-Bound Interest Rate Problem. Ph.D. Thesis, Charles University, Prague, Czech Republic, 2013.
34. Gorman, J.D.; Hero, A.O. Lower Bounds For Parametric Estimation with Constraints. *IEEE Trans. Inf. Theory* **1990**, *36*, 1285–1301. [CrossRef] [CrossRef]
35. Terrence Joseph Moore, J. A Theory of Cramer-Rao Bounds for Constrained Parametric Models. Ph.D. Thesis, University of Maryland, College Park, MD, USA, 2010.
36. King, B.F.; Wolter, K.M. *Introduction to Variance Estimation;* Springer: Berlin/Heidelberg, Germany, 1987; Volume 82, p. 688. [CrossRef]

sensors

MDPI

Article

Classification of Unmanned Aerial Vehicles Based on Acoustic Signals Obtained in External Environmental Conditions

Marzena Męsikowska

Faculty of Mechatronics and Mechanical Engineering, Kielce University of Technology, 25-314 Kielce, Poland; marzena@tu.kielce.pl; Tel.: +48-730-930-962

Abstract: Detection of unmanned aerial vehicles (UAVs) and their classification on the basis of acoustic signals recorded in the presence of UAVs is a very important source of information. Such information can be the basis of certain decisions. It can support the autonomy of drones and their decision-making system, enabling them to cooperate in a swarm. The aim of this study was to classify acoustic signals recorded in the presence of 17 drones while they hovered individually at a height of 8 m above the recording equipment. The signals were obtained for the drones one at a time in external environmental conditions. Mel-frequency cepstral coefficients (MFCCs) were evaluated from the recorded signals. A discriminant analysis was performed based on 12 MFCCs. The grouping factor was the drone model. The result of the classification is a score of 98.8%. This means that on the basis of acoustic signals recorded in the presence of a drone, it is possible not only to detect the object but also to classify its model.

Keywords: unmanned aerial vehicle; discriminant analysis; drone classification

1. Introduction

Drones, also known as unmanned aerial vehicles (UAVs), have become one of the most dynamically developing areas of aviation technology in recent years. Their versatility and ability to perform a variety of tasks have contributed to a wide range of applications in many fields and areas of the economy [1–14]. From agriculture and rescue operations to infrastructure inspection and transportation, drones contribute significantly to the efficiency and safety of many processes. Drones are widely used in precision agriculture [1,2]. Using drones, farmers can monitor the conditions of crops, identify farm areas that require irrigation or fertilization, and assess plant health [3]. Drones generate high-resolution images and maps, which enable optimization and efficient management of farms, thereby increasing crop yield. Drones are used to inspect hard-to-reach or dangerous locations such as high-voltage power lines, pipelines, telecommunication towers, and bridges [4–7]. In rescue operations, drones can quickly reach disaster sites, detect fires, provide first aid, and monitor situations in real time [8]. In security, drones are used for surveillance and border patrol [9,10]. Drones can be used in transport, especially medical transport [11–13], thus speeding up the delivery process, especially in urban areas with heavy traffic. However, to take full advantage of drones, it is necessary to effectively address issues related to autonomy, regulation, security and privacy protection [14].

Drones are equipped with a number of advanced technologies. These solutions not only increase their functionality but enable their autonomous or remotely controlled operation [15,16]. They include navigation systems for precise position tracking and flight stability, data analysis and mission planning, sensors and cameras for collecting visual and topographic data, and real-time data transmission via radio and satellite systems [17]. Artificial intelligence and machine learning can be applied to boost the autonomy of drones [18]. Machine learning, particularly deep learning, is the foundation of autonomous systems. Neural networks enable real-time analysis of large sensory data sets, allowing for object recognition, navigation in complex environments, and decision-making. The integration of sensors and the fusion of data obtained from these sensors allow for information to be obtained, inferences to be made, and specific decisions to be reached, making it possible for drones to independently map out optimal routes without obstacles [19]. Machine learning algorithms analyze power consumption patterns and adjust flight parameters to minimize battery consumption. Drones equipped with appropriate image processing and data analysis algorithms can autonomously map areas and inspect infrastructure such as bridges [20], power lines, and buildings without human intervention. Detection and classification of unmanned aerial vehicles based on acoustic signals could play a key role in the development of autonomous systems for drones.

A very important direction in the development of drones is the cooperation of drones, or the so-called "work in a swarm". Here, it is important to avoid collision, recognize neighboring objects, detect the directions of approaching objects, achieve full autonomy of flight, and map out the path in real time. Acoustic systems can be useful for this particular application, as they can boost the autonomy of the drone. Using acoustic signals, it is possible to detect drones in natural environments even at a distance of 1 km and to indicate their exact distances, thus enhancing drone detection systems [21]. If it is possible to detect the direction of an incoming object acoustically, which is still being researched, acoustic sensors could resolve collision problems and enable drones to cooperate in a swarm. In addition, the classification of environmental signals in the presence of drones could improve their decision-making and autonomy.

When the drone is the carrier of the acoustic sensor, its noise may constitute a problem. To deal with this problem, it is necessary to discard the redundant components of the carrier from the signals and thus obtain the signals from the environment. This study carried out an acoustic analysis based on the acoustic signals obtained in the presence of various drones in order to classify the drones according to their UAV models. Following this classification, it may be possible to separate the noisy components of the drones from the useful signals. This, in turn, will make it easier to obtain signals from the environment, which will not only enable more accurate applications of drones in various fields—e.g., ecology (listening to birds), precision agriculture (acoustic observation of plantations), rescue systems, voice control of drones [22–24]—but will allow drones to cooperate.

Acoustic classification of UAVs can be valuable in Unmanned Ground Vehicle (UGV)-UAV cooperation in scenarios where direct communication is not possible or in environments where GNSS (Global Navigation Satellite System) signals are unavailable [25]. In such cases, the ability of UGVs to identify and interact with UAVs using acoustic signals would be a robust alternative, enhancing operational effectiveness in challenging conditions.

Despite the many advantages of drones, their use also comes with some challenges, such as regulation, security, and privacy protection [26–28]. The dynamic development of drone technology requires appropriate and safe legal regulations. The widespread use of drones can lead to privacy violations; thus, it requires proper regulation and protective measures. Detection of drones plays a very important role in security. A variety of sensing techniques have been proposed for drone detection, including acoustic, optical, radar detection systems and passive radiofrequency sensing [29]. Detection of small-sized drones can be very challenging [30]. Deep learning techniques, particularly the You Only Look Once (YOLO) algorithm, have been extensively explored and have shown promising results in UAV detection [31]. Privacy protection may be provided by acoustic sensors that can detect and classify objects at different heights and distances [21,32–34]. Acoustic systems for drone detection and classification may significantly boost security and privacy protection as well as the autonomy of drones.

The aim of this paper is to perform acoustic analysis and discriminant function analysis of acoustic signals recorded in the presence of UAVs hovering at a height of 8 m above the recording equipment in external environmental conditions. Seventeen different UAVs were used in the experiment. The acoustic analysis included the analysis of the characteristic frequencies of the background sound levels in the presence of the UAVs. Discriminant function analysis was used to investigate differences between the UAV models based on the acoustic signals recorded in the presence of each UAV. This research provides information on the classification accuracies of UAV models based on acoustic signals.

Drone detection and classification can significantly enhance security, privacy protection, and the autonomy of drones. This work investigates how acoustic signals acquired in presence of unmanned aerial vehicles can be classified. The analysis will demonstrate whether sound signals obtained in the drone regions show significant differences. The remainder of this article is organized as follows: Section 2 presents the materials and methods used in this study, the results are shown and discussed in Sections 3 and 4, respectively, and the conclusions and future steps are presented in Section 5.

2. Materials and Methods

The materials and methods used in this experiment are described in the following subsections.

2.1. UAVs Used in the Experiment

Seventeen UAVs were used for the experiment. Their structures and models are presented in Table 1.

Table 1. The UAVs used in the experiment.

UAV Number	UAV Structure	UAV Model
D1	X4	MATRICE 300
D2	X4	Mavic 3
D3	X4	Mavic Air 2S
D4	X4	Mavic Air 2
D5	X4	Mavic Mini 2
D6	X4	Mavic 2 Pro
D7	X4	Mavic 2 Pro
D8	X4	Mavic 3
D9	X4	Phantom 4
D10	X4	Mavic 2 Zoom
D11	X4	Mavic Mini 2
D12	X6	Yuneec H520
D13	X6	Yuneec H520E RTK
D14	X6	S900 [1]
D15	X6	X6D [1]
D16	X6	Y6 [1]
D17	X4	Phantom 4

[1] Non-commercial construction of UAV.

Twelve of the drones used in the experiment have an X4 structure (four rotating propellers) while five have an X6 structure (six rotating propellers). Several drones of the same models were used in the experiment. Drones D5 and D11 are two different drones of the model Mavic Mini 2, drones D6 and D7 are two different drones of the model Mavic 2 Pro, drones D9 and D17 are two different drones of the model Phantom 4, and drones D2 and D8 are two different drones of the model Mavic 3. Each drone was observed separately while hovering at a height of 8 m above the recording equipment. The X4 UAVs are presented in Figure 1. The X6 UAVs are presented in Figure 2.

Figure 1. The X4 UAVs used in the experiment (**a**) D1; (**b**) D2; (**c**) D3; (**d**) D4; (**e**) D5; (**f**) D6; (**g**) D7; (**h**) D8; (**i**) D9; (**j**) D10; (**k**) D11; (**l**) D17.

Figure 2. The X6 UAVs used in the experiment (**a**) D12; (**b**) D13; (**c**) D14; (**d**) D15; (**e**) D16.

2.2. Measurement and Recording of Acoustic Signals

Recordings of acoustic signals in the presence of UAVs took place in four different places: in two Polish cities, Kielce and Gdańsk, and in two places in the vicinity of the city of Gdańsk. The recordings were taken separately for the seventeen UAVs. During the recording, the UAV hovered at a height of 8 m directly over the recording equipment, as shown in Figure 3.

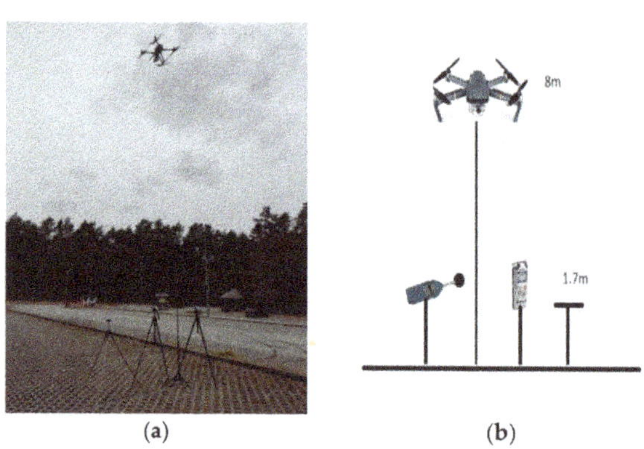

Figure 3. Measurement of acoustic signal. (**a**) Real environmental conditions; (**b**) illustration.

Recordings were taken with Olympus LS-11 digital recorder and Norsonic 140 sound analyzer. The recording equipment was placed 1.7 m above the ground. For each UAV, five (5) one-minute-long recordings were taken with Olympus LS-11 at a frequency of 44.1 kHz. Five recordings were also taken for each UAV using Norsonic 140 sound analyzer.

The measurement schedule, including dates, places, weather conditions, and the UAVs recorded that day, is presented in Table 2.

Table 2. Measurement schedule: places, dates, weather conditions, and the UAVs [35].

Day	Date	Place	Conditions	UAVs
Day 1	15 March 2023	Kielce	Temperature: 5 °C Air Pressure: 1014 hPa Humidity: 51% Wind: 22 km/h	D1, D2
Day 2	15 April 2023	Gdańsk	Temperature: 10 °C Air Pressure: 1015 hPa Humidity: 78% Wind: 25 km/h	D3, D4, D5, D6
Day 3	16 April 2023	Dębogórze, vicinity of Gdańsk	Temperature: 6 °C Air Pressure: 1022 hPa Humidity: 93% Wind: 18 km/h	D7, D8, D9
Day 4	17 April 2023	Dębogórze, vicinity of Gdańsk	Temperature: 7 °C Air Pressure: 1030 hPa Humidity: 80% Wind: 22 km/h	D10, D11, D12, D13
Day 5	18 April 2023	Łapalice, vicinity of Gdańsk	Temperature: 8 °C Air Pressure: 1033 hPa Humidity: 90% Wind: 25 km/h	D14, D15, D16, D17

2.3. Acoustic Analysis of Signals

Acoustic analysis of signals obtained using a Norsonic 140 sound analyzer consisted of frequency analysis of characteristic high background sound levels (peaks) and analysis of A-weighted sound levels obtained in the presence of the unmanned aerial vehicles.

2.4. MFCC Extraction from Recordings

Twelve Mel-Frequency Cepstral Coefficients (MFCCs) were extracted from recordings of signals in the presence of UAVs obtained with Olympus LS-11 recorder. MFCCs were used because of their efficient classification in previous experiments in which discriminant function analysis was applied to analyze sounds recorded in the presence of UAV [33]. These coefficients are also efficient in recognition systems where they provide high recognition accuracy.

2.5. Discriminant Analysis of MFCC

Discriminant function analysis of the 12 MFCCs was performed to investigate the differences between the UAV models. The UAV models were taken as the grouping variables and the MFCCs as the independent variables.

The discriminant analysis consisted of the discrimination stage and the classification stage. It was performed using STATISTICA software version 13.3 [36]. In the discrimination stage, the maximum number of discriminant functions evaluated was equal to the number of discriminant variables minus one. A canonical analysis was used to determine the successive functions and their canonical roots. The standardized coefficients were estimated for each discriminant function. The contribution of the variable to the discrimination between groups becomes greater as the standardized coefficients become larger. Chi-square tests with successive roots removed were investigated. The coefficient of the canonical correlation (canonical-R), which ranges between 0 (no association) and 1 (very high association), is a measure of the association between the i-canonical discriminant function and the group. Wilks' lambda statistic, which ranges between 0 (excellent discrimination) and 1 (no discrimination), is used to determine the statistical significance of discrimination.

The classification stage followed the determination of the variables that discriminate the UAV groups. Because there were thirteen model groups, thirteen classification functions were created according to Equation (1), viz.:

$$K_i(h) = c_{i0} + w_{i1}mfcc_1 + w_{i2}mfcc_2 + \ldots + w_{i12}mfcc_{12} \tag{1}$$

where h is the UAV considered as a group (mavic2zoom, mavicmini2, phantom4, matrice300, mavic3, mavicair2s, mavicair2, mavic2pro, yuneech520, yuneech520ertk, s900, x6d, y6), the subscript i denotes the respective group, c_{i0} is a constant for the i-th group, w_{ij} is the weight of the j-th variable in the computation of the classification score for the i-th group, and $mfcc_j$ is the observed Mel-frequency cepstral value for the respective case. The classification functions were used to determine to which group each case most likely belongs. A case was classified as belonging to the group for which it had the highest classification score, or more precisely, for which $K_i(h)$ assumed the highest value. The classification matrix was used to present the number of cases that were correctly classified and the number that were misclassified.

3. Results

The following results of the acoustic analysis and discriminant function analysis of signals detected in the presence of the UAVs in external environmental conditions were obtained in the experiment.

3.1. Results of Acoustic Analysis

The A-weighted sound levels of the UAVs obtained with the Norsonic 140 sound analyzer in external environmental conditions are presented in Figure 4:

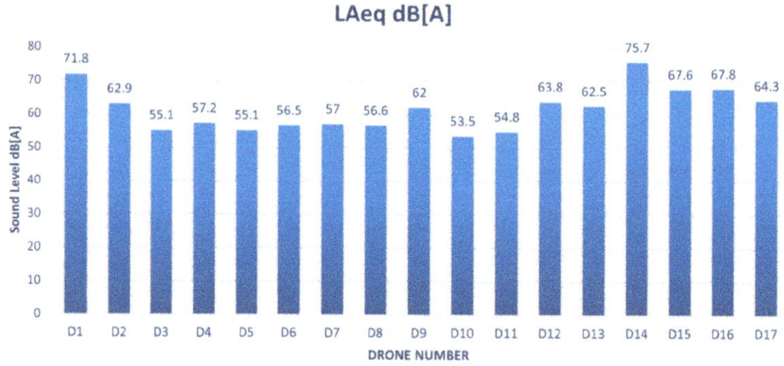

Figure 4. The A-weighted sound levels obtained for drones.

The background sound levels of the UAVs recorded with the Norsonic 140 sound analyzer in external environmental conditions are presented in Figure 5.

In Figure 5, the background sound levels obtained in the absence of UAVs in environmental conditions of the city of Kielce (BG) are also presented. The characteristic peak of the BG appeared at 25 Hz. The acoustic analysis showed that the X4 model resulted in smaller A-weighted and background sound levels than the X6 model. Characteristic peaks of the UAVs and their frequencies, according to Figure 5, are presented in Table 3.

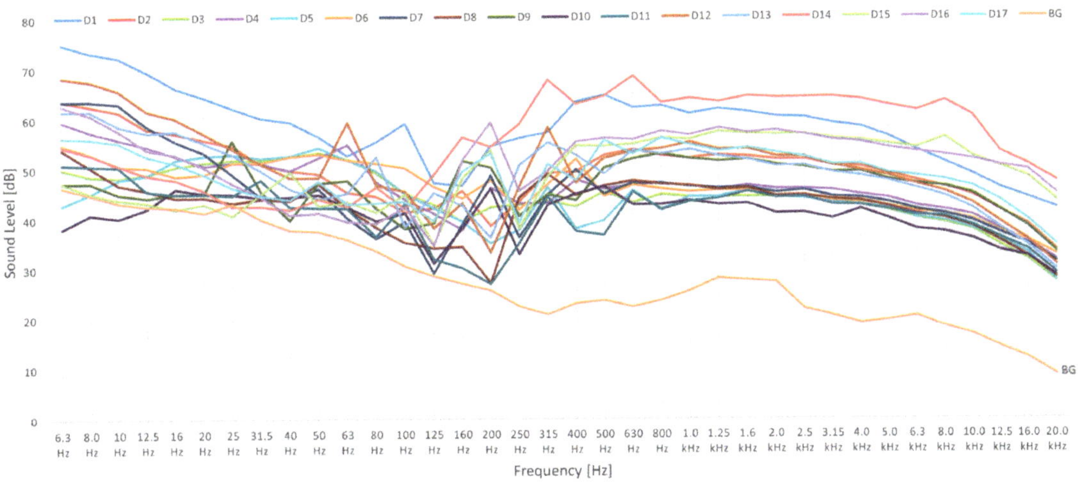

Figure 5. The background sound levels obtained for UAVs.

Table 3. The characteristic frequencies of peaks (^—normal, ^—high) of the UAVs.

UAV	D1	D2	D3	D4	D5	D6	D7	D8	D9	D10	D11	D12	D13	D14	D15	D16	D17
12.5 Hz						^											
16 Hz									^	^							
20 Hz							^	^							^		
25 Hz		^				^			^								
31.5 Hz				^				^			^						
40 Hz	^														^		
50 Hz		^	^		^	^	^	^	^	^				^		^	^
63 Hz				^					^		^	^			^		
80 Hz		^											^	^			
100 Hz	^					^	^			^	^	^			^	^	^
125 Hz													^				
160 Hz		^						^	^			^	^	^	^	^	^
200 Hz	^		^	^		^	^		^	^					^	^	^
250 Hz								^					^				
315 Hz		^	^		^		^	^	^	^	^	^	^	^			^
400 Hz	^			^		^	^								^		
500 Hz	^	^	^						^	^		^				^	^
630 Hz		^			^	^	^	^			^		^	^			
800 Hz	^		^	^					^						^	^	^
1 kHz					^					^		^	^	^			
1.25 kHz		^													^	^	
1.6 kHz				^	^	^	^	^	^	^	^			^			^
2.5 kHz		^	^		^		^	^		^					^		
4 kHz			^		^			^		^							^

According to Table 3, the presented peaks and their associated frequencies characterize the acoustic background obtained in the presence of the drones. The background sound levels are a combination of the UAV signals and the surrounding sounds. Maximum large peaks, marked in red in Table 3, show very high and easily visible peaks. Same models of drones showed similar characteristic frequencies. For example, the pair D2 and D8 showed characteristic peaks at 50 Hz, 160 Hz, 315 Hz, and 630 Hz; the pair D5 and D11 showed characteristic peaks at 315 Hz and 630 Hz; the pair D9 and D17 showed characteristic peaks at 50 Hz, 160 Hz, 200 Hz, 315 Hz, 500 Hz, and 800 Hz; and the pair D6 and D7 showed characteristic peaks at 50 Hz, 100 Hz, 200 Hz, 400 Hz, and 630 Hz. The most common characteristic frequencies for the UAVs were 50 Hz, 200 Hz, and 315 Hz.

3.2. Results of Discriminant Function Analysis

Discriminant function analysis was performed with 12 MFCCs as the independent variables and the UAV models as the grouping variables. The analysis showed significant main effects used in the model (Wilks' lambda: 0.0000009; approx. $F(144, 540) = 15.34$; $p < 0.00001$). Eleven discriminant functions (Root0, Root1, Root2, Root3, Root4, Root5, Root6, Root7, Root8, Root9, and Root10) were created. Chi-square tests performed at the canonical stage with successive roots removed are presented in Table 4.

Table 4. Chi-square tests with successive roots removed.

Roots Removed	Canonical R	Wilks' Lambda	Chi-Square	*p*-Value
0	0.984	0.0000	995.17	0.00000
1	0.968	0.0000	749.98	0.00000
2	0.948	0.0004	551.43	0.00000
3	0.907	0.0044	387.49	0.00000
4	0.868	0.0251	263.53	0.00000
5	0.774	0.1015	163.55	0.00000
6	0.695	0.2537	98.08	0.00000
7	0.543	0.4903	50.96	0.00162
8	0.475	0.6951	26.01	0.05395
9	0.304	0.8971	7.76	0.55816
10	0.106	0.9884	0.83	0.93388

According to Table 4, chi-square tests with successive roots removed were significant for all discriminant functions used in the model ($R = 0.984$; Wilks' lambda = 0.0000; $p < 0.00000$). The removal of the first discriminant function resulted in a high canonical-R between groups and discriminant functions ($R = 0.968$). The removal of the second, third, fourth, fifth, sixth, seventh, and eight discriminant functions also resulted in a high canonical-R.

After the canonical stage and derivation of discriminant functions with 12 MFCC features that mostly discriminate between groups, the classification stage followed. The coefficients of the classification functions were determined. The classification functions were used to establish to which group each case most likely belongs. The classification matrix was obtained to show the number of cases that were correctly classified and those that were misclassified.

The coefficients of the classification functions obtained for the groups are presented in Table 5.

The results of classification of the UAV model groups using the classification functions K(h) are presented in Table 6.

The value five in Table 6 means that for five considered records of the UAV model, five were correctly classified as belonging to the considered group using the respective classification function K(h). The value 10 means that for 10 considered records of the UAV model, 10 were correctly classified. The value 9 means that for 10 considered records, 9 were correctly classified and 1 was misclassified. The value zero (0) means that no record was classified as belonging to the considered group using the function K(h). The Total row (besides the first column) contains the number of all cases classified under the given function K(h). The value 11 means that for 10 considered records of the UAV model, 10 were correctly classified and 1 record was additional and misclassified. The percentage values are the average values of correctly classified cases.

Table 5. The coefficients of classification functions.

ci	K(Mavic 2 Zoom)	K(Mavic Mini 2)	K(Phantom 4)	K(Matrice 300)	K(Mavic 3)	K(Mavic Air 2S)	K(Mavic Air 2)	K(Mavic 2 Pro)	K(Yuneec H520)	K(Yuneec H520E RTK)	K(S900)	K(X6D)	K(Y6)
wi1	78.68	72.36	125.46	131.42	116.42	83.64	83.56	84.21	130.48	134.29	128.93	72.83	85.57
wi2	−27.42	−52.07	−64.47	−70.11	−48.30	−45.95	−24.60	−33.13	−70.16	−69.48	−60.50	−64.23	−45.68
wi3	136.73	101.03	96.68	70.82	119.72	116.44	139.50	140.17	103.73	96.45	147.69	84.23	85.31
wi4	−140.46	−208.73	−176.03	−94.83	−150.06	−200.34	−152.54	−188.17	−183.64	−182.19	−151.99	−160.93	−139.35
wi5	188.23	246.56	279.98	253.03	247.08	234.93	187.79	209.62	303.74	303.28	321.99	185.94	208.90
wi6	−66.20	109.61	−123.52	−221.46	−151.74	38.25	−78.20	−3.50	−93.66	−107.68	−183.84	15.65	−76.86
wi7	−67.45	−112.55	−99.88	−43.51	−74.61	−103.67	−65.22	−81.71	−92.31	−103.93	−135.60	−29.11	−56.70
wi8	55.20	−20.30	125.69	122.23	94.47	45.06	58.55	44.72	133.24	153.70	118.96	55.67	58.10
wi9	−129.30	−109.27	−123.04	−135.49	−152.85	−97.02	−125.01	−162.71	−132.91	−114.35	−211.17	−77.91	−98.97
wi10	−52.32	57.16	−96.49	−147.66	−105.03	5.49	−57.61	−33.90	−100.15	−98.34	−95.28	−35.94	−44.19
wi11	304.85	340.06	359.10	268.74	372.54	318.73	314.04	367.75	337.62	324.41	409.87	129.89	197.51
wi12	−168.27	−371.09	−120.45	36.67	−66.37	−306.44	−119.78	−186.34	−208.51	−206.46	−119.42	−137.74	−74.97
ci0	−379.01	−443.49	−579.10	−496.55	−526.51	−459.50	−403.43	−466.95	−639.39	−635.76	−734.34	−379.69	−352.66

Table 6. The classification matrix.

Group	%	K(Mavic 2 Zoom)	K(Mavic Mini 2)	K(Phantom 4)	K(Matrice 300)	K(Mavic 3)	K(Mavic Air 2S)	K(Mavic Air 2)	K(Mavic 2 Pro)	K(Yuneec H520)	K(Yuneec H520E RTK)	K(S900)	K(X6D)	K(Y6)
Mavic 2 Zoom	100.0	5	0	0	0	0	0	0	0	0	0	0	0	0
Mavic Mini 2	100.0	0	10	0	0	0	0	0	0	0	0	0	0	0
Phantom 4	90.0	0	0	9	0	1	0	0	0	0	0	0	0	0
Matrice 300	100.0	0	0	0	5	0	0	0	0	0	0	0	0	0
Mavic 3	100.0	0	0	0	0	10	0	0	0	0	0	0	0	0
Mavic Air 2S	100.0	0	0	0	0	0	5	0	0	0	0	0	0	0
Mavic Air 2	100.0	0	0	0	0	0	0	5	0	0	0	0	0	0
Mavic 2 Pro	100.0	0	0	0	0	0	0	0	10	0	0	0	0	0
Yuneec H520	100.0	0	0	0	0	0	0	0	0	5	0	0	0	0
Yuneec H520E RTK	100.0	0	0	0	0	0	0	0	0	0	5	0	0	0
S900	100.0	0	0	0	0	0	0	0	0	0	0	5	0	0
X6D	100.0	0	0	0	0	0	0	0	0	0	0	0	5	0
Y6	100.0	0	0	0	0	0	0	0	0	0	0	0	0	5
Total	98.8	5	10	9	5	11	5	5	10	5	5	5	5	5

Classification of the UAV models was very accurate, as shown by the 100% value obtained for correctly classified cases, except for the Phantom 4 model, whose accuracy percentage was 90%. One record from the ten Phantom 4 drones was misclassified as a Mavic 3 model.

According to Table 6, the classification was accurate (98.8%). Discriminant analysis showed significant differences between drones of different models but no significant differences between those of the same models.

4. Discussion

The acoustic analysis yielded higher A-weighted sound levels and background sound levels for the X6 UAVs than for X4 UAVs. The A-weighted sound levels of the drones with an X4 structure were above 50 dB(A), while the A-weighted sound levels of the X6 drones were above 60 dB(A). The highest A-weighted sound level of 75.7 dB(A) was exhibited by the D14-X6-S900 model. The D1-X4-Matrice 300 model, also showed an A-weighted sound level above 70 dB(A) value. The background sound levels presented in Figure 5 resulted in peaks that could be characteristic for UAVs hovering at 8 m over the recording equipment, but also for other sounds in the surroundings. When investigating the BG, the surrounding factors may have a minor effect on the recordings of UAVs in this experiment. The characteristic peaks presented in Table 3 were similar for drone pairs of the same models, viz.: D2 and D8, D5 and D11, D9 and D17, and D6 and D7. The most common frequencies, which were obtained for almost all the UAV models, were 50 Hz, 200 Hz, and

315 Hz. To specify the characteristic peaks, more data records of same UAV models need to be analyzed.

Discriminant analysis based on MFCC showed significant differences between the different UAV model groups, but no significant differences between UAVs of the same models. One out of ten records of the Phantom 4 model was incorrectly classified as a Mavic 3 model, resulting in a 90% classification accuracy for the Phantom 4 group. In general, the mean classification accuracy for all of the UAV models was 98.8%. This high classification accuracy shows that UAV models can be classified based on acoustic signals. An acoustic system can serve as an additive system for other systems, e.g., vision and radar systems, to detect and classify drones. Previous research on drones shows that acoustic systems can accurately detect drones even from a distance of 1 km. Some acoustic systems can detect drones as well as the drone models. Such systems work even at night, enhancing privacy area protection. In the current study, the drones were observed and analyzed at four different places, but this had no influence on the classification accuracies.

Future research should focus on extracting features that will provide more accurate information about drones and obtaining classification scores of UAV models from other altitudes and distances. The surrounding factors may affect the accuracy of classification when increasing the distance between the UAV and recording equipment, which will be the subject of further research. Previous research on drones has shown that the acoustic signals of selected drones can be used to determine the altitudes and distances at which the drones are hovering [33]. Other information that can be obtained from drone acoustic signals may include the structure of the drone (X4 vs. X6 vs. X8 vs. X3) and its loading. Initial listening tests showed that information about the loading of a model can be obtained from the acoustic signal. The sound of the same drone with and without loading shows differences in the sound signal during listening tests. Future research should aim to numerically confirm the listening tests and obtain information about the structure of the drone and loading from acoustic signals. Such information obtained from an acoustic signal may allow us to detect an object, classify it, determine its loading, distance and height, and understand the nature of the signal. This, in turn, makes it possible to develop an acoustic sensor for an unmanned acoustic system that can perform the above activities directly from the unmanned platform. It will therefore be necessary to reject the components of the sensor carrier and acquire environmental signals. Such an operation is possible only after the nature of the drone signals has been understood, thus allowing the rejection of carrier components and the acquisition of environmental signals.

5. Conclusions

The aim of this study was to perform acoustic analysis and discriminant function analysis of acoustic signals recorded in the presence of UAVs hovering at a height of 8 m above the recording equipment in external environmental conditions. Seventeen different UAVs were used in the experiment.

Acoustic analysis was based on A-weighted sound levels and background sound levels in the presence of the UAVs. The acoustic analysis showed that drones of X4 model yielded smaller A-weighted and background sound levels than those of X6 model. The most common frequencies of background sound levels (peaks) obtained for almost every UAV model were 50 Hz, 200 Hz, and 315 Hz.

Discriminant function analysis showed significant differences between different UAV models, but no significant differences between the same UAV models. Classification of the UAV models was 98.8% accurate. Discriminant analysis and MFCC features showed very accurate classification results for the models.

Future research should evaluate the impact of other hovering distances of UAVs from the recording equipment on the efficiency of classification and concentrate on the classification of the structure of the drone (X4 vs. X6 vs. X8 vs. X3).

Funding: This research was funded in whole by National Science Centre, Poland, Grant number: 2022/06/X/ST7/00586. For the purpose of open access, the author has applied a CC-BY public copyright license to any Author Accepted Manuscript (AAM) version arising from this submission.

Institutional Review Board Statement: Not applicable.

Informed Consent Statement: Not applicable.

Data Availability Statement: The data presented in this study are available on the website https://t47-marzena.s3.kielce.pl/index.html accessed on 27 August 2024 under the terms and conditions of the Creative Commons Attribution (CC BY) license (https://creativecommons.org/licenses/by/4.0/ accessed on 27 August 2024). Author: Marzena Mięsikowska.

Conflicts of Interest: The author declares no conflicts of interest.

References

1. Mogili, U.R.; Deepak, B.B.V.L. Review on Application of Drone Systems in Precision Agriculture. *Procedia Comput. Sci.* **2018**, *133*, 502–509. [CrossRef]
2. Abbas, A.; Zhang, Z.; Zheng, H.; Alami, M.M.; Alrefaei, A.F.; Abbas, Q.; Naqvi, S.A.H.; Rao, M.J.; Mosa, W.F.A.; Abbas, Q.; et al. Drones in Plant Disease Assessment, Efficient Monitoring, and Detection: A Way Forward to Smart Agriculture. *Agronomy* **2023**, *13*, 1524. [CrossRef]
3. Montes, O.; Castro, R.; Villanueva, C.; Pérez, M.; Uribe, M.; Jimenez, C. Fast analysis of plant health, plant count and nitrogen status using drones. *Res. Crops* **2020**, *21*, 82–87. [CrossRef]
4. PA, O.; SJ, O.; CC, I.; GH, K. Unmanned Aerial Vehicle for Pipeline Surveillance: A Review. *J. Energy Res. Rev.* **2024**, *16*, 47–61. [CrossRef]
5. Isa, M.F.M.; Rahim, N.Z.A.; Fathi, M.S. It's a bird... It's a plane... It's a drone...: Telecommunication Tower Inspection Using Drone. In Proceedings of the 2019 6th International Conference on Research and Innovation in Information Systems (ICRIIS), Johor Bahru, Malaysia, 2–3 December 2019; pp. 1–5.
6. Jacobsen, R.H.; Matlekovic, L.; Shi, L.; Malle, N.; Ayoub, N.; Hageman, K.; Hansen, S.; Nyboe, F.F.; Ebeid, E. Design of an Autonomous Cooperative Drone Swarm for Inspections of Safety Critical Infrastructure. *Appl. Sci.* **2023**, *13*, 1256. [CrossRef]
7. Pafitis, M.; Savva, A.; Kyrkou, C.; Kolios, P.; Theocharides, T. MELETI: A Machine-Learning-Based Embedded System Architecture for Infrastructure Inspection with UAVs. In Proceedings of the Embedded Machine Learning for Cyber-Physical, IoT, and Edge Computing: Use Cases and Emerging Challenges, Stanford, CA, USA, 25–27 March 2024; pp. 285–311. [CrossRef]
8. Simoes, D.; Rodrigues, A.; Reis, A.B.; Sargento, S. Forest Fire Monitoring through a Network of Aerial Drones and Sensors. In Proceedings of the 2020 IEEE International Conference on Pervasive Computing and Communications Workshops, PerCom Workshops, Austin, TX, USA, 23–27 March 2020. [CrossRef]
9. Ahmadian, N.; Lim, G.J.; Torabbeigi, M.; Kim, S.J. Smart border patrol using drones and wireless charging system under budget limitation. *Comput. Ind. Eng.* **2022**, *164*, 107891. [CrossRef]
10. Ho, Y.-H.; Tsai, Y.-J. Open Collaborative Platform for Multi-Drones to Support Search and Rescue Operations. *Drones* **2022**, *6*, 132. [CrossRef]
11. Enayati, S.; Li, H.; Campbell, J.F.; Pan, D. Multimodal Vaccine Distribution Network Design with Drones. *Transp. Sci.* **2023**, *57*, 1069–1095. [CrossRef]
12. Rosser, J.C., Jr.; Vignesh, V.; Terwilliger, B.A.; Parker, B.C. Surgical and Medical Applications of Drones: A Comprehensive Review. *JSLS J. Soc. Laparoendosc. Surg.* **2018**, *22*, e2018.00018. [CrossRef]
13. Elmeseiry, N.; Alshaer, N.; Ismail, T. A Detailed Survey and Future Directions of Unmanned Aerial Vehicles (UAVs) with Potential Applications. *Aerospace* **2021**, *8*, 363. [CrossRef]
14. Chamola, V.; Kotesh, P.; Agarwal, A.; Gupta, N.; Guizani, M. A Comprehensive Review of Unmanned Aerial Vehicle Attacks and Neutralization Techniques. *Ad Hoc Netw.* **2021**, *111*, 102324. [CrossRef]
15. Lee, M.-T.; Chuang, M.-L.; Kuo, S.-T.; Chen, Y.-R. UAV Swarm Real-Time Rerouting by Edge Computing D* Lite Algorithm. *Appl. Sci.* **2022**, *12*, 1056. [CrossRef]
16. Lonklang, A.; Botzheim, J. Improved Rapidly Exploring Random Tree with Bacterial Mutation and Node Deletion for Offline Path Planning of Mobile Robot. *Electronics* **2022**, *11*, 1459. [CrossRef]
17. Hong, S.J.; Han, Y.; Kim, S.Y.; Lee, A.Y.; Kim, G. Application of deep-learning methods to bird detection using unmanned aerial vehicle imagery. *Sensors* **2019**, *19*, 1651. [CrossRef] [PubMed]
18. Lee, T.; Mckeever, S.; Courtney, J. Flying Free: A Research Overview of Deep Learning in Drone Navigation Autonomy. *Drones* **2021**, *5*, 52. [CrossRef]
19. Gupta, A.; Fernando, X. Simultaneous Localization and Mapping (SLAM) and Data Fusion in Unmanned Aerial Vehicles: Recent Advances and Challenges. *Drones* **2022**, *6*, 85. [CrossRef]
20. Mandirola, M.; Casarotti, C.; Peloso, S.; Lanese, I.; Brunesi, E.; Senaldi, I. Use of UAS for damage inspection and assessment of bridge infrastructures. *Int. J. Disaster Risk Reduct.* **2022**, *72*, 102824. [CrossRef]

21. Międsikowska, M.; Nowakowski, M.; Lorenc, W.; Chodnicki, M. Signal Frequency Analysis of X4 Unmanned Aerial Vehicle. Mechanics in Aviation ML-XVII 2016, Volume II. Available online: https://ptmts.org.pl/pliki/wydawnictwa/ML-2016-t-II-WAT.pdf (accessed on 22 August 2024).
22. Mięsikowska, M. Analysis of sound levels and speech intelligibility in the presence of X4 Unmanned Aerial Vehicle in External Environmental Conditions. In Proceedings of the 2023 Signal Processing: Algorithms, Architectures, Arrangements, and Applications (SPA), Poznan, Poland, 20–22 September 2023; pp. 31–35, ISBN 979-8-3503-0498-5. [CrossRef]
23. Mięsikowska, M. Discriminant analysis of voice commands in the presence of unmanned aerial vehicle. *Information* **2021**, *12*, 23. [CrossRef]
24. Mięsikowska, M. Speech Intelligibility in the presence of X4 Unmanned Aerial Vehicle. In Proceedings of the 2018 Signal Processing: Algorithms, Architectures, Arrangements, and Applications (SPA), Poznan, Poland, 19–21 September 2018; pp. 310–314. [CrossRef]
25. Sivaneri, V.O.; Gross, J.N. UGV-to-UAV cooperative ranging for robust navigation in GNSS-challenged environments. *Aerosp. Sci. Technol.* **2017**, *71*, 245–255. [CrossRef]
26. Hlavata, R.; Kamencay, P.; Sykora, P.; Hudec, R.; Radilova, M. Recognition of Dangerous Objects using Deep Learning. In Proceedings of the 2024 34th International Conference Radioelektronika (RADIOELEKTRONIKA), Zilina, Slovakia, 17–18 April 2024; pp. 1–4. [CrossRef]
27. Dumitrescu, C.; Minea, M.; Costea, I.M. Development of an Acoustic System for UAV Discovery and Tracking Employing Concurrent Neural Networks. 2020. Available online: https://videleaf.com/wp-content/uploads/2021/01/Development-of-an-Acoustic-System-for-UAV-discovery-and-tracking-employing-Concurrent-Neural-Networks.pdf (accessed on 22 August 2024).
28. Brighente, A.; Conti, M.; Peruzzi, G.; Pozzebon, A. ADASS: Anti-Drone Audio Surveillance Sentinel via Embedded Machine Learning. In Proceedings of the 2023 IEEE Sensors Applications Symposium (SAS), Ottawa, ON, Canada, 18–20 July 2023; pp. 1–6. [CrossRef]
29. Flak, P.; Czyba, R. RF Drone Detection System Based on a Distributed Sensor Grid With Remote Hardware-Accelerated Signal Processing. *IEEE Access* **2023**, *11*, 138759–138772. [CrossRef]
30. Liu, M.; Wang, X.; Zhou, A.; Fu, X.; Ma, Y.; Piao, C. UAV-YOLO: Small Object Detection on Unmanned Aerial Vehicle Perspective. *Sensors* **2020**, *20*, 2238. [CrossRef] [PubMed]
31. Fang, A.; Feng, S.; Liang, B.; Jiang, J. Real-Time Detection of Unauthorized Unmanned Aerial Vehicles Using SEB-YOLOv8s. *Sensors* **2024**, *24*, 3915. [CrossRef] [PubMed]
32. Dumitrescu, C.; Minea, M.; Costea, I.M.; Cosmin Chiva, I.; Semenescu, A. Development of an Acoustic System for UAV Detection. *Sensors* **2020**, *20*, 4870. [CrossRef] [PubMed]
33. Mięsikowska, M. Analysis of signal of X8 unmanned aerial vehicle. In Proceedings of the Signal Processing-Algorithms, Architectures, Arrangements, and Applications Conference Proceedings, SPA, Poznan, Poland, 20–22 September 2017; pp. 69–72. [CrossRef]
34. Satish, A.; Medda, A. Acoustic UAV Detection using Spherical Array Beamforming. Conference Record. In Proceedings of the Asilomar Conference on Signals, Systems and Computers, Pacific Grove, CA, USA, 31 October–2 November 2022; pp. 446–450. [CrossRef]
35. Mięsikowska, M. Project Website, UAVs Recordings. Available online: https://t47-marzena.s3.kielce.pl/index.html (accessed on 23 August 2024).
36. Discriminant function analysis–STATSoft Electronic Documentation. Available online: https://statisticasoftware.wordpress.com/2012/06/25/discriminant-function-analysis/ (accessed on 21 August 2024).

sensors

Review

A Review of Unmanned Aerial Vehicle Based Antenna and Propagation Measurements

Venkat R. Kandregula [1,*], Zaharias D. Zaharis [2], Qasim Z. Ahmed [1], Faheem A. Khan [1], Tian Hong Loh [3], Jason Schreiber [4], Alexandre Jean René Serres [5] and Pavlos I. Lazaridis [1]

1 School of Computing and Engineering, University of Huddersfield, Huddersfield HD1 3DH, UK; q.ahmed@hud.ac.uk (Q.Z.A.); f.khan@hud.ac.uk (F.A.K.); p.lazaridis@hud.ac.uk (P.I.L.)
2 School of Electrical and Computer Engineering, Aristotle University of Thessaloniki, 54124 Thessaloniki, Greece; zaharis@auth.gr
3 Electromagnetic & Electrochemical Technologies Department, National Physical Laboratory, Teddington TW11 0LW, UK; tian.loh@npl.co.uk
4 Sixarms, West Burleigh, Burleigh Heads 4219, Australia; jason@sixarms.com
5 Center of Electrical Engineering and Informatics, Universidade Federal de Campina Grande, Campina Grande 58708-110, Brazil; alexandreserres@dee.ufcg.edu.br
* Correspondence: v.r.kandregula@hud.ac.uk

Abstract: This paper presents a comprehensive survey of state-of-the-art UAV–based antennas and propagation measurements. Unmanned aerial vehicles (UAVs) have emerged as powerful tools for in situ electromagnetic field assessments due to their flexibility, cost-effectiveness, and ability to operate in challenging environments. This paper highlights various UAV applications, from testing large–scale antenna arrays, such as those used in the square kilometer array (SKA), to evaluating channel models for 5G/6G networks. Additionally, the review discusses technical challenges, such as positioning accuracy and antenna alignment, and it provides insights into the latest advancements in portable measurement systems and antenna designs tailored for UAV use. During the UAV–based antenna measurements, key contributors to the relatively small inaccuracies of around 0.5 to 1 dB are identified. In addition to factors such as GPS positioning errors and UAV vibrations, ground reflections can significantly contribute to inaccuracies, leading to variations in the measured radiation patterns of the antenna. By minimizing ground reflections during UAV–based antenna measurements, errors in key measured antenna parameters, such as HPBW, realized gain, and the front-to-back ratio, can be effectively mitigated. To understand the source of propagation losses in a UAV to ground link, simulations were conducted in CST. These simulations identified scattering effects caused by surrounding buildings. Additionally, by simulating a UAV with a horn antenna, potential sources of electromagnetic coupling between the antenna and the UAV body were detected. The survey concludes by identifying key areas for future research and emphasizing the potential of UAVs to revolutionize antenna and propagation measurement practices to avoid the inaccuracies of the antenna parameters measured by the UAV.

Keywords: absorbers; broadcasting systems; knife edge diffraction (KED); parabolic reflector; path loss; propagation measurements; square kilometer array (SKA); UAV-based measurements; unmanned aerial vehicles (UAVs)

1. Introduction

Unmanned aerial vehicles (UAVs) or remotely piloted aircraft systems, simply known as drones, are in high demand for in situ measurements because of their mobility, low cost, hovering capability, and low maintenance expenses. Advances in technology, such as software–defined radios (SDRs), have facilitated the utilization of UAVs for antenna and propagation measurement [1,2]. UAVs are an effective antenna measurement solution for projects such as the square kilometer array (SKA) [3], broadcasting systems [4], large biconical antennas [5], such as those of the French aerospace research center Office National d'Etudes et de Recherches Aérospatiales (ONERA) [6], and parabolic reflector antennas [7], because of their in situ electromagnetic field measurement capabilities. UAV measurements can also help to identify multipath–related propagation losses. In radar applications that use antenna arrays, UAV measurements are useful for the final calibration after the antennas are enclosed in radomes and deployed. This is particularly crucial in cases in which the RF system is large. Climatic conditions, such as rain, snow, or harsh weather, degrade the antenna radiation performance in terms of side–lobe level (SLL) and ripples in the beam peak. The degradation of these antenna parameters can eventually lead to a deterioration of the antenna gain and cross-polarization levels [8]. Conventional methods may not be feasible in such situations, making UAV–based measurements a viable alternative.

In practice, antenna measurements with UAVs can be performed either in the Fresnel region (also known as the radiative near field) or in the Fraunhofer region (also known as the far field). For example, in shortwave communications, an antenna operating at 20–30 MHz can achieve very long communication distances owing to ionospheric propagation. Nevertheless, in this frequency range, the antenna size is large. To avoid large separation distances between the antenna under test (AUT) and the UAV, near–field measurements are an appropriate and practical alternative, whereby the near–field measurements are mathematically transformed into far–field radiation patterns. In this paper, the utilization of UAVs to perform near–field measurements is presented and involves the UAV following a specific spatial path [9–11] to collect measurements of the AUT using an SDR.

Measuring antenna performance using the traditional outdoor method [12] involves measuring the antenna in the far–field region. Accordingly, these measurements require several pieces of expensive and heavy equipment. The AUT is in the receiving mode and is connected to a mixer and microwave receiver [13], which can retrieve the amplitude and phase information of the AUT. Supported by model towers, the AUT is placed on a multi-axis rotary positioner, such as an azimuth–over–elevation-over-azimuth positioner [14]. These positioners can rotate the AUT by 360° in the azimuth plane and provide limited motion in the elevation plane.

Unlike traditional outdoor measurements, UAV–based measurements do not rely on large or heavy equipment. For example, instead of a heavy spectrum analyzer, an SDR, which is compact and lightweight, can be used. In addition, these SDRs can be controlled from ground level. Taking advantage of UAVs' capability to hover and perform circular trajectories would make them free from the positioners and controllers used in the traditional measurement methods. In addition to the UAV–based antenna measurements, UAVs can also be used for propagation measurements. UAVs equipped with SDRs can be used to perform propagation measurements, which involve estimating signal strength as a function of frequency and time. These measurements are then post–processed to extract the channel coefficients, such as path loss and angle of arrival, which are used to model the channel. The channel model can then be used to design an air–to–ground (ATG) communication system [15] and evaluate its performance under different operating conditions. UAVs are used to evaluate the performance of existing ATG communication systems and to identify potential sources of interference or signal degradation [16]. Propagation measurements can be conducted by mounting a transmitting antenna and lightweight transmitter with a built–in battery on a UAV and a standard calibrated antenna acting as a receiver connected to a spectrum analyzer at the ground level. A critical issue is airframe shadowing, which obstructs the line–of–sight (LOS) path between the transmitter and receiver caused by the UAV body. In this regard, it is important to verify that the transmitting antenna mounted on the UAV is in direct LOS with the receiver before performing any propagation measurements [17].

This article describes how UAVs equipped with an antenna and an SDR configured as a receiver can be used to measure antenna and propagation characteristics in various practical cases. In contrast, previous studies have emphasized only one specific case each time. The main contributions of this study are in the following topics:

- A comprehensive review of the latest advancements in UAV–based antenna and propagation measurements encompassing various techniques and applications.
- Implementation of both near–field and far–field measurement techniques, emphasizing the practical advantages, including a discussion on the use of SDRs in UAV–based measurements.
- Exploration of a wide range of applications, from the SKA and large–scale biconical antennas to 5G/6G network evaluations.
- Illustration of practical implementations and effectiveness of UAV–based measurement systems, presented using real–world test cases, such as measurements of parabolic reflector systems and large–scale propagation channel effects.

The remainder of this paper is organized as shown in Figure 1. Section 2 focuses on UAV–based measurements. In this section, we provide design recommendations for UAV–based measurements. By following the design recommendations, we describe how UAV–based measurements can be performed in the far– and near–field regions. For UAV–based antenna measurements in the far–field region, we describe the procedure for measuring base station antennas (BASTAs) and digital television (DTV) stations deployed in the field. To measure antennas operating at lower frequencies, a larger distance between the AUT and UAV is required, owing to the large dimensions of the AUT. To overcome these large separation distances, antenna measurements can be performed in the near–field region.

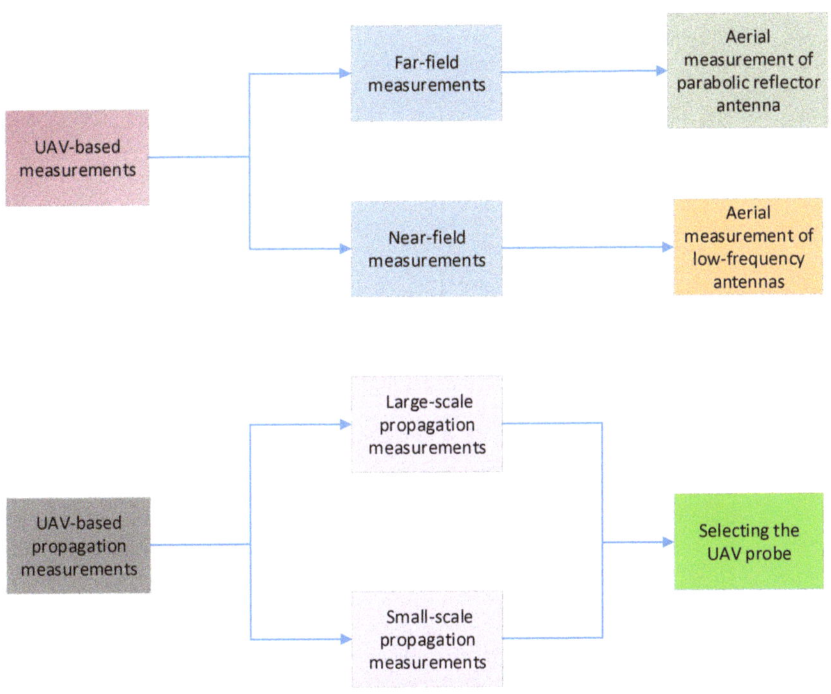

Figure 1. Organization of the paper.

We explain UAV–based near–field measurements for structurally large antennas, such as low–frequency aperture arrays (LFAAs) and biconical antennas. UAV–based propagation measurements are presented in Section 3. In this section, a detailed explanation of the simulation results for large–scale and small–scale propagation measurements is presented. Based on the UAV–based antenna and propagation measurements, we explain the design considerations for selecting the UAV probe, and the conclusions are provided in Section 4.

2. UAV–Based Measurements

Characteristics such as the matching impedance, polarization, radiation efficiency, directivity, gain, and radiation patterns are used to ascertain the antenna performance. For radar systems with stringent specifications, such as polarimetric weather radar, calibration of the systems after deployment in the field is crucial. Figure 2 describes the methodology adopted for UAV–based measurements, in which the UAV is equipped with an antenna at a height h_{uav} that serves as a signal source, while the AUT is at a height h_{aut} from the ground plane. An antenna mounted on a UAV can act as either a transmitting (TX) or a receiving (RX) antenna. As shown in Figure 2, the AUT and the TX antenna mounted on the UAV were separated by distance R. Here, α denotes the half–power beam width (HPBW) of the TX antenna mounted on the UAV, and α_h denotes the plane angle subtended at the antenna mounted on the UAV by the AUT height [18].

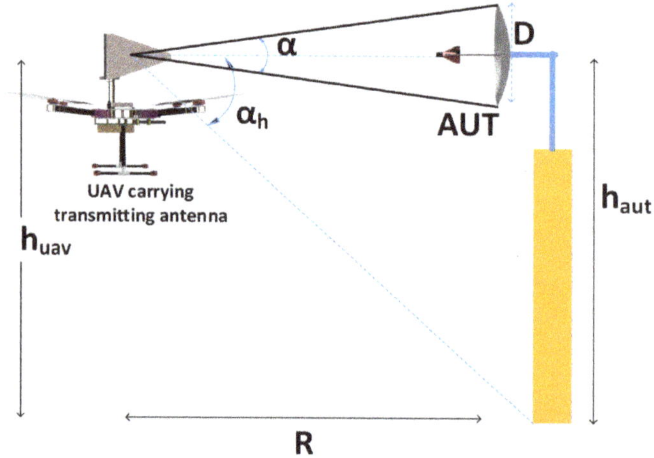

Figure 2. Measurement configuration of the UAV system.

In the process of measuring the radiation pattern of an AUT using a UAV, several factors, such as the phase curvature of the incident wave, ground reflections, and amplitude taper of the source antenna, affect the measurement accuracy. To avoid measurement inaccuracies, caution should be exercised regarding the variations in the phase and amplitude of the incident field and the interference from ground reflections. For UAV–based measurements, as a rule of thumb, the TX antenna is selected such that it has a wider HPBW than the AUT to prevent measurement errors. Similarly, the phase curvature of the incident field on the AUT affects the accuracy of the measured SLL. To overcome these errors, the phase deviation over the planar test aperture is maintained below 22.5°. To achieve this phase deviation, it is necessary to have a separation distance greater than $2D^2/\lambda$, where D is the maximum dimension of the antenna and λ is the operating wavelength [19]. The other factors that affect the accuracy of the measurements are ground reflections.

Ground reflections can often cause signals to be added constructively/destructively, resulting in apparent gain values that are higher or lower than expected. Also, maintaining a sufficient distance between the TX and RX antennas prevents the distortion of the patterns caused by ground reflections. Different approaches have been used to reduce ground reflections, including diffraction screening and absorbers between the TX and RX antennas. The use of a TX antenna with a low SLL can also prevent ground reflections [20]. At the same time, aligning the beam peak of the TX antenna with the AUT is essential. This can be achieved in a couple of iterations, which involve measuring the power received at the AUT when a UAV is carrying the TX antenna that flies at different altitudes. At a particular height, the UAV can further vary its altitude slightly to detect the position corresponding to the maximum received power of the AUT. At this height, the beam peak from the TX is appropriately aligned with the AUT. With the TX aligned with the AUT and by maintaining a low SLL, ground reflections can be avoided to a certain extent.

2.1. UAV–Based Far–Field Measurements

To perform far–field measurements, it is essential to maintain a minimum distance of $2D^2/\lambda$, as discussed previously. A UAV equipped with high–precision controller boards and RF measurement equipment, such as an SDR [21] with directional antennas, can be used for far–field measurements. For instance, in the inspection of reflector systems [7] and structurally large antennas, UAV–based measurements are cost–effective and reliable. The measurement techniques proposed here have several advantages over conventional methods, such as the use of helicopters equipped with RF payloads [22], in terms of cost and maneuverability. A compact and lightweight design allows UAVs to easily reach any location for measurements. They can also hover at a specific location, which enhances their ability to conduct RF measurements, with improved results. Various techniques, such as fast Fourier transform (FFT) [23], angular deconvolution [24], spatial mode filtering [25], frequency impulse response, and Hilbert transform, can be used to filter noise when measurements are conducted outdoors in a noisy environment. Considering that the above methods are not generic and cannot be applied to all environmental conditions, [26] proposed a filtering technique referred to as locally weighted regression and dispersion smoothing, which can be used to filter out high–frequency noise.

To validate this methodology, measurements were conducted in an anechoic chamber as well as an outdoor environment using the proposed filtering technique on a Yagi–Uda antenna operating at 2.4 GHz and a horn antenna at 5.3 GHz. An analysis of the HPBW measured in an anechoic chamber and an outdoor environment showed that they were in good agreement, with a difference of only 1°. In the far–field measurements, the far–field patterns on a sphere of a constant radius were estimated. The elevation and azimuth angles, denoted by θ and ψ, were the variables used to identify the location on the sphere. The phase information of the AUT was obtained using a vector network analyzer (VNA), and the two-dimensional amplitude information may be calculated using the total electric field:

$$|E| = \sqrt{(E_\theta)^2 + (E_\psi)^2}, \tag{1}$$

where E_θ, E_ψ are the electric field in the elevation and azimuth planes, respectively.

In UAV–based far–field measurements, the AUT is placed on a tripod and the UAV follows a vertical and horizontal path around the AUT. The UAV carries an RX antenna while moving and collecting data points. The AUT is stationary in this scenario, unlike the conventional outdoor ranges, which can be either elevated or elevated slant ranges, and requires the AUT to rotate by means of positioning commands operated by a computer.An example of a conventional elevated slant range is shown in Figure 3, in which a TX antenna, such as a quad–ridged horn, is mounted on a tall structure, and an AUT, such as an offset–fed parabolic reflector system, is mounted on an azimuth positioner. To eliminate errors caused by ground reflection, absorbers are placed between the source antenna and the AUT. Generally, a TX antenna is designed to have a low SLL, and the height at which the source antenna is mounted should be selected such that the main beam illuminates the AUT. Traditional measurements are performed using a stationary source antenna, and the radiation pattern can be obtained in both the elevation and azimuth cuts by rotating the AUT placed on the positioner. However, in UAV–based measurements, the UAV carrying the antenna follows a trajectory and the AUT is stationary. The advantage of this setup is that it does not require heavy and expensive positioners to obtain AUT radiation patterns.

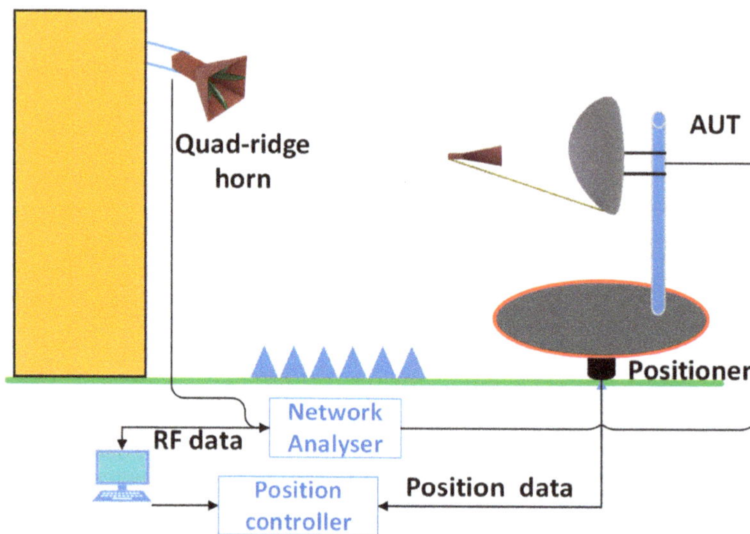

Figure 3. Conventional elevated slant test range.

As shown in Figure 4, a UAV with a quad-ridged dual–polarized horn operating at 6–24 GHz was used as a far–field transmitting antenna [27] to measure the microwave vision group (MVG) SR40 parabolic reflector system. During the outdoor measurements, the UAV and AUT were maintained at 350 and 750 m, respectively. The TX antenna mounted on the UAV was supported by a gimbal, which was used to detect radiation patterns in the elevation plane, and the rotation of the UAV around the parabolic antenna was performed to measure the radiation patterns in the azimuth plane. All the measurements were performed at 14.5 GHz. During UAV measurements, errors may occur because of external winds, reflections from surfaces, and misalignment between the probe and the AUT caused by UAV propeller vibrations. To minimize these errors, various methods are employed, such as measurement of the AUT under additional conditions, such as rotating the device at 180°, conducting measurements with different separation distances, and taking multiple measurements at a time in one cut and averaging them.

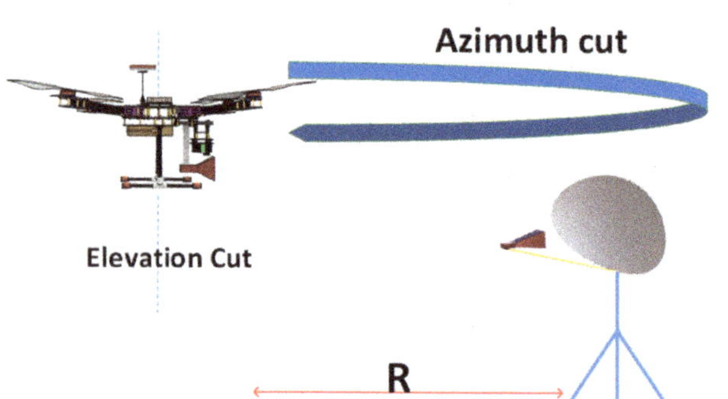

Figure 4. UAV–based in situ measurement for a parabolic reflector antenna system.

Figure 5 illustrates how UAVs can measure the performance of antennas mounted on ships. In [28], a UAV carrying a vertically polarized ground plane monopole antenna and three vertical radials was used as an RX antenna to measure an X–band vertically polarized reflector antenna mounted on a ship that resonates at 9.5 GHz. Throughout the UAV flight, the antenna mounted on the UAV is directed towards the AUT; however, the variance in the pitch and roll axes while the UAV is in motion creates a polarization mismatch. Based on the experiments conducted in [28], a loss of 0.2 dB is observed in the measured radiation pattern due to polarization mismatch. Traditionally, to measure the performance of an antenna installed on a ship, an RX antenna connected to a spectrum analyzer is required to collect the signals. The RX antenna is placed at the ground level on the shore. The reflector antenna, which is the AUT installed on the ship, continuously transmits signals. To measure the radiation pattern of the AUT in the azimuth plane, the ship carrying the reflector antenna should follow a circular trajectory in the sea, while the receiver collecting the signals is static at ground level. Conversely, for UAV–based measurements, an SDR mounted on the UAV is used to receive the signals. A reflector antenna installed on the ship transmits signals. Here, the ship carrying the TX is static, and the UAV carrying the RX antenna follows a circular trajectory with a constant radius around the ship situated at the center of the circle. When measuring an antenna installed on a ship, it is essential to meet specific criteria to ensure that there are no losses due to polarization mismatches [29] or multipath reflections from seawater [30]. Considering that the AUT has a maximum diameter of 1 m and operates at 9.5 GHz, it is imperative to maintain a far–field distance (Fraunhofer distance) greater than 60 m, and no obstacles should block the first Fresnel zone between the ship and the UAV.

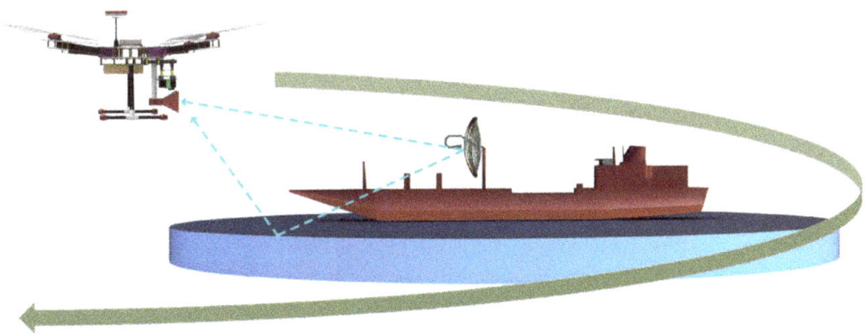

Figure 5. UAV–based measurement for a parabolic reflector antenna system placed on a ship.

2.1.1. Aerial Measurement of Base Station Antennas

To establish communication, a mobile BASTA [31] is essential and serves as a communication hub for wireless devices. Owing to the exponential increase in the number of devices connected to wireless networks, BASTAs are being deployed at an unprecedented rate to provide connectivity to users. Certain errors may occur when BASTAs are deployed in the field. These errors include undesired antenna twists, antenna tilts, errors in antenna alignments, and the effects of adjacent objects and towers on radiation patterns. In such cases, performing an in–situ measurement allows one to identify faults and repair the system, thereby improving its performance. In traditional airborne measurements, a helicopter is used to measure the radiation pattern. However, these methods are expensive and require heavy equipment. Advancements in UAVs and the miniaturization of RF components, such as portable spectrum analyzers, have enabled the measurement of mobile BASTA systems using UAVs [32].

UAV–based measurements for BASTAs involve measuring vertical and horizontal radiation patterns. The vertical radiation pattern is determined using the procedure shown in Figure 6. To ensure optimal reception from the AUT, the UAV should be positioned at an appropriate height to maintain LOS with the BASTA. Once the optimal vertical location is determined, the UAV follows a vertical path and reconstructs the radiation pattern. Typically, a BASTA consists of several antenna elements, such as dipoles, arranged in an array. Each of these elements has its own radiation pattern. The resultant radiation pattern is formed by combining all the elements in an array. Therefore, the UAV must follow a vertical path to reconstruct its vertical radiation pattern. The electrical down–tilt and null fill can be determined by obtaining the vertical radiation pattern of the BASTA. Electrical down–tilt [33] and null fills are significant parameters that affect the base station coverage area. Ground users experience maximum signal strength when the main lobe is directed towards their area, whereas they do not receive any signal when a null fill is directed towards them. In [31], a UAV programmed with a mask R–CNN was used to automatically determine the base station orientation. R–CNN is an object detection algorithm used to detect specific regions in an image. The proposed method in [31] involves creating a database named UAV–antenna, which consists of 19,715 communication BASTA images. This is achieved by the capturing of BASTA images by UAVs. Secondly, mask R–CNN applies a selective search scheme to identify the pixel coordinates of the BASTA. These pixel coordinates are used to measure the BASTA's tilt angle. Based on the proposed method, after completing the measurements it was found that the actual tilt angle of the antenna system deviated by 1°–2° from the intended tilt angle.

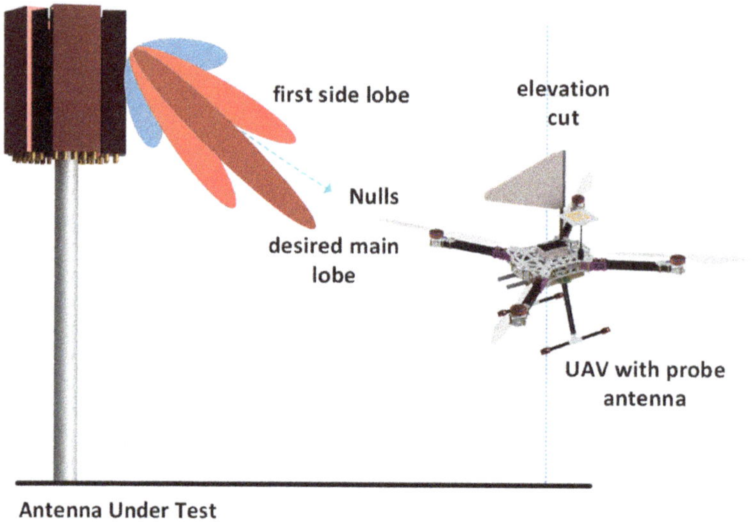

Figure 6. Vertical radiation pattern of a BASTA using a UAV.

Figure 7 illustrates the procedure followed to obtain the horizontal radiation pattern of a BASTA deployed in the field, which was measured using a UAV with an RX antenna. Based on the optimum height, at which the RX antenna mounted on the UAV receives the maximum power from the AUT, which is determined during the vertical pattern measurements, the UAV follows a trajectory in a circular path around the AUT. Obtaining a horizontal radiation pattern enables one to determine azimuth HPBW, sector power ratio (SPR), and front-to-back ratio (FBR). SPR is the ratio of power outside the desired sector to power inside the desired sector. This helps to improve the antenna design, which requires the SPR to be as low as possible to achieve lower co–channel interference and better call quality. Ideally, a BASTA should have an SPR less of than 3% and an FBR greater than 25 dB [34].

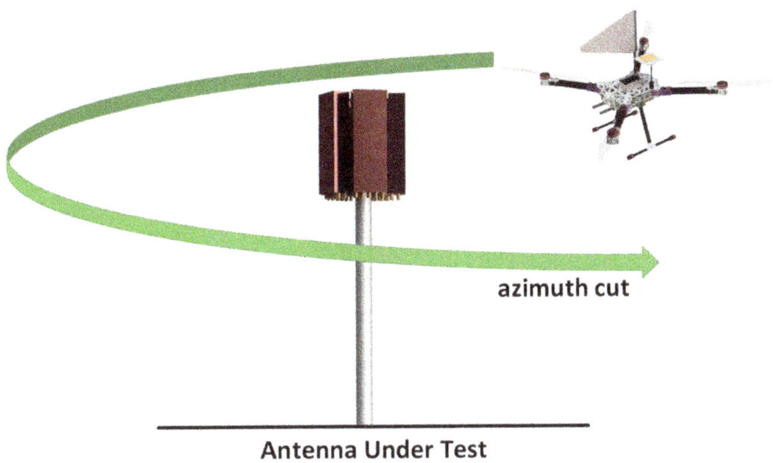

Figure 7. Horizontal radiation pattern of a BASTA using a UAV.

2.1.2. Aerial Measurement of Broadcasting Antennas

In [35], a SixArms custom-built hexacopter with a log–periodic antenna was used to perform broadcasting antenna measurements in the far-field region. It was used to measure the effective radiated power (ERP) and horizontal radiation pattern (HRP), as well as the vertical radiation pattern (VRP). As shown in Figure 8, high–power broadcast antennas can experience certain deviations in their performance, leading to a degradation in the overall coverage area. The feeding mechanism of a broadcast antenna [36] plays a significant role, and these systems are vulnerable to changes. Upon performing the in–situ measurements and comparing the measured VRP with the design specifications, it was observed that there was a 1° deviation in the tilt in the test case of [32]. Similarly, a change of 0.5° in the electrical tilt was observed owing to the change in the mechanical lean of the broadcast antennas. By measuring the HRP, other common errors, such as incorrect panel orientation and inverted panels, could be identified and eliminated, thereby enhancing the overall performance of the broadcast antenna [37]. By comparing the HRP of the broadcasting antenna measured with the UAV with the design specifications, it became apparent that a 10° deviation arose in this test case from an incorrect panel orientation after the panel was installed on the tower. Similarly, measuring the HRP of the broadcast antenna, when there was a taller tower at just 650 ft, showed a 3 dB notch in the plots, which implies that the adjacent tower impacts the measured system.

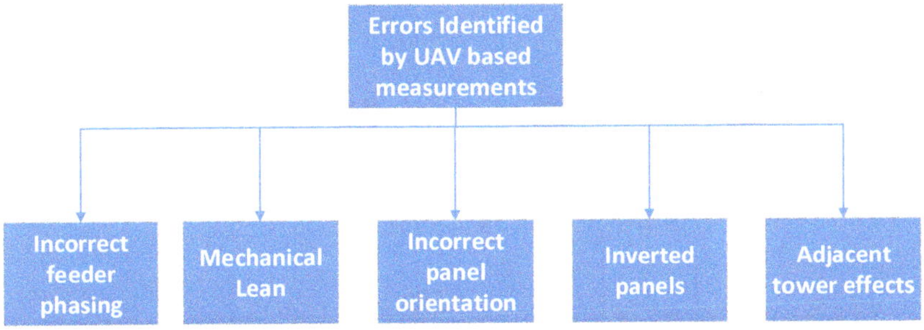

Figure 8. Common errors in broadcasting systems.

Ideally, when amplitude–only measurements are performed, UAVs carrying transmitters/receivers are in the far–field region (the Fraunhofer region) to measure the radiation pattern [20]. To validate the theoretical concept of far–field regions, the SixArms Airborne Radio Measurement Systems (ARMS-RFX) UAV was used to measure a DTV station at 720 m and 2025 m from the AUT. DTV broadcasting antennas are composed of antenna arrays formed by similar elements. The total height of a DTV broadcasting antenna with all the elements in an array is typically 20 m, with a maximum antenna dimension of 20 m and a frequency of operation at 515 MHz (UHF channel 21 in the USA). Far–field measurements with UAVs can be performed by maintaining at least 1450 m from the AUT. As shown in Figure 9, ARMS–RFX UAVs equipped with an ARMS receiver comprising a real–time spectrum analyzer and an embedded PC [38] were used to measure the DTV transmitter station at 720 m and 2025 m from the AUT. Log–periodic antennas (LPDA) mounted on top of the UAV received signals from the AUT. The radiation pattern of the AUT was calculated instantly for every 0.1°, as the UAV took a vertical path. Using a telemetry link, the measured radiation patterns were transmitted to the ground user for quick verification.

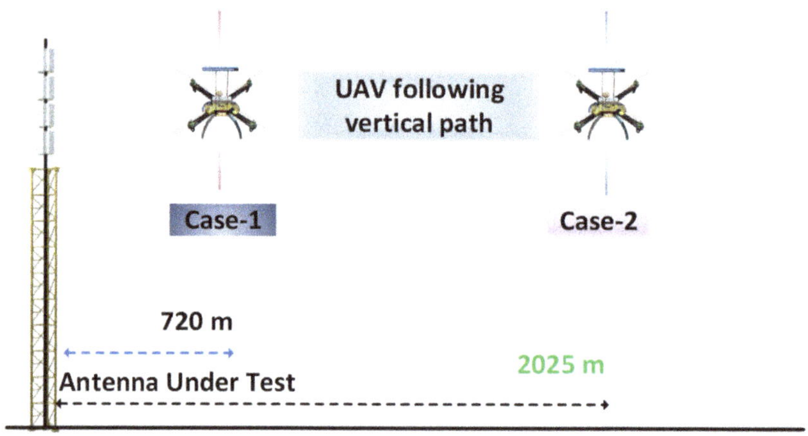

Figure 9. SixArms airborne measurements for broadcasting systems.

The elevation patterns taken at 720 m, which should have been 1450 m according to the theoretical far–field distance calculations, and at a far–field distance of 2025 m from the AUT, are depicted in Figure 10. In Figure 10a, the red dashed lines represent the elevation pattern of the AUT according to the manufacturer's data sheet, and the solid black line represents the measurements performed with the UAV at 720 m. Similarly, Figure 10b represents the measurements performed at 2025 m from the AUT. From the elevation patterns measured at 720 and 2025 m, it was observed that the measurements do not always have to be in the far–field region. UAV–based measurements can be performed closer to the AUT and are still valid. From the two cases depicted in Figure 10, it is evident that when measurements are performed at 720 m from the AUT, the null fill and null depths vary slightly compared to measurements performed at 2025 m from the AUT. In the following sections, we discuss UAV–based antenna measurements in near–field regions. This technique enables the assessment of antennas that are significantly large. Additionally, conducting measurements in the near–field offers the advantage of reducing the flight times required for UAV operations. In Table 1, we present an overview of the reflector and broadcasting antennas measured in far–field regions using UAVs.

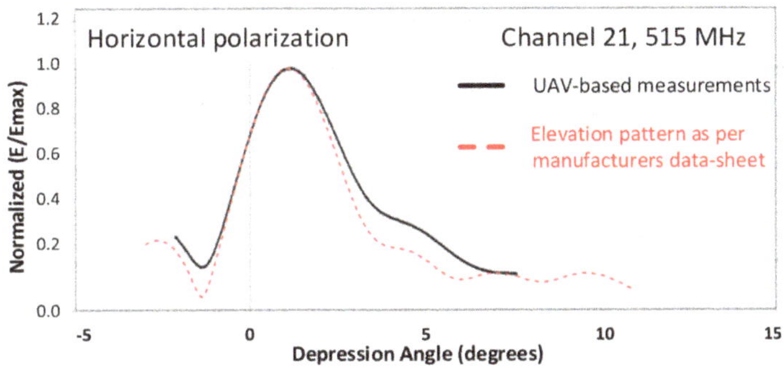

(**a**)

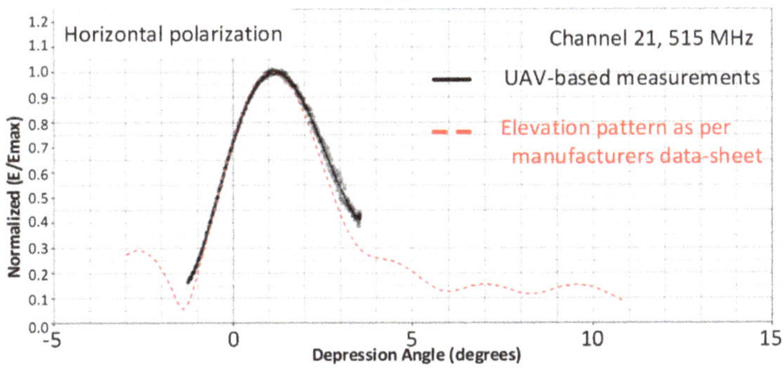

(**b**)

Figure 10. (**a**) UAV–based measurements at 720 m and (**b**) UAV–based measurements at 2025 m.

Table 1. Far–field measurements.

Reference	Frequency of Operation	Far–Field Distance	AUT
[27]	14.5 GHz	350 m	Parabolic reflector.
[28]	9.5 GHz	60 m	Reflector mounted on ship.
[35]	515 MHz	2025 m	Broadcasting antenna.

2.2. *UAV–Based Near–Field Measurements*

To meet the far–field criteria, low–frequency antennas require a considerable distance between the AUT and UAV carrying the antenna. Such long distances may result in excessive free–space path loss, which can reduce overall system accuracy. In such cases, near–field measurements can be useful for EM wave measurements in the radiative near–field region. It is easy and quick to conduct near–field measurements with a UAV and does not require heavy and complicated equipment. In [39], a biconical antenna operating at 110 MHz was measured using a UAV under indoor conditions. Time domain gating techniques were applied to avoid ground reflections and UAV motion. Similarly, in [40], navigation systems were measured in the near–field region. The measurement results from [39,40] prove that UAV–based near–field measurements can be performed accurately with low–cost equipment.

As shown in Figure 11, UAV–based measurements were performed at radiating near–field regions beginning at 2.5 m for a grid reflector with a maximum diameter of 1 m that operates at 4.65 GHz [41]. Because the accuracy of UAV–based near–field measurements depends on UAV coordinates, [42] used dual–band real–time kinematics (RTK), which made it possible to obtain UAV coordinates within 10 arcseconds. The UAV was equipped with two monopoles operating in the 4–7 GHz frequency band that were separated from each other by a space of 80 cm. This setup measured a reflector system operating at 4.65 GHz fed by a circularly polarized helix antenna. The monopoles were mounted on a UAV using three–dimensional (3D) printed components, which are highly resistant to mechanical vibrations and transparent to EM waves. Near–field measurements were performed by obtaining the equivalent current distribution over the surface of the AUT. A phase–less retrieval technique was utilized to plot the radiation pattern based on the integral equation method [43], in which the simulated AUT was modeled with equivalent electric and magnetic fields on a closed surface. After obtaining the radiation patterns for the reflector system deployed in the field, measurements were conducted in anechoic chambers. Comparisons between the results obtained from the UAV–based near–field measurements and the measurements in anechoic chambers indicate that when the reflector system is deployed in the field, there is an offset in the beam position and the main beam is widened owing to misalignment errors and ground reflections.

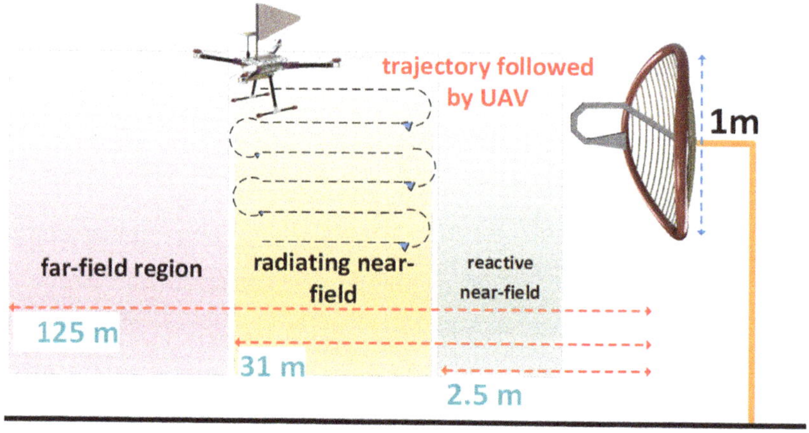

Figure 11. UAV–based measurements in radiating near field.

In Figure 12, a UAV carrying a monopole is used to characterize an antenna operating at 3–30 MHz. The AUT is a Nostradamus ONERA system [44] consisting of a set of 288 biconical antennas arranged along a branch separated by 120°. Biconicals are omnidirectional, and each has a height of 7 m and a width of 6 m. To validate the accuracy and functionality of UAV–based near–field measurements for characterizing the high–frequency (HF) antenna by ONERA, ref. [6] used a Dà-Jiāng Innovations (DJI) Matrice 600 Pro carrying an antenna to measure a monopole that was 6 m high when placed on a ground plane. The measured data were compared with simulated data, and good agreement was achieved, thus proving that UAV–based near–field measurements are a cost–effective solution for characterizing HF antenna systems.

When selecting the appropriate material for mounting the antenna on the UAV, it is important to ensure that the UAV body does not degrade the performance of the antenna. In [45], a Mikrokopter equipped with a signal generator and a short monopole was employed to measure a 6 m parabolic dish fed by a dual–polarized LPDA operating in the frequency range of 300 MHz to 3 GHz. In this UAV model, the ground plate was made of aluminum, and to minimize the impact of the UAV body on the antenna, a mesh structure was placed between the frame of the UAV and the antenna. Similarly, the UAV propellers can generate harmonics that are influenced by the propellers' rotations per minute (RPM) and their dimensions. The Doppler spectrum and harmonics [46] generated by the propellers can be studied using the double–edge diffraction model [47]. It was observed that when an antenna mounted on a UAV transmits signals while the propellers rotate, the signal received by the AUT exhibits a Doppler effect. The Doppler effect, harmonics, and scattering experienced by EM waves from an antenna lead to a drop in the power received by the AUT, which affects the radiation pattern of the AUT. The effects of propellers can be avoided by placing the antenna on the UAV at a location which is far from the propellers. Other approaches, such as using fiber glass material instead of carbon fiber for the propellers, are validated in [48,49]. All the measurements of the radiation patterns in this case were performed at 328.5 MHz.

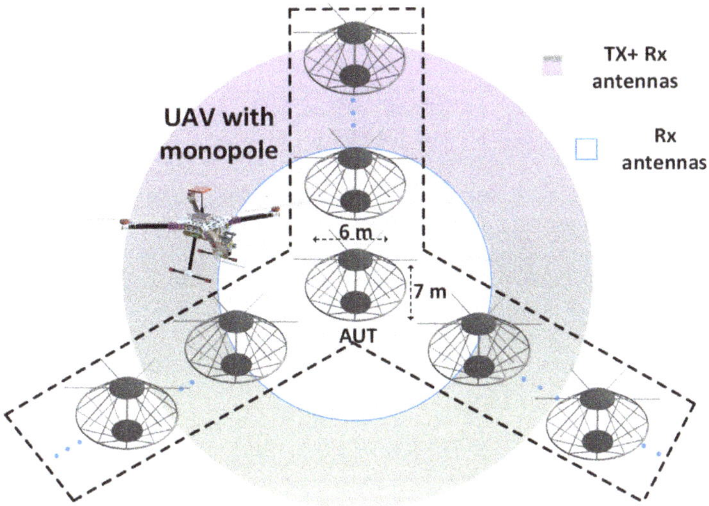

Figure 12. UAV–based measurements for an array of HF wire biconical antennas.

Aerial Measurements of Low–Frequency Antennas

The SKA is an array of telescopes that operates based on the principle of aperture synthesis and is designed for excellent spatial and angular resolution. A square kilometer array log–periodic antenna (SKALA) is a very large structure, and measuring such a large structure in the far–field region requires a large measurement distance. Performing UAV–based far–field measurements for these antennas is not economical owing to the battery limitations of UAVs. Sometimes, the UAV altitude needs to be more than 120 m, which is not possible owing to UAV flying guidelines. In such cases, the antennas are measured in near–field regions. One such case is the SKALA; it consists of 256 LPDAs with a diameter of 38 m. In the SKALA, each LPDA comprises a bowtie dipole for impedance matching. In [50], a pre–aperture array verification system for SKA was measured with UAVs in the near field. The AUT consisted of 16 active elements. All the elements were designed to feature a dual–polarization operation in the frequency range of 50–350 MHz, a minimum directivity of 8 dBi, and an intrinsic cross–polarization ratio exceeding 15 dB. An inter–element spacing of $\lambda/2$ was maintained to achieve better control over beam steering. However, maintaining an inter–element spacing of $\lambda/2$ is subject to mutual coupling [3].

The pre–aperture array verification system of SKA, which is the AUT, has an overall size of 9.2 m over a 16 m ground plane mesh. The metallic grid ground plane improved the overall directivity of the system and provided protection from humidity, weather, and terrain conditions. The UAV was equipped with a portable signal generator and a dipole resonating at 175 MHz. The main challenge in these measurements is obtaining accurate phase values. In [50], to address this problem, an additional reference antenna with a known phase [51] was used. This reference antenna was placed 12 λ (20 m) from the center of the array, as shown in Figure 13. The UAV was equipped with a dipole and followed a quasi–planar trajectory at an altitude of 24 m. With an average speed of 3 m/s, the UAV took 15 min to complete the trajectory.

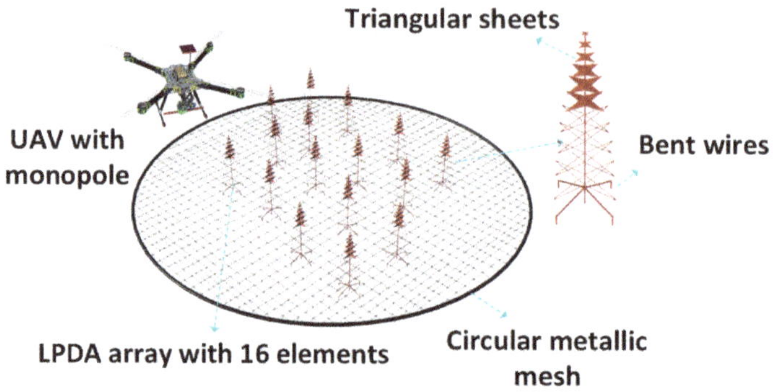

Figure 13. UAV with monopole flying over the LPDA array.

When the UAV follows the trajectory, the receiver system connected to the AUT at the ground level acquires voltages corresponding to the horizontal and vertical directions. The time stamps of the global navigation satellite systems (GNSSs) on the UAV and the time stamps of the GNSS at the ground level are synchronized at the receiver connected to the AUT. Finally, with the complex voltages and phase values, the embedded element patterns are reconstructed by performing a near–field to far–field transformation. The measured data from the UAV were compared with the simulated data, and the simulations were performed in the CST studio suite. From these comparisons, it was observed that the UAV–based near–field measurements were accurate. A deviation of 1 dB was observed in the amplitude. Thus, UAVs can be used to measure large structures such as the SKA in the near–field region. Thus, UAV–based measurements for the SKALA can help to identify areas where design improvements are required to improve the efficiency of the entire system. Table 2 presents a summary of large antennas, such as reflector antennas, ground plane antennas, ONERA biconicals, and SKALA LPDAs measured by UAV near–field techniques.

Table 2. Near–field measurements.

Reference	Frequency of Operation	Near–Field Distance	AUT
[41]	4.65 GHz	3.4 m	Offset reflector
[52]	20 MHz	4 m	Ground plane antenna
[50,51]	175 MHz	15 m	SKALA

3. UAV–Based Propagation Measurements

With advancements in technology and the demand for wireless connectivity, especially for UAVs and other applications, there is a great demand for wireless networks with low latency. These networks require a latency as low as 1 ms, which is a significant improvement over the 40 ms latency of fourth–generation (4G) networks. Transmission needs to be moved to millimeter wave (mmWave) or even terahertz (THz) frequencies to achieve such low latency. EM waves experience higher propagation losses at such high frequencies, owing to diffraction [53] and scattering from rough surfaces. Therefore, understanding the propagation environment through propagation measurements [54] is essential. The propagation of EM waves in an environment can be evaluated by using UAV–based propagation measurements. In addition, UAV–based propagation measurements enable the measurement of key performance indicators (KPIs), such as reference signal received quality (RSRQ) and reference signal received power (RSRP). In [55], a hexacopter carrying a smartphone, sensors, and guided autonomous flight paths was used to measure RSRP and RSRQ. During the transmission, the signal was attenuated by various factors before it reached the receiver. The most common reasons for signal attenuation are path loss, shadowing, and multipaths. Shadowing and multipath components, such as reflection, refraction, diffraction, and scattering, are primarily caused by obstacles. In sub-Section 3.1, we discuss large–scale and small–scale propagation in detail.

3.1. *UAV–Based Large–Scale Propagation Measurements*

UAVs can be used to measure large–scale propagation effects that occur mainly owing to path loss and shadowing. To estimate the path loss, the commonly used models are the free–space path loss (FSPL), two–ray, basic log–distance, and modified log–distance models. For the path loss calculation, the FSPL requires information about the transmitter antenna gain (G_t), receiver antenna gain (G_r), operating wavelength (λ), and separation (d) between the UAV and receiver. On the other hand, when the UAV is at a lower altitude and ground reflections are present between the UAV and receiver, two–ray models can be utilized to estimate the path loss. In [56], a DJI Mavic 2 Enterprise UAV equipped with a LoRa sleeve dipole operating at 868 MHz was used for propagation measurements between a UAV and a wireless sensor network (WSN) and between a UAV and an unmanned surface vehicle (USV). The results of the UAV–based propagation measurements for a scenario in which the UAV moved vertically up to 30 m, and another scenario in which the UAV moved horizontally away from the receiver (WSN or USV), were compared with path loss estimations from two–ray models. Based on the comparison, the two–ray model underestimated the path loss, resulting in a mean difference of 6.45 dB between the UAV and USV and 15.5 dB between the UAV and WSN. These findings demonstrate the necessity of improving the two–ray model to increase the accuracy of path loss measurements. A higher level of precision is required for mmWave [57–59] and for situations in urban areas [60], which are surrounded by multiple buildings and obstacles.

Log–distance path loss models are more general and appropriate for calculating path loss [61]. In evaluating the path loss for a channel between a UAV and a receiver, the log–distance method considers the path loss exponent (α). An improvement in the log–distance model is the modified log–distance model. According to the modified log–distance model [62,63], we can determine the path loss by

$$PL(\mathrm{dB}) = PL_0(\mathrm{dB}) + 10\alpha log_{10}\left(\frac{d}{d_0}\right) - 10log_{10}\left(\frac{\Delta h}{h_{opt}}\right) + C_p + 10log_{10}\left(1 + \frac{\Delta f}{f_c}\right), \quad (2)$$

When estimating the path loss, the modified log–distance model considers an additional parameter known as the height of the UAV from the ground (h_{gnd}), the minimum height of the UAV (h_{opt}) that provides the lowest path loss, and a constant loss factor C_p, representing the losses due to the antenna orientation on the UAV and carrier frequency f_c. PL_0 (dB) is the path loss at the reference distance d_0, Δf is the Doppler variation in frequency, and Δh is the difference between h_{gnd} and h_{opt}.

To determine the extent to which the various factors discussed above attenuate the signals, we created a scenario with a UAV carrying the transmitting antenna. The propagation measurements [64], as described in Figure 14, consisted of a car and horn antenna that was identical to that mounted on a UAV. The ground–level antenna worked in the receive mode and was mounted on a 2 m mast in front of a 10 m building. The horn antenna was mounted on the UAV hovering 12 m above ground level. Modeling and simulations were performed using commercially available EM software, CST Studio suite 2023 [65] at 5 GHz. Part of the signal was diffracted by the corners of the buildings at a height of 10 m and by metallic components of masts and obstacles, such as cars, before reaching the receiver at ground level.

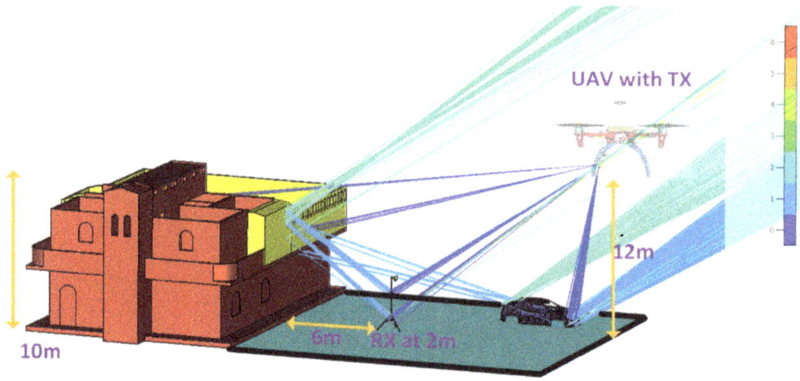

Figure 14. Scattering effect in a semi–urban area simulated in CST.

In [66], a UAV equipped with a dipole and scanner capable of measuring RSRP was utilized to calculate the path loss, with the UAV's altitude varying from 1.5 m to 120 m above ground level. To understand the influence of LOS and non–line–of–sight (NLOS) conditions on the path loss calculations in [67], the path loss was measured in an area of 500 m × 500 m. As part of this experiment, propagation measurements were performed in an urban area in Greece, which consists of buildings and trees. Measurements were conducted at 2.12 GHz for both LOS and NLOS scenarios, with the UAVs operating at an altitude of 6 to 200 m above ground level. The attenuation caused by reflections and diffractions from buildings and obstacles was also considered for the path loss calculations using the log–distance path loss model. The calculations were based on the uniform theory of diffraction (UTD) and geometric optics (GO). The path loss calculations varied for the LOS and NLOS conditions. The UAV flew at 100 m above the ground; for the LOS condition, α was as low as 2.6; however, at the same altitude, for the NLOS condition, α was 7.2, indicating that the path loss values differed depending on the test conditions.

The polarization mismatch of the antenna mounted on the UAV also affects the accuracy of the path loss measurements. In [68], a DJI Phantom 4 UAV with a vertically polarized dipole working in the 3.1–4.8 GHz range was used for propagation measurements in three different scenarios. Initial measurements were performed between the UAV and receiver, assuming that no obstacles were present between the UAV and receiver. During this experiment, the receiver antenna at ground level was vertically polarized in one case and horizontally polarized in the other. By always leaving the transmitter located on the UAV vertically polarized, the VV and VH cases were produced. For the VV condition, a path loss of 72 dB and a path loss of 80 dB were observed for the VH condition. This indicates that there was an additional path loss resulting from the polarization mismatch. For ATG propagation measurements, EM waves are attenuated by multipath components, such as reflections, diffractions, and scattering. To estimate these losses, using the FSPL, two–ray [69], and log–distance path loss models may not be accurate in certain scenarios. Alternatively, empirical models, such as multi–slope log–distance path loss models [70], height–dependent two–ray models [71], and excess path loss models [72], are more reliable.

The models that have been addressed so far are all deterministic. These models do not consider the dielectric properties of obstacles that attenuate the signals. In such instances, statistical models such as log–normal shadowing are used to calculate the attenuation of signals due to random variations. There are two crucial variables in a log–normal shadowing expression, $\mu_{\psi B}$: the mean of the random variable, and $\sigma_{\psi B}$: its standard deviation. A DJI N3 UAV, in combination with a $\lambda/4$ monopole, was used in [73] to measure path loss while the UAV moved from 0 to 24 m in height. In the context of path loss measurements, [73] proposed an altitude–dependent propagation loss model based on a zero-mean-behavior random variable. Under NLOS conditions, after performing propagation measurements at 1 GHz and 4 GHz, the $\sigma_{\psi B}$ [74] value increased with the frequency and distance of the UAV from the receiver. The typical range of $\sigma_{\psi B}$ is 5–12 dB for terrestrial macrocells and 4–13 dB for terrestrial microcells. For aerial wireless channels, $\sigma_{\psi B}$ ranges from 1.2 to 5.24 dB, and it is observed that when UAVs fly at high altitudes, $\sigma_{\psi B}$ can be as low as 1.2 dB [75].

3.2. UAV–Based Small–Scale Propagation Measurements

In [76], to investigate the dependence of small–scale fading on the altitude of a UAV, a hexacopter was equipped with a circularly polarized cloverleaf wire antenna, as shown in Figure 15. The receiving system consisted of a magnetic mount wideband high–frequency (MGRM–WHF) antenna, which is independent of the ground plane and was installed on a mast at the ground station, 1.5 m above the ground. The test environment consisted of multiple buildings and metal containers with the UAV taking a vertical path ranging from 0 to 100 m in height and a horizontal path maintaining 20–60 m from the receiver at ground level. The path loss exponents (PLEs) were estimated by varying the height of the UAVs to determine the relationship between small–scale fading and the UAV altitude. Furthermore, small–scale fading calculations are categorized into LOS and NLOS conditions. The Rician-K factor was utilized in the LOS case to explain the fading behavior. Adding the height parameter of a UAV to the Rician–K factor provided a better understanding of the small–scale fading. At lower altitudes, multipath components [77] from buildings and metallic containers combined vectorially at the receiver, causing fading. The cumulative distribution functions (CDF) [78] estimated small–scale fading in both the LOS and NLOS conditions.

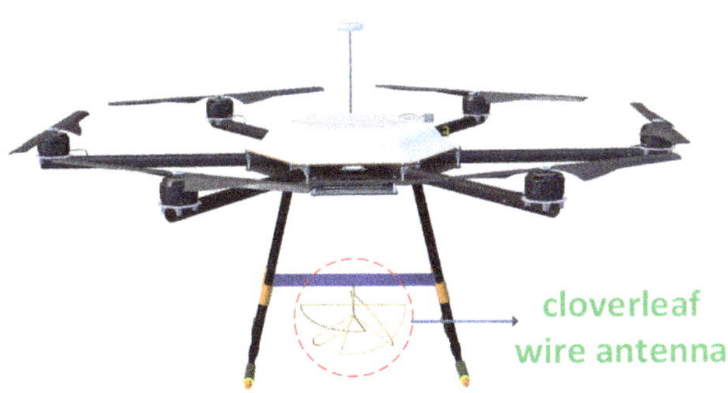

Figure 15. Hexacopter carrying cloverleaf wire antenna.

In [75], a DJI N3 six–rotor UAV, equipped with a $\lambda/4$ monopole, was used to determine the fading depth, using UAVs by varying their vertical paths from 0 to 24 m in height, with a receiving station positioned 25 m away from the UAV. A $\lambda/4$ monopole with a ground plane and gain of 5.2 dBi, connected to a portable signal generator, enabled the continuous transmission of signals. We note from the measurements that the fading depth was independent of the operation frequency, which was more evident for the LOS conditions than for the NLOS conditions. A distribution function, such as the Nakagami, Rayleigh, Weibull, or Gaussian function, can describe the fading amplitude. By maintaining a root mean square error (RMSE) as low as 0.02 dB for both the LOS and NLOS scenarios, the log–logistic function [79] is the best distribution function among the available distribution functions.

In [80], to investigate the scattering effect of the buildings, ATG propagation measurements were conducted using a custom–built UAV equipped with a mmWave conical horn antenna (operating in the 26–40 GHz range) configured as the receiver and a planar elliptical dipole ultra–wideband (UWB) antenna (operating in the 3.1–5.3 GHz range) configured as the transmitter. In contrast, the ground station featured an mmWave conical horn antenna as the transmitter and a UWB antenna as the receiver. These ATG propagation measurements were instrumental in understanding the propagation characteristics of urban environments. This study presented power angle profiles for ATG propagation, which showed that in urban areas the presence of building rooftops causes a reduction in signal strength due to scattering. Additionally, the power elevation profile results indicated that when the UAV was at a higher altitude (50 m), the dominant propagation mechanism was due to reflections from buildings located behind the ground station. Measurements of outdoor–to–indoor coverage, conducted with the UAV hovering outside a building and the ground station positioned inside the building, revealed significant losses as the signals propagated through the building walls at both mmWave and UWB frequencies.

In ATG channels, which consist of a wireless channel between the UAV and the ground system, knife–edge diffraction (KED) is a commonly employed method for estimating the signal strength attenuation caused by diffraction. In KED, the EM wave diffracted by the building corners is determined by considering the obstacles to be thin and perfectly absorbing. The magnitude of the diffraction losses is calculated using mathematical formulas that consider Fresnel diffraction parameters. According to the UTD, diffraction losses are estimated using wedge geometry, which involves the wedge angle and reflection coefficient of obstacles and empirical models such as the linear regression model and the creeping wave linear model [81]. In [82], to understand the accuracy of KED and the empirical models, diffraction loss measurements were performed over a roof top in urban environments at 28 GHz. The measurement setup consisted of a transmitter antenna with a beamwidth of 10° and a receiver antenna with a beamwidth of 30°. The measurements were conducted at two sites to understand the influence of the TX distance from the LOS boundary and the RX distance from the LOS/NLOS boundary. The study found that the diffraction losses increased when the distance from the diffraction edge increased and decreased when the distance between the TX and the building decreased. The loss was shown to be proportional to the diffraction angle.

Using UAV–based propagation measurements, we can estimate the attenuation of the signal when the transmitter follows vertical and horizontal paths. In contrast, conventional methods fail to evaluate diffraction losses and multipath components from the corners and edges of buildings, which are typically between 10 m and 25 m in height. In [83], horn antennas with a gain of 20 dBi and an HPBW of 17° were used indoors and outdoors for propagation measurements. The indoor measurements were analyzed using three types of wall construction: plastic boards, wooden walls, and dry walls. During the measurements, the receiving and transmitting horns were placed at a 1.4 m height above the corner of the wall. The measurements were performed at 10 GHz, 20 GHz, and 26 GHz. The measured data were compared with the theoretical estimates using the KED model. The practical measurements were in good agreement with the theoretical calculations for a dry wall. However, the KED overestimated losses by 2–4 dB in the case of wooden walls and plastic boards. For outdoor measurements, it was found that KED accurately calculated the diffraction losses for sharp edges, whereas linear models using a minimum mean square error (MMSE) linear fit derived from actual measured data were more accurate for rounded edges.

To investigate the scattering effects of buildings, we created an ATG propagation scenario. In this setup, two buildings with heights of 10 m and 20 m were modeled using CST Studio Suite. Building 1 was modeled with a height of 20 m while Building 2 was modeled with a height of 10 m. A horn antenna configured as a receiver was mounted on a mast placed on top of Building 2 to identify potential scattering regions. In the simulation environment, we modeled a UAV equipped with a horn antenna flying at a height of 17 m above ground level. The complete simulation setup, including the scattered rays, is shown in Figure 16. The two buildings were positioned 30 m apart in this scenario. The structure was analyzed using an asymptotic solver based on the shooting and bouncing ray (SBR) technique, which allowed us to observe how signals were diffracted at the corners of the buildings. The SBR technique provides an initial estimation of ATG propagation. However, to accurately understand scattering effects in real–world scenarios, practical UAV–based propagation measurements are necessary. In such measurements, a UAV equipped with a transmitter antenna and a portable signal generator would be used. On the ground, a receiver setup consisting of a horn antenna mounted on a mast and a spectrum analyzer connected to the antenna would be used to calculate the power levels of the received signal.

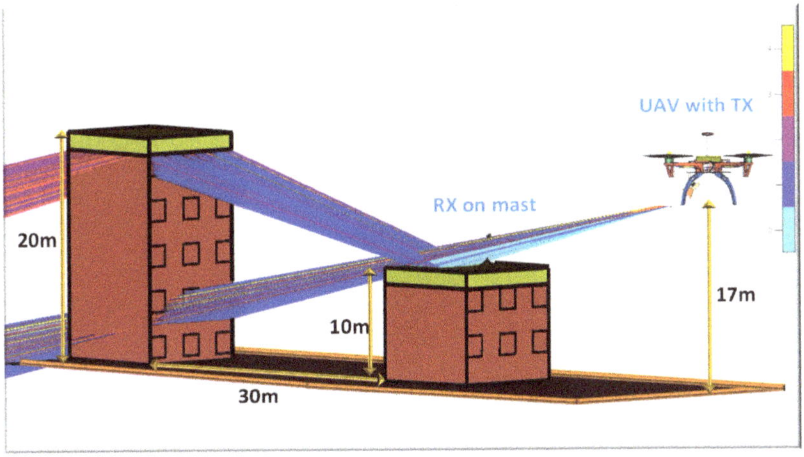

Figure 16. Scattering effect in an urban area simulated in CST Studio Suite.

Several propagation measurements were carried out in [84] using a hexacopter equipped with a narrowband antenna resonating at 440 MHz and a wideband antenna operating between 1 GHz and 6 GHz. The measurements were performed in a suburban area at 440 MHz and 1 GHz. The UAV flew in a vertical path with an altitude of 0–25 m over two buildings of 15 m and 25 m in height. Although there were other obstacles, such as trees and cars, in addition to the two main buildings, the diffraction losses owing to other obstacles were minimal at high altitudes. A comparison was made between UAV–based propagation measurements and theoretical modeling, such as the KED model. In general, the measurements by UAVs and the theoretical calculations are in good agreement at lower frequencies; however, at higher frequencies, the diffraction losses are more significant, and the theoretical calculations underestimate these losses.

3.3. Selecting the UAV Antenna

Choosing an appropriate antenna for UAV–based measurements is essential before conducting measurements. To ensure that the UAV–based in situ measurements are accurate, it is important to calibrate the antenna before mounting it on the UAV. Several factors are considered when selecting a UAV antenna: it should be compact, lightweight, mechanically stable, unaffected by wind, and electromagnetically insensitive to the structure of the UAV. Because of the several metallic components on the body of the UAV, directional antennas are likely to experience EM coupling with the UAV body, which can degrade its performance.

To understand the EM behavior of the antenna [85] mounted on the UAV, we simulated a complete UAV structure using the CST Studio Suite. Figure 17 depicts a DJI F450 UAV equipped with a pyramidal horn antenna simulated at 8 GHz using the SBR technique. Apart from the main beam from the horn, there is a portion of signals scattered from the UAV body, which can create errors in antenna measurements. This explains the necessity for care to be taken before selecting an antenna and understanding its behavior after mounting [86] it on the UAV [87]. There are several ways to mitigate the effects of scattering, including changing the antenna design [87,88], optimizing the antenna location [89], and using RF absorbers in areas where the UAV body exhibits potential reflections. Whenever we choose antennas for UAVs, there is always a tradeoff between an antenna with a narrow beamwidth and an antenna with wider beamwidth. Antennas with a wider beamwidth cover wide angles, which means that small deviations in the alignment of the UAV relative to the AUT have a less pronounced effect. The signal remains closer to the intended polarization, minimizing the introduction of unwanted cross–polarized components. However, with an antenna with a wider beamwidth, there will be high scattering from UAVs, affecting the co–polarization and cross–polarization patterns.

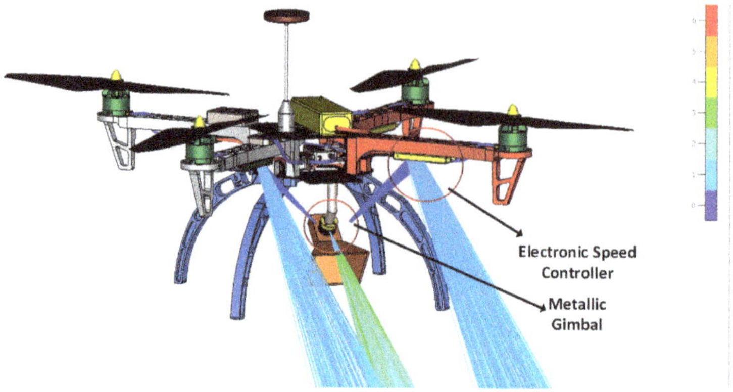

Figure 17. Fields scattered by the UAV body.

On the other hand, antennas with narrow beam widths have an advantage in terms of low scattering from the UAV body. However, they have some limitations as well. The major challenge for these antennas is alignment between the antenna mounted on the UAV and the AUT. In the case of an antenna with a narrow beamwidth, vibrations from the UAV body can create a misalignment between the antenna mounted on the UAV and the boresight of the AUT. To minimize misalignment errors, additional efforts must be made to maintain the antenna's beam peak at the AUT's boresight throughout its trajectory.

Table 3 presents the different antennas used in the literature. An omnidirectional or directional antenna was used depending on the area of application. These antennas are specially designed for UAV applications, considering beamwidth and radiation pattern constraints. In cases such as dipole [90] and helix [91] antennas, the ground plane is included as part of the antenna. The commonly used antennas with directional or omnidirectional patterns have limitations. Directional antennas are prone to misalignment errors; hence, additional precautionary steps are required to overcome them. Omnidirectional antennas have the limitations of low gain. On the other hand, in [92], an array of half-bowtie antennas was designed to cover all hemispherical regions. This design has better coverage with an HPBW of 240° in the azimuth and 98.6° in elevation, similar to an omnidirectional antenna; with a gain of around 5.9 dBi.

Table 3. State–of–the–art antennas for UAV applications.

Reference	Antenna	Frequency Range (GHz)	Radiation Pattern
[21]	Micro–Strip Patch	1.8–2.7	Directional
[87]	Micro–Strip Patch	2.4–5.2	Directional
[93]	Log–periodic meandered dipole array	0.85–2.2	Directional
[94]	Vivaldi	1.5–4.5	Directional
[95]	Horn	7.5–18	Directional
[90]	Dipole	0.55–1.6	Omnidirectional
[96]	Surface Wave	6.1–18	Directional
[91]	Helix	0.6–1.1	Omnidirectional
[97]	Quasi–Yagi	23–28.5	Directional
[92]	Half-bowtie antenna	4.1–5.6	Directional

3.4. Accuracy Analysis of UAV–Based Antenna Measurement

The accuracy of the UAV-based antenna measurements mainly depends on the accuracy of the RF equipment mounted on the UAV, the amount of vibration experienced by the UAV, the accuracy of the GPS positioning, and the external environmental conditions. To understand how these aspects affect the measurements, Table 4 describes the variations in the radiation patterns measured by the UAV. All these measurements were performed by a UAV to characterize an antenna installed outdoors. After characterizing the antenna in terms of the radiation pattern, the same antenna was measured in anechoic chambers, and in some cases, it was simulated using commercially available EM solvers. As described in Table 4, it can be understood that due to the UAV vibrations, external environmental conditions, and drifting in the UAV positions, a maximum error of 1 dB in the peak amplitude is noticed. These results indicate that UAV measurements are an accurate and a reliable solution for characterizing an antenna.

Table 4. Accuracy of UAV–based antenna measurements.

Reference	Frequency (GHz)	Difference Between UAV and Anechoic Chamber Measurements	Difference Between UAV and Simulation Results
[1]	0.75	NA	0.5 dB in peak amplitude
[4]	0.47 to 0.7	NA	0.6 to 1 dB in peak amplitude
[7]	0.7 to 0.8	NA	0.5 to 1 dB in peak amplitude
[27]	14.5	0.38 dB in peak amplitude	NA
[42]	4.65	Widening in radiation pattens	NA
[50]	0.175	NA	<1 dB in peak amplitude
[98]	0.05 to 0.35	NA	0.5 to 1 dB in peak amplitude
[99]	8 to 12	0.5 dB in peak amplitude, 0.06° in HPBW	NA
[100]	0.05 to 0.32	NA	<0.1 dB in peak amplitude and 1 dB in SLL
[101]	44	1 dB in peak amplitude	NA
[102]	4 to 6	3 dB in peak amplitude	NA

In UAV–based antenna measurements, the radiation patterns of the antenna in either the azimuth or elevation plane were obtained by following a predefined trajectory. For the UAV to follow this predefined trajectory, the UAV path planning is achieved through software tools such as QGroundControl v1.3.8 [103]. The accuracy with which the GPS follows the defined waypoints depends on the accuracy of the GPS used on the UAV. Based on the frequency of operation, GPS systems such as differential RTK (D–RTK), RTK, differential GNSS, and real–time differential GPS are used. These high–precision systems are particularly used for UAV measurement applications, which can provide centimeter and sub–meter accuracy. Based on the UAV–based antenna measurements conducted in [27], it is evident that a total deviation of 0.38 dB between the UAV–based measurements and the measurements in the anechoic chamber is observed. Out of the 0.38 dB variation in the peak amplitude of the radiation patterns, 0.36 dB is due to the external environmental conditions; small inaccuracies in D-RTK positioning result in a 0.01 dB variation, and variations in RF component behavior due to outdoor temperature result in a 0.02 dB variation. In [7], the horizontal deviation between the trajectory followed by the UAV and the trajectory planned in the software was less than 2 m. These deviations were due to environmental conditions such as wind; in any case, these effects resulted in an angular deviation of less than 0.38° in the UAV measurements. In [101], small deviations in the trajectory followed by the UAV resulted in a deviation of 0.02 dB, and the variations in the relative orientation of the

UAV could produce an uncertainty of ±2%, producing a variation of 0.005 dB in the UAV measured results. Similarly, in [50], differential GNSS, which can provide an accuracy of only sub–meter level accuracy, was used. Although a centimeter–level accuracy GPS such as D-RTK was not used here, a deviation of 0.03° was observed in the UAV measurements due to the deviations in the UAV positioning. This is because the measurements were performed at 175 MHz; in this case, such low frequencies do not demand centimeter–level accuracy in UAV positioning. In [99], where UAV measurements were performed at 8–12 GHz, D–RTK was used for UAV positioning, and deviations in the UAV positioning resulted in a variation of ±0.01° in the UAV measurements.

From these analyses of the accuracy of UAV–based antenna measurements, it is evident that factors such as GPS positioning, vibrations in UAV, and changes in UAV alignment due to external environmental conditions result in relatively small deviations in the radiation pattern of the antenna. However, in [42,102], major deviations in UAV measurements and measurements from anechoic chambers were observed. This is due to the ground reflections that account for deviations in the antenna parameters, such as side-lobe level and HPBW, and, in certain scenarios, in the peak gains. Ground reflections create multipath interference, where signals add constructively or destructively before reaching the receiver antenna [104]. Based on the constructive or destructive interference, ground reflections lead to either an increase or decrease in the measured antenna gain.

To mitigate the effects of ground reflections, strategies such as those using radiation-absorbent material or diffraction fencing can be employed. Radiation–absorbent material, simply known as an absorber, is used in anechoic chambers, which helps to reduce interferences due to reflection from the ground. However, several absorbers would be required for outdoor measurements to mitigate the ground reflections; this is impractical and costly. Another approach is to use a metallic diffraction fence to block the ground–reflected waves. This approach is applicable and used in elevated slant measurement ranges [105].

To overcome ground reflections, we propose the use of two UAVs, which allows the AUT and receiving antenna mounted on the UAV to maintain higher altitudes from the ground level. In the two proposed UAV antenna measurements, one of the UAVs is configured as a transmitter consisting of a portable signal generator and an AUT. Similarly, the second UAV, which is configured as a receiver, consists of a real–time spectrum analyzer and an antenna to receive signals from the transmitter UAV. The proposed solution enables non–tethered UAV operation, allowing UAVs to maintain a higher altitude from the ground level, thus avoiding the effects of ground reflections. By adopting two UAV antenna measurements and placing the antennas at appropriate locations on the UAV to avoid electromagnetic coupling, antenna measurements with good accuracy can be performed in outdoor environments.

4. Conclusions

In this article, we presented a comprehensive review of UAV–based antenna and propagation measurements, offering a detailed analysis of the various factors influencing these measurements. The study provides a set of guidelines for selecting the UAV antennas, ensuring higher levels of measurement accuracy. Additionally, we compared traditional slant–range methods with innovative, low–cost UAV test setups and explored the extension of path loss models by incorporating UAV altitude as a critical parameter.

The discussion included several practical test cases, such as the use of parabolic reflector systems on ships, BASTAs, LFAAs, one of the world's largest radio telescopes, and ONERA's Nostradamus system, which features 288 biconical antennas operating in the HF range (3–30 MHz). We also examined propagation measurements for both large–scale and small–scale channel effects.

The findings suggest that advancements in portable devices like SDRs, high–precision positioning systems with centimeter–level accuracy, custom antenna designs, and UAVs constructed from lightweight and durable materials such as carbon fiber have significantly expanded the potential for UAV–based antenna and propagation measurements. For applications in 5G/6G, where accuracy is paramount, UAV–based test setups have emerged as the preferred measurement solution.

In conclusion, we anticipate that this review will serve as a valuable reference for the further development of UAV–based measurement solutions, driving innovation and precision in the field.

Author Contributions: Conceptualization, V.R.K., P.I.L. and Z.D.Z.; methodology, V.R.K., P.I.L. and Z.D.Z.; formal analysis, V.R.K.; investigation, P.I.L.; writing—original draft, V.R.K. and P.I.L.; writing—review and editing, V.R.K., P.I.L., Z.D.Z., Q.Z.A., F.A.K., J.S. and T.H.L.; supervision, P.I.L., Z.D.Z., Q.Z.A., F.A.K. and A.J.R.S.; funding acquisition, P.I.L. All authors have read and agreed to the published version of the manuscript.

Funding: This research was supported by the European Union through the Horizon 2020 Marie Skłodowska-Curie Innovative Training Networks Programme "Mobility and Training for beyond 5G Ecosystems (MOTOR5G)" under grant agreement no. 861219.

Informed Consent Statement: Not applicable.

Data Availability Statement: Not applicable.

Conflicts of Interest: Author Jason Schreiber was employed by the company Sixarms. The remaining authors declare that the research was conducted in the absence of any commercial or financial relationships that could be construed as a potential conflict of interest.

References

1. Paonessa, F.; Ciorba, L.; Righero, M.; Giordanengo, G.; Virone, G.; Vecchi, G. SDR-Enabled Near-Field Measurements with UAVs. In Proceedings of the IEEE Conference on Antenna Measurements and Applications (CAMA), Genoa, Italy, 15–17 December 2023; pp. 159–160.
2. Tena Sanchez, R.; Sierra Castaner, M.; Foged, L.J. A Referenceless Antenna Measurement System Based on Software-Defined Radio. *IEEE Antennas Propag. Mag.* **2020**, *62*, 108–118. [CrossRef]
3. Paonessa, F.; Ciorba, L.; Kyriakou, G.; Bolli, P.; Virone, G. UAV-Based Measurement of Sharp Spectral Resonances in Mutually Coupled SKA-Low Elements. *IEEE Antennas Wirel. Propag. Lett.* **2023**, *22*, 2735–2739. [CrossRef]
4. Cadavid, A.N.; Aristizabal, J.; Alhucema, M.D.P. Antenna pattern verification for Digital TV Broadcast systems in Andean countries based on UAV's. In Proceedings of the IEEE-APS Topical Conference on Antennas and Propagation in Wireless Communications (APWC), Cartagena, Colombia, 10–14 September 2018; pp. 858–861.
5. Jovićević, S.; Jovanović, A. The Analysis of the Biconical Antenna by the Least-Squares Boundary Residual Method. *Int. J. Electron. Commun.* **2003**, *57*, 415–419. [CrossRef]
6. Herbette, Q.; Saillant, S.; Menelle, M.; Urbani, B.; Bourey, N.; Darces, M.; Helier, M. HF Radar antenna near field assessment using a UAV. In Proceedings of the IEEE International Radar Conference (RADAR), Toulon, France, 23–27 September 2019; pp. 1–4.
7. Zhang, J.; Liu, J.; Wu, F.; Chen, X.; Li, J.; Timbie, P.T.; Das, S.; Yan, R.; He, J.; Calvin, O. Beam Measurements of the Tianlai Dish Radio Telescope Using an Unmanned Aerial Vehicle. *IEEE Antennas Propag. Mag.* **2021**, *63*, 98–109.
8. Oguchi, T. Electromagnetic wave propagation and scattering in rain and other hydrometeors. *Proc. IEEE* **1983**, *71*, 1029–1078. [CrossRef]
9. Avolio, V.; Capozzoli, A.; Celentano, L.; Curcio, C.; Liseno, A.; Savarese, S. Reducing the Scanning Time in Near-Field Measurements with an Optimized Sampling and an Optimized Controller on Arduino Due. In Proceedings of the AMTA Proceedings, Williamsburg, VA, USA, 4–9 November 2018; pp. 1–6.
10. Zhou, H.; Xiong, H.-L.; Liu, Y.; Tan, N.-D.; Chen, L. Trajectory Planning Algorithm of UAV Based on System Positioning Accuracy Constraints. *Electronics* **2020**, *9*, 250. [CrossRef]
11. Fritzel, T.; Steiner, H.-J.; Straus, R. Laser tracker metrology for UAV-based antenna measurements. In Proceedings of the IEEE Conference on Antenna Measurements & Applications (CAMA), Västerås, Sweden, 3–6 September 2018; pp. 1–3.
12. Barry, E.; Betjes, P.; Kim, E. Measurement Uncertainties in Outdoor Far-field Antenna Ranges. In Proceedings of the Antenna Measurement Techniques Association Symposium (AMTA), Renton, WA, USA, 8–13 October 2023; pp. 1–6.
13. Mezieres, N.; Mattes, M.; Fuchs, B. Antenna Characterization From a Small Number of Far-Field Measurements via Reduced-Order Models. *IEEE Trans. Antennas Propag.* **2022**, *70*, 2422–2430. [CrossRef]
14. MVG. Microwave Vision Group (MVG) Positioners and Controllers Overview. Available online: https://www.mvg-world.com/en/resources/product-catalogs/positioners-and-controllers-overview (accessed on 1 August 2024).
15. Jiang, H.; Zhang, Z.; Wang, C.-X.; Zhang, J.; Dang, J.; Wu, L.; Zhang, H. A Novel 3D UAV Channel Model for A2G Communication Environments Using AoD and AoA Estimation Algorithms. *IEEE Trans. Commun.* **2020**, *68*, 7232–7246. [CrossRef]
16. Mao, K.; Zhu, Q.; Qiu, Y.; Liu, X.; Song, M.; Fan, W.; Kokkeler, A.B.J.; Miao, Y. A UAV-Aided Real-Time Channel Sounder for Highly Dynamic Nonstationary A2G Scenarios. *IEEE Trans. Instrum. Meas.* **2023**, *72*, 1–15. [CrossRef]

17. Jiang, H.; Xiong, B.; Zhang, H.; Basar, E. Physics-Based 3D End-to-End Modeling for Double-RIS Assisted Non-Stationary UAV-to-Ground Communication Channels. *IEEE Trans. Commun.* **2023**, *71*, 4247–4261. [CrossRef]
18. *ANSI/IEEE Std 149-1979*; IEEE Standard Test Procedures for Antennas. The Institute of Electrical and Electronics Engineers, Inc.: New York, NY, USA, 30 November 1979; pp. 1–144.
19. Skulkin, S.P.; Turchin, V.I.; Kascheev, N.I. Range Distance Requirements for Large Antenna Measurements for Square Aperture With Uniform Field Distribution. *IEEE Antennas Wirel. Propag. Lett.* **2018**, *17*, 1257–1260. [CrossRef]
20. Umeyama, A.Y.; Salazar-Cerreno, J.L.; Fulton, C.J. UAV-Based Far-Field Antenna Pattern Measurement Method for Polarimetric Weather Radars: Simulation and Error Analysis. *IEEE Access* **2020**, *8*, 191124–191137. [CrossRef]
21. Mauermayer, R.A.M.; Kornprobst, J. A Cost-Effective Tethered-UAV-Based Coherent Near-Field Antenna Measurement System. *IEEE Open J. Antennas Propag.* **2022**, *3*, 984–1002. [CrossRef]
22. Bredemeyer, J.; Schrader, T.; Kleine-Ostmann, T. Quasi-stationary Signal-in-Space Measurements using Traceable Antennas. In Proceedings of the 17th International Flight Inspection Symposium (IFIS), Braunschweig, Germany, 4–8 June 2012.
23. Jinhwan, K.; De, A.; Sarkar, T.K.; Hongsik, M.; Weixin, Z.; Salazar-Palma, M. Free Space Radiation Pattern Reconstruction from Non-Anechoic Measurements Using an Impulse Response of the Environment. *IEEE Trans. Antennas Propag.* **2012**, *60*, 821–831. [CrossRef]
24. Spirlet, M.; Geuzaine, C.; Beauvois, V. Experimental Correction of Radiation Patterns Between Electromagnetic Environments. *IEEE Trans. Antennas Propag.* **2017**, *65*, 1330–1338. [CrossRef]
25. Knapp, J.; Eibert, T.F. Near-Field Antenna Pattern Measurements in Highly Reflective Environments. *IEEE Trans. Antennas Propag.* **2019**, *67*, 6159–6169. [CrossRef]
26. Froes, S.M.; Corral, P.; Novo, M.S.; Aljaro, M.; Lima, A.C.D.C. Antenna Radiation Pattern Measurement in a Nonanechoic Chamber. *IEEE Antennas Wirel. Propag. Lett.* **2019**, *18*, 383–386. [CrossRef]
27. Culotta-Lopez, C.; Gregson, S.; Buchi, A.; Rizzo, C.; Trifon, D.; Skeidsvoll, S.; Barbary, I.; Espeland, J. On the Uncertainty Sources of Drone-Based Outdoor Far-Field Antenna Measurements. In Proceedings of the IEEE Antenna Measurement Techniques Association Symposium (AMTA), Daytona Beach, FL, USA, 24–29 October 2021; pp. 1–6.
28. Núñez, J.; Orgeira-Crespo, P.; Ulloa, C.; García-Tuñón, I. Analysis of the operating conditions for UAV-based on-board antenna radiation pattern measurement systems. *PLoS ONE* **2021**, *16*, e0245004. [CrossRef]
29. Friedlander, B. Polarization Sensitivity of Antenna Arrays. *IEEE Trans. Signal Process.* **2019**, *67*, 234–244. [CrossRef]
30. Li, X.; Song, H.; Liu, C. Path Loss Modeling for Wireless Network Deployment in Water Surface Environments. *IEEE Antennas Wirel. Propag. Lett.* **2022**, *21*, 1090–1094. [CrossRef]
31. Zhai, Y.; Ke, Q.; Xu, Y.; Deng, W.; Gan, J.; Zeng, J.; Zhou, W.; Scotti, F.; Labati, R.D.; Piuri, V. Mobile Communication Base Station Antenna Measurement Using Unmanned Aerial Vehicle. *IEEE Access* **2019**, *7*, 119892–119903. [CrossRef]
32. Fernandez, M.G.; Lopez, Y.A.; Andres, F.L.H. On the Use of Unmanned Aerial Vehicles for Antenna and Coverage Diagnostics in Mobile Networks. *IEEE Commun. Mag.* **2018**, *56*, 72–78. [CrossRef]
33. Yang, J.; Ding, M.; Mao, G.; Lin, Z.; Zhang, D.-G.; Luan, T.H. Optimal Base Station Antenna Downtilt in Downlink Cellular Networks. *IEEE Trans. Wirel. Commun.* **2019**, *18*, 1779–1791. [CrossRef]
34. Li, R.; Huang, Y.; Du, J.; Cui, Y. Front-to-back ratio improvement of a compact base-station antenna for 5G NR applications. *Microw. Opt. Technol. Lett.* **2024**, *66*, e34063. [CrossRef]
35. Schreiber, J. Antenna pattem reconstitution using unmanned aerial vehicles (UAVs). In Proceedings of the IEEE Conference on Antenna Measurements & Applications (CAMA), Syracuse, NY, USA, 23–27 October 2016; pp. 1–3.
36. Smart-Drones. Colibrex Drone-Based Antenna Measurements. Available online: https://www.colibrex.com/en/references-case-studies/antenna-measurement (accessed on 1 August 2024).
37. Chaurasia, K.K.; Sahu, R.; Tripathi, S.; Gupta, A.K. Coverage Analysis of Broadcast Networks With Users Having Heterogeneous Content/Advertisement Preferences. *IEEE Trans. Commun.* **2022**, *70*, 2057–2071. [CrossRef]
38. Sixarms. Airborne Radio Measurement Systems 9 KHz to 12 GHz. Available online: https://www.sixarms.com (accessed on 1 August 2024).
39. Geise, R.; Vaske, A.; Fritzel, T.; Strauß, R. Indoor Antenna Calibration with UAVs—On Time Domain Gating and the Influence of Motion. In Proceedings of the IEEE 17th European Conference on Antennas and Propagation (EuCAP), Florence, Italy, 26–31 March 2023; pp. 1–5.
40. Geise, R.; Weiss, A.; Neubauer, B.; Alkar, A. A Field Test for Phaseless Measurements for Nearfield Inspections of Navigation Systems with UAVs. In Proceedings of the 2021 USNC-URSI Radio Science Meeting (USCN-URSI RSM), Honolulu, HI, USA, 9–13 August 2021; pp. 3–7.
41. Fernández, M.G.; López, Y.Á.; Andrés, F.L.H. Advances in Antenna Measurement and Characterization Using Unmanned Aerial Vehicles. In Proceedings of the IEEE 13th European Conference on Antennas and Propagation (EuCAP), Krakow, Poland, 31 March–5 April 2019; pp. 1–5.
42. García Fernández, M.; Álvarez López, Y.; Las-Heras, F. Dual-Probe Near-Field Phaseless Antenna Measurement System on Board a UAV. *Sensors* **2019**, *19*, 4663. [CrossRef]
43. Bod, M.; Moradi, G.; Sarraf-Shirazi, R. Phaseless near-field to far-field transformation based on source current reconstruction and signal subspace optimization. *Int. J. RF Microw. Comput.-Aided Eng.* **2022**, *32*, e22913. [CrossRef]

44. Bazin, V.; Molinie, J.P.; Munoz, J.; Dorey, P.; Saillant, S.; Auffray, G.; Rannou, V.; Lesturgie, M. Nostradamus: An OTH Radar. *IEEE Aerosp. Electron. Syst. Mag.* **2006**, *21*, 3–11. [CrossRef]
45. Martinez Picar, A.; Marque, C.; Anciaux, M.; Lamy, H.; Ranvier, S. Antenna pattern calibration of radio telescopes using an UAV-based device. In Proceedings of the IEEE International Conference on Electromagnetics in Advanced Applications (ICEAA), Turin, Italy, 7–11 September 2015; pp. 981–984.
46. Birtcher, C.R.; Balanis, C.A.; Decarlo, D. Rotor-blade modulation on antenna amplitude pattern and polarization: Predictions and measurements. *IEEE Trans. Electromagn. Compat.* **1999**, *41*, 384–393. [CrossRef]
47. Hou, H.-A.; Wang, L.-C.; Lin, H.-P. Micro-Doppler Shift and Its Estimation in Rotary-Wing UAV Sub-6 GHz Communications. *IEEE Wirel. Commun. Lett.* **2021**, *10*, 2185–2189. [CrossRef]
48. Virone, G.; Paonessa, F.; Ciorba, L.; Lumia, M.; Addamo, G.; Peverini, O.A.; Bolli, P. Effect of Conductive Propellers on VHF UAV-based Antenna Measurements: Experimental Results. In Proceedings of the IEEE Conference on Antenna Measurements & Applications (CAMA), Antibes Juan-les-Pins, France, 15–17 November 2021; pp. 46–47.
49. Faul, F.T.; Korthauer, D.; Eibert, T.F. Impact of Rotor Blade Rotation of UAVs on Electromagnetic Field Measurements. *IEEE Trans. Instrum. Meas.* **2021**, *70*, 1–9. [CrossRef]
50. Ciorba, L.; Virone, G.; Paonessa, F.; Righero, M.; Acedo, E.D.L.; Matteoli, S.; Beltran, E.C.; Bolli, P.; Giordanengo, G.; Vecchi, G.; et al. Large Horizontal Near-Field Scanner Based on a Non-Tethered Unmanned Aerial Vehicle. *IEEE Open J. Antennas Propag.* **2022**, *3*, 568–582. [CrossRef]
51. Gueuning, Q.; Cavillot, J.; Gregson, S.; Craeye, C.; De Lera Acedo, E.; Brown, A.K.; Parini, C. Plane-wave spectrum methods for the near-field characterization of very large structures using UAVs: The SKA radio telescope case. In Proceedings of the 15th European Conference on Antennas and Propagation (EuCAP), Dusseldorf, Germany, 22–26 March 2021; pp. 1–5.
52. Steiner, H.J.; Straus, R.; Lamm, M.; Henkel, P. UAV-Based Near-Field Antenna Measurements at 20 MHz of a Reference Ground-Plane Antenna. In Proceedings of the IEEE Conference on Antenna Measurements & Applications (CAMA), Antibes Juan-les-Pins, France, 15–17 November 2021; pp. 48–50.
53. Vitucci, E.M.; Chen, J.; Degli-Esposti, V.; Lu, J.S.; Bertoni, H.L.; Yin, X. Analyzing Radio Scattering Caused by Various Building Elements Using Millimeter-Wave Scale Model Measurements and Ray Tracing. *IEEE Trans. Antennas Propag.* **2019**, *67*, 665–669. [CrossRef]
54. Guven, D.; Jamroz, B.F.; Chuang, J.; Gentile, C.; Horansky, R.D.; Remley, K.A.; Williams, D.F.; Quimby, J.T.; Weiss, A.J.; Leonhardt, R. Methodology for Measuring the Frequency Dependence of Multipath Channels Across the Millimeter-Wave Spectrum. *IEEE Open J. Antennas Propag.* **2022**, *3*, 461–474. [CrossRef]
55. Platzgummer, V.; Raida, V.; Krainz, G.; Svoboda, P.; Lerch, M.; Rupp, M. UAV-Based Coverage Measurement Method for 5G. In Proceedings of the IEEE 90th Vehicular Technology Conference (VTC2019-Fall), Honolulu, HI, USA, 22–25 September 2019; pp. 1–6.
56. Zamfirescu, C.; Halunga, S.; Marcu, I.; Badescu, A. Towards joint WSN-UVs networks supporting IoT-applications in harsh environments a measurement-campaign and mathematical models. *IEEE Internet Things J.* **2022**.
57. Zhu, Q.; Yao, M.; Bai, F.; Chen, X.; Zhong, W.; Hua, B.; Ye, X. A general altitude-dependent path loss model for UAV-to-ground millimeter-wave communications. *Front. Inf. Technol. Electron. Eng.* **2021**, *22*, 767–776. [CrossRef]
58. Rappaport, T.S. Wireless Communications--Principles and Practice, (The Book End). *Microw. J.* **2002**, *45*, 128–129.
59. Khan, S.K.; Naseem, U.; Siraj, H.; Razzak, I.; Imran, M. The role of unmanned aerial vehicles and mmWave in 5G: Recent advances and challenges. *Trans. Emerg. Telecommun. Technol.* **2021**, *32*, e4241. [CrossRef]
60. Yang, Z.; Zhou, L.; Zhao, G.; Zhou, S. Channel Model in the Urban Environment for Unmanned Aerial Vehicle Communications. In Proceedings of the 12th European Conference on Antennas and Propagation (EuCAP 2018), London, UK, 9–13 April 2018; pp. 1–5.
61. El Khaled, Z.; Ajib, W.; McHeick, H. Log Distance Path Loss Model: Application and Improvement for Sub 5 GHz Rural Fixed Wireless Networks. *IEEE Access* **2022**, *10*, 52020–52029. [CrossRef]
62. Khawaja, W.; Guvenc, I.; Matolak, D. UWB Channel Sounding and Modeling for UAV Air-to-Ground Propagation Channels. In Proceedings of the IEEE Global Communications Conference (GLOBECOM), Washington, DC, USA, 4–8 December 2016; pp. 1–7.
63. Saad, W.; Bennis, M.; Mozaffari, M.; Lin, X. *Wireless Communications and Networking for Unmanned Aerial Vehicles*; Cambridge University Press: Cambridge, UK, 2020.
64. Rodriguez-Pineiro, J.; Dominguez-Bolano, T.; Cai, X.; Huang, Z.; Yin, X. Air-to-Ground Channel Characterization for Low-Height UAVs in Realistic Network Deployments. *IEEE Trans. Antennas Propag.* **2021**, *69*, 992–1006. [CrossRef]
65. CST. Electromagnetic Field Simulation Software for Designing, Analyzing and Optimizing Electromagnetic Components. Available online: https://www.3ds.com/products-services/simulia/products/cst-studio-suite (accessed on 1 August 2024).
66. Amorim, R.; Nguyen, H.; Mogensen, P.; Kovacs, I.Z.; Wigard, J.; Sorensen, T.B. Radio Channel Modeling for UAV Communication Over Cellular Networks. *IEEE Wirel. Commun. Lett.* **2017**, *6*, 514–517. [CrossRef]
67. Athanasiadou, G.E.; Tsoulos, G.V. Path Loss characteristics for UAV-to-Ground Wireless Channels. In Proceedings of the IEEE 13th European Conference on Antennas and Propagation (EuCAP), Krakow, Poland, 31 March–5 April 2019; pp. 1–4.
68. Khawaja, W.; Ozdemir, O.; Erden, F.; Guvenc, I.; Matolak, D.W. UWB Air-to-Ground Propagation Channel Measurements and Modeling Using UAVs. In Proceedings of the IEEE Aerospace Conference, Big Sky, MT, USA, 2–9 March 2019; pp. 1–10.

69. Jo, S.-W.; Jang, J.H.; Shim, W.-S. An analysis of path loss models of LTE-Maritime for mobile communication system in maritime environments. In Proceedings of the IEEE 22nd International Symposium on Wireless Personal Multimedia Communications (WPMC), Lisbon, Portugal, 24–27 November 2019; pp. 1–5.
70. Cai, X.; Gonzalez-Plaza, A.; Alonso, D.; Zhang, L.; Rodriguez, C.B.; Yuste, A.P.; Yin, X. Low altitude UAV propagation channel modelling. In Proceedings of the IEEE 11th European Conference on Antennas and Propagation (EUCAP), Paris, France, 19–24 March 2017; pp. 1443–1447.
71. Goddemeier, N.; Daniel, K.; Wietfeld, C. Role-Based Connectivity Management with Realistic Air-to-Ground Channels for Cooperative UAVs. *IEEE J. Sel. Areas Commun.* **2012**, *30*, 951–963. [CrossRef]
72. Simunek, M.; Pechac, P.; Perez Fontan, F. Excess Loss Model for Low Elevation Links in Urban Areas for UAVs. *Radioengineering* **2011**, *20*, 561–568.
73. Briso, C.; Calvo, C.; Cui, Z.; Zhang, L.; Xu, Y. Propagation Measurements and Modeling for Low Altitude UAVs From 1 to 24 GHz. *IEEE Trans. Veh. Technol.* **2020**, *69*, 3439–3443. [CrossRef]
74. Qixing, F.; McGeehan, J.; Tameh, E.K.; Nix, A.R. Path Loss Models for Air-to-Ground Radio Channels in Urban Environments. In Proceedings of the IEEE 63rd Vehicular Technology Conference, Melbourne, VIC, Australia, 7–10 May 2006; pp. 2901–2905.
75. Tu, H.D.; Shimamoto, S. A Proposal of Wide-Band Air-to-Ground Communication at Airports Employing 5-GHz Band. In Proceedings of the IEEE Wireless Communications and Networking Conference, Budapest, Hungary, 5–8 April 2009; pp. 1–6.
76. Qiu, Z.; Chu, X.; Calvo-Ramirez, C.; Briso, C.; Yin, X. Low Altitude UAV Air-to-Ground Channel Measurement and Modeling in Semiurban Environments. *Wirel. Commun. Mob. Comput.* **2017**, *2017*, 1587412. [CrossRef]
77. Xi, C.; Briso, C.; Danping, H.; Xuefeng, Y.; Jianwu, D. Channel Modeling for Low-Altitude UAV in Suburban Environments Based on Ray Tracer. In Proceedings of the 12th European Conference on Antennas and Propagation (EuCAP), London, UK, 9–13 April 2018; pp. 1–5.
78. Khuwaja, A.A.; Chen, Y.; Zheng, G. Effect of User Mobility and Channel Fading on the Outage Performance of UAV Communications. *IEEE Wirel. Commun. Lett.* **2020**, *9*, 367–370. [CrossRef]
79. Cui, Z.; Briso-Rodriguez, C.; Guan, K.; Zhong, Z.; Quitin, F. Multi-Frequency Air-to-Ground Channel Measurements and Analysis for UAV Communication Systems. *IEEE Access* **2020**, *8*, 110565–110574. [CrossRef]
80. Fuschini, F.; Barbiroli, M.; Vitucci, E.M.; Semkin, V.; Oestges, C.; Strano, B.; Degli-Esposti, V. An UAV-Based Experimental Setup for Propagation Characterization in Urban Environment. *IEEE Trans. Instrum. Meas.* **2021**, *70*, 1–11. [CrossRef]
81. Rappaport, T.S.; Maccartney, G.R.; Sun, S.; Yan, H.; Deng, S. Small-Scale, Local Area, and Transitional Millimeter Wave Propagation for 5G Communications. *IEEE Trans. Antennas Propag.* **2017**, *65*, 6474–6490. [CrossRef]
82. Kim, K.-W.; Kim, M.-D.; Park, J.-J.; Lee, J.; Liang, J.; Lee, K.-C. Diffraction Loss Model Based on 28 GHz Over-Rooftop Propagation Measurements. In Proceedings of the IEEE 86th Vehicular Technology Conference (VTC-Fall), Toronto, ON, Canada, 24–27 September 2017; pp. 1–5.
83. Deng, S.; Maccartney, G.R.; Rappaport, T.S. Indoor and Outdoor 5G Diffraction Measurements and Models at 10, 20, and 26 GHz. In Proceedings of the IEEE Global Communications Conference (GLOBECOM), Washington, DC, USA, 4–8 December 2016; pp. 1–7.
84. Ramirez, C.C.; Briso-Rodríguez, C. Diffraction Path Losses Measurements for Low Altitude UAVs. In Proceedings of the 13th European Conference on Antennas and Propagation (EuCAP), Krakow, Poland, 31 March–5 April 2019; pp. 1–3.
85. Kandregula, V.R.; Lazaridis, P.I.; Zaharis, Z.D.; Ahmed, Q.Z.; Khan, F.A.; Hafeez, M.; Chochliouros, I.P. Simulation Analysis of a Wideband Antenna on a Drone. In Proceedings of the IEEE International Black Sea Conference on Communications and Networking (BlackSeaCom), Sofia, Bulgaria, 6–9 June 2022; pp. 288–292.
86. Hamdalla, M.Z.M.; Caruso, A.N.; Hassan, A.M. Electromagnetic Compatibility Analysis of Quadcopter UAVs Using the Equivalent Circuit Approach. *IEEE Open J. Antennas Propag.* **2022**, *3*, 1090–1101. [CrossRef]
87. Akhter, Z.; Bilal, R.M.; Shamim, A. A Dual Mode, Thin and Wideband MIMO Antenna System for Seamless Integration on UAV. *IEEE Open J. Antennas Propag.* **2021**, *2*, 991–1000. [CrossRef]
88. Peng, J.-J.; Qu, S.-W.; Xia, M.; Yang, S. Conformal Phased Array Antenna for Unmanned Aerial Vehicle With ±70° Scanning Range. *IEEE Trans. Antennas Propag.* **2021**, *69*, 4580–4587. [CrossRef]
89. Kandregula, V.R.; Lazaridis, P.; Zaharis, Z.; Mihovska, A.; Zeeshan, Q.; Khan, F.A.; Hafeez, M.; Antonopoulos, C.S. Optimization of a printed log-periodic antenna position on a UAV. In Proceedings of the 11th International Conference on Computation in Electromagnetics, Cannes, France, 11–14 April 2023.
90. Scorrano, L.; Di Rosa, A.; Orobello, B.; Manna, A.; Trotta, F. Experimental Investigations of a Novel Lightweight Blade Antenna for UAV Applications. *IEEE Antennas Propag. Mag.* **2017**, *59*, 108–178. [CrossRef]
91. Fernandez Gonzalez, J.-M.; Padilla, P.; Valenzuela-Valdes, J.F.; Padilla, J.-L.; Sierra-Perez, M. An Embedded Lightweight Folded Printed Quadrifilar Helix Antenna: UAV telemetry and remote control systems. *IEEE Antennas Propag. Mag.* **2017**, *59*, 69–76. [CrossRef]
92. Jeong, Y.-Y.; Lee, W.-S. Wideband Printed Half Bow-Tie Antenna Array Based on a Quad-Mode Reconfigurable Feeding Network for UAV Communications. *IEEE Open J. Antennas Propag.* **2021**, *2*, 238–248. [CrossRef]
93. Kwon, O.H.; Lee, S.; Lee, J.M.; Hwang, K.C. A compact, low-profile log-periodic meandered dipole array antenna with an artificial magnetic conductor. *Int. J. Antennas Propag. Hindawi* **2018**, *2018*, 7261076. [CrossRef]

94. Azhar, A.; Eibert, T.F. A Dual-Polarized Tapered Slot Antenna for UAV-Based Collection of Locally Coherent Field Data. In Proceedings of the IEEE 15th European Conference on Antennas and Propagation (EuCAP), Dusseldorf, Germany, 22–26 March 2021; pp. 1–4.
95. Hoel, K.V.; Kristoffersen, S.; Moen, J.; Holm, G.; Lande, T.S. Characterization of a 3D printed wideband waveguide and horn antenna structure embedded in a UAV wing. In Proceedings of the IEEE 10th European Conference on Antennas and Propagation (EuCAP), Davos, Switzerland, 10–15 April 2016; pp. 1–4.
96. Chen, Z.; Shen, Z. Wideband Flush-Mounted Surface Wave Antenna of Very Low Profile. *IEEE Trans. Antennas Propag.* **2015**, *63*, 2430–2438. [CrossRef]
97. Gong, Z.; Ge, S.; Guo, T.; Zhang, Q.; Chen, Y. A compact planar 24GHz quasi-Yagi antenna for unmanned aerial vehicle radar applications. In Proceedings of the IEEE International Conference on Computational Electromagnetics (ICCEM), Kumamoto, Japan, 8–10 March 2017; pp. 104–105.
98. De Lera Acedo, E.; Bolli, P.; Paonessa, F.; Virone, G.; Colin-Beltran, E.; Razavi-Ghods, N.; Aicardi, I.; Lingua, A.; Maschio, P.; Monari, J.; et al. SKA aperture array verification system: Electromagnetic modeling and beam pattern measurements using a micro UAV. *Springer Sci. Bus. Media LLC Exp. Astron.* **2018**, *45*, 1–20. [CrossRef]
99. Segales, A.R.; Schvartzman, D.; Burdi, K.; Salazar-Cerreño, J.L.; Fulton, C.; Palmer, R.D. Far-Field Antenna Pattern Measurements Using a UAS-Based Dynamic Gimbal Scan System. *IEEE Trans. Antennas Propag.* **2024**, *72*, 2368–2381. [CrossRef]
100. Paonessa, F.; Ciorba, L.; Virone, G.; Bolli, P.; Magro, A.; McPhail, A.; Minchin, D.; Bhushan, R. First Results on the Experimental Validation of the SKA-low Prototypes Deployed in Australia Using an Airborne Test Source. In Proceedings of the 2020 XXXIIIrd General Assembly and Scientific Symposium of the International Union of Radio Science, Rome, Italy, 29 August–5 September 2020; pp. 1–3.
101. Paonessa, F.; Virone, G.; Ciorba, L.; Addamo, G.; Lumia, M.; Dassano, G.; Zannoni, M.; Franceschet, C.; Peverini, O.A. Design and Verification of a Q-Band Test Source for UAV-Based Radiation Pattern Measurements. *IEEE Trans. Instrum. Meas.* **2020**, *69*, 9366–9370. [CrossRef]
102. Garcia-Fernandez, M.; Lopez, Y.A.; Andres, F.L.H. Unmanned aerial system for antenna measurement and diagnosis: Evaluation and testing. *IET Microw. Antennas Amp; Propag.* **2019**, *13*, 2224–2231. [CrossRef]
103. Dronecode. QGroundControl. Available online: https://qgroundcontrol.com/ (accessed on 1 August 2024).
104. Li, H.; Ding, L.; Wang, Y.; Wang, Z. Air-to-Ground Channel Modeling and Performance Analysis for Cellular-Connected UAV Swarm. *IEEE Commun. Lett.* **2023**, *27*, 2172–2176. [CrossRef]
105. Balanis, C.A.; Birtcher, C.R. Antenna Measurements. In *Modern Antenna Handbook*; John Wiley & Sons: Hoboken, NJ, USA, 2008; pp. 977–1033.

sensors

MDPI

Article

Web Real-Time Communications-Based Unmanned-Aerial-Vehicle-Borne Internet of Things and Stringent Time Sensitivity: A Case Study

Agnieszka Chodorek [1] and Robert Ryszard Chodorek [2,*]

[1] Department of Applied Computer Science, Faculty of Electrical Engineering, Automatic Control and Computer Science, Kielce University of Technology, Al. 1000-lecia P.P. 7, 25-314 Kielce, Poland

[2] Institute of Telecommunications, Faculty of Computer Science, Electronics and Telecommunications, AGH University of Krakow, Al. Mickiewicza 30, 30-059 Krakow, Poland

* Correspondence: chodorek@agh.edu.pl; Tel.: +48-12-617-4803

Abstract: The currently observed development of time-sensitive applications also affects wireless communication with the IoT carried by UAVs. Although research on wireless low-latency networks has matured, there are still issues to solve at the transport layer. Since there is a general agreement that classical transport solutions are not able to achieve end-to-end delays in the single-digit millisecond range, in this paper, the use of WebRTC is proposed as a potential solution to this problem. This article examines UAV-borne WebRTC-based IoT in an outdoor environment. The results of field experiments conducted under various network conditions show that, in highly reliable networks, UAV and WebRTC-based IoT achieved stable end-to-end delays well below 10 ms during error-free air-to-ground transmissions, and below 10 ms in the immediate vicinity of the retransmitted packet. The significant advantage of the WebRTC data channel over the classic WebSocket is also demonstrated.

Keywords: IEEE 802.11ac; internet of things; low latency; real-time transmissions; unmanned aerial vehicle; WebRTC; WebSocket

1. Introduction

Unmanned aerial vehicles (UAVs) are currently one of the fastest developing multi-role carrier technologies. These ubiquitous devices now have a multitude of economic, commercial, leisure, military, and academic uses [1], and their uses range from individuals flying them for recreation to large commercial package and medical supply companies [2]. They can be used to transport parcels and people between locations [3–5]. Equipped with on-board cameras and Internet of Things (IoT) systems, UAVs are used to monitor pollution [6,7], weather [8,9], road traffic [10], and crop production [11,12]. An important part of these UAV applications is the communication and computing support, including a flying range extender [13], a flying router [13,14], and a flying computer for aerial mobile edge computing (AMEC) purposes [15].

These and other UAV applications can be time-sensitive in a broad sense, i.e., they may have arbitrary time constraints imposed. These can be relatively large if they are related to the delivery of parcels or people. Such deliveries may have to be completed within a specific time window [3,4] or as soon as possible. It is estimated that, in a large city, the time needed for transporting parcels or people by air may be a few dozen percent shorter than the time for land transport [5]. Time constraints may also be relatively small when they concern the provision of real-time or near-real-time information: either to detect and locate the source of pollution [6,7], or for disaster response purposes [13,14]. The need for real-time information may also arise in the case of observations of weather [8,9], crop production [11,12], and road traffic [10], etc., if data are sent from the UAV to a ground station, where they are used for analysis, real-time visualization, and decision-making.

Time-sensitive UAV-IoT applications may also involve collecting and processing data from sensors located in a given area. For example, the integration of UAVs with the internet of medical things (IoMT) was reported in [15]. This UAV-enabled system implements AMEC functionality. In the proposed solution, communication delays were reduced from a range of 17 ms to 30 ms to a range of 12.5 ms to 24 ms. The freshness of data is expressed in the so-called age of information (AoI), i.e., the time that has passed since the generation of the most recently received data [16]. AoIs known from the literature include UAV flight time, hover time, and maintenance time, and range from less than 600 s to less than 1800 s [16] or from almost 1400 s to less than 2200 s [17], with low-AoI systems starting with an AoI of 70–80 s [18,19]. AoI improvement methods are based on optimizing the UAV trajectory [16–19], and the transmission delay for such large times is negligible.

Currently, the main challenge in the field of wireless communication with UAVs is time-sensitive applications that require low latency, defined as end-to-end delays measured in single-digit milliseconds at the application level. Examples of such applications are presented in Table 1. The traffic generated by these applications is deterministic, meaning hard real-time with no jitter, or non-deterministic, where low jitter can be observed. A high reliability of transmission is required, and in the case of deterministic traffic, ultra-high reliability is needed. This approach breaks with the classic division of telecommunications traffic into elastic and inelastic, where only inelastic traffic had to meet stringent time requirements and only elastic traffic had to be characterized by a high transmission reliability [20].

Table 1. Selected time-sensitive applications with stringent time constraints related to both UAVs and IoT that can be carried by UAVs.

Application	Time Constraint	Paper
Connecting autonomous vehicles	below 1 ms	[21]
Transport industry	3 or 7 ms	[22]
Intelligent transportation system	5 to 10 ms	[23]
Internet of drones (remote control)	5 to 50 ms	[23]
Approaching autonomous navigation infrastructure	10 ms	[24]
Mobile robots: video-operated remote control	10 to 100 ms	[25]
Command and control of UAV networks	10, 40, or 140 ms	[22]

It is important to note here that using a low-latency network does not guarantee low end-to-end delays at the application level. This state of affairs is blamed on the mechanisms of classic transport protocols, which are unable to effectively meet the requirements of low delays [23]. Another problem is the socket application programming interface (API) for these protocols, which is too low-level, simple, and inflexible [23]. The authors believe that a solution to the above problems, at least in the case of time-sensitive UAV-borne IoT, could be the use of web real-time communications (WebRTC), which, as the name suggests, provides native real-time communication on the Web. The World Wide Web Consortium (W3C) in the document in [26] announced the general need for building a WebRTC-based IoT. Requirement N15 included in [26] states that a WebRTC-based IoT should be able to provide low and consistent latency under varying network conditions.

Main Contributions and Organization of This Paper

In our previous paper, we proposed a WebRTC-based application capable of operating like the classic IoT [27], intended for use in UAV-borne monitoring systems. This application was a part of our UAV- and WebRTC-based open universal framework [28]. In this paper, we present the results of field experiments aimed at verifying whether, and to what extent, a UAV-borne IoT based on the current WebRTC standard is able to provide low and consistent latency under varying network conditions. The main contributions of this paper are as follows:

- Supplementing the application in [27], working as an element of the framework [28], with high-resolution time measurement and timer synchronization procedures.
- Carrying out delay measurements at the level of the transport protocol and at the level of the web logical channel during air-to-ground IoT transmissions under varying network conditions, and then performing a statistical analysis of these delays.
- For the completeness of the results, a comparison of the obtained results with the results obtained for IoT transmission via a classic web logical channel, i.e., WebSocket, in the same circumstances.

The rest of this paper is organized as follows: Section 2 analyzes related work. Section 3 discusses the materials and methods used during the experiments. Section 4 describes the field experiments, including post-selection of the measurement series for further analysis. Section 5 presents and discusses the measures of the location of the selected series of end-to-end delays, while Section 6 compares the transmissions carried out with the use of WebRTC and WebSocket in terms of the measures of location, as well as the measures of variation derived from these measures of location. Section 7 summarizes our experiences.

2. Related Work

While Section 1 provides a broad background, Section 2 discusses both real-time alternatives and the authors' prior application solutions that formed the basis of this paper. The review of existing solutions covers time-sensitive applications that generate non-deterministic traffic. Although real-time transmission is usually associated with multimedia streaming (as are IoT real-time transmissions [29]), the paper only discusses the transmission of non-media data, usually data coming from sensors. The discussion focuses on aspects of the transport layer, i.e., transport protocols and interfaces. The criterion for selecting literature was the various non-media real-time transmission techniques found in the literature, preferring papers that explicitly provided transmission times in a local area network. Most references concern IoT communication between UAVs and ground stations.

Non-deterministic traffic generated by time-sensitive applications is most often transmitted air-to-ground using wireless local networks (WLANs), usually built using the Institute of Electrical and Electronics Engineers (IEEE) 802.11 standard [27,28,30–40], but also using standards for broadband cellular networks: the long-term evolution (LTE) standard [40,41], also known as the fourth-generation (4G) technology standard, and the fifth-generation (5G) technology standard [15,30,40]. Among IEEE 802.11 [42] networks, popular versions of the physical layer are used, such as 802.11g [31,32] and 802.11n [33–35,40], as well as the 802.11p version intended for the vehicular environment [30]. The works [27,28] employed the 802.11ac version, which is able to provide latencies below 10 ms, including providing handover latencies below 10 ms thanks to the fast roaming service [43].

The 802.11 standard was also used in [44], where a time-sensitive application, intended to work on board a UAV, was tested using a laptop and an unmanned ground vehicle (UGV). Another time-sensitive application in which the sensor system was carried on board a UGV was presented in [45]. The network used in [45] was based on software-defined radio (SDR) working with software-defined networking (SDN). In [46], a stationary robot communicated via both 802.11 WLAN and evolved high-speed packet access (HSPA+), also known as the 3.75G technology standard. As a side note, refs. [44,46] also tested transmissions between stationary end systems over longer distances using the public infrastructure of an Internet service provider (ISP). These are not the subject of this article, because current networks, including those built using 5G technology [47], are only able to provide low-latency services locally, using dedicated low-latency solutions with limited range.

Time-sensitive applications typically send sensor data using the classic transmission control protocol (TCP), which provides reliable congestion- and flow-controlled transmission. Applications use the TCP transport protocol directly [41] or via the WebSocket web logical channel [30,34–36,38,39,44–46]. The works in [31–33] used a multipath version of the TCP protocol, i.e., the multipath transmission control protocol (multipath TCP or MPTCP), which allows multi-homed senders to increase transmission efficiency. The use of MPTCP has been shown to reduce latency under certain conditions [48], although the protocol is sensitive to path asymmetry, especially if the paths are built with different technologies (e.g., 4G and 802.11) [49]. In the works in [31,32], MPTCP was modified to meet the requirements of time-sensitive networking (TNS). In [33], to deal with network stability in the face of high UAV mobility, the MPTCP scheduling algorithm was modified.

While the TCP protocol has been typically used for transmission of non-media data, another classic transport protocol, i.e., the user datagram protocol (UDP), is used for audio/video transmission, due to its simple structure and mechanisms reduced to an absolute minimum, including the lack of window-based control. Currently, the UDP is most often used as an underlay protocol for other transport protocols, where it occupies a lower sublayer of the transport layer. The most popular solution is to use the real-time transport protocol (RTP) in the upper sublayer. A RTP/UDP protocol stack is typically used for multimedia communications. This solution is also used in the WebRTC video channel. In [27,28,34,35,46], a WebRTC video channel was used to transmit video from a camera. In [44], it was used to transmit data from lidar. WebRTC uses the RTP protocol implemented in a WebRTC-capable browser and the UDP protocol implemented in an operating system.

UDP can also be used as a transport protocol for transmitting non-media data. In [41], a low-latency reliable transmission (LRT) application layer protocol operating directly over UDP was proposed to reduce the delay during ground-to-UAV transmission through a cellular network. In the abovementioned work [31], control data were sent via the UDP protocol, while the remaining data were sent via the MPTCP protocol. The WebRTC data channel used the stream control transmission protocol (SCTP) over UDP. To ensure reliable transmission, the SCTP uses error control and congestion control mechanisms similar to those of TCP. In the works in [27,28,46], the SCTP protocol was used to transmit data from sensors. Similarly to the RTP, the SCTP is always implemented in a WebRTC-capable browser, regardless of the operating system's implementation of the SCTP. This is due to the need to use the new version of the SCTP standard intended for WebRTC [50]. However, the implementation of the SCTP in the operating system can also be used for IoT data transmission [37]. The UDP transport protocol implemented in the operating system is also the basis for the quick UDP internet connections (QUIC) protocol [51], initially intended for web applications and now proposed for low-latency communication in the next-generation IoT [52]. The papers in [38,39] presented the results of evaluations of a real IoT transmitted using the QUIC in an emulated [38] or simulated [39] wireless environment.

Applications, including web browsers, use the TCP and UDP transport protocols implemented in the operating system, communicating with them via the socket interface. In [31–33], applications communicated with the MPTCP protocol in the operating system via a classic stream socket. The WebSocket web logical channel uses the TCP protocol in the operating system, also communicating with it via a stream socket. Applications that send data via WebSocket, such as [30,34,44,45], use the high-level WebSocket API. WebRTC offers separate web logical channels for media and non-media transmission, each of which is associated with a separate high-level API used by WebRTC applications such as [27,28,34,35,44,46]. The RTP and SCTP protocols are implemented in WebRTC-capable browsers and communicate with the UDP protocol in the operating system via classic datagram sockets. WebRTC does not use the SCTP in the operating system and therefore does not use an SCTP socket.

A comparison of related work is presented in Table 2. Our WebRTC-based UAV-borne IoT application is presented in [27,28]. In this work, data were not transmitted in the web of things (WoT) architecture [53], using an intermediate server, but in the classic IoT manner, using a peer-to-peer WebRTC architecture. In [44], transmissions of lidar data via the WebRTC video channel and via Websocket were compared. In [34,35], IoT data were transmitted via Websocket, and only video was transmitted via WebRTC. In [46], WoT data were transferred from a robot to a WoT server over WebSockets, and then from the server to the recipient over WebRTC. In [40], WebRTC data channel was used to control a UAV and transmit telemetry. The remaining papers did not use WebRTC. In [30,36,45], only a WebSocket logical channel was used, while, in [15,31–33,37–39,41], a web logical channel was not used at all.

Table 2. Related work.

Paper	Carrier	Network Technology	Transport Protocol	Web Logical Channel	API
[15]	UAV	5G	n/a [1,5]	n/a [1,5]	n/a [1,5]
[27,28]	UAV	802.11ac	RTP, SCTP	WebRTC	WebRTC
[30]	UAV	5G, 802.11p	TCP [4]	WebSocket	n/a [1,5]
[31,32]	UAV	802.11g	MPTCP	n/a [1]	socket
[33]	UAV	802.11n	MPTCP	n/a [1]	socket
[34,35]	UAV	802.11n	RTP [4], TCP [4]	WebRTC, WebSocket	WebRTC, WebSocket
[36]	UAV	802.11	TCP [4]	WebSocket	WebSocket
[41]	UAV	4G	UDP	n/a [1]	n/a [1,5]
			TCP	n/a [1]	n/a [1,5]
[44]	n/a [1,3], UGV [3]	802.11	TCP [4]	WebSocket	WebSocket
			RTP [4]	WebRTC	WebRTC
[45]	UGV	SDR	TCP [4]	WebSocket	WebSocket
[46]	n/a [1,2]	802.11, 3.75G	TCP [4], RTP [4], SCTP [4]	WebSocket, WebRTC	WebSocket, WebRTC
[38,39]	n/a [1]	802.11, Cellular Satellite	QUIC ,TCP	n/a [1,5]	socket
[37]	n/a [1]	802.11	SCTP, TCP	n/a [1,5]	n/a [1,5]
[40]	UAV	802.11n, 4G, 5G	SCTP	WebRTC	WebRTC
this paper	UAV	802.11ac	SCTP	WebRTC	WebRTC
			TCP	WebSocket	WebSocket

[1] not applicable, [2] stationary robot, [3] the target application is UAV, [4] stated implicitly, [5] simulation.

WebRTC applications are web-based equivalents of classic, standalone multimedia applications based on the session initiation protocol (SIP). The management plane protocol stack [54] and the production plane protocol stack for media streaming [55] are similar to their legacy SIP architecture counterparts. WebRTC applications are loaded from web servers as part of web pages and use web browsers as run-time environments. This makes them highly portable and secure, as detected browser vulnerabilities are eliminated on an ongoing basis. What distinguishes WebRTC from other web techniques, such as WebSocket, is its dual protocol stack, the idea of which was taken from the SIP architecture. As an effect, WebRTC can be used to transmit both media streams and non-media flows. Streams and flows are cryptographically protected and congestion-controlled. Since both the media stream and non-media flow use TCP-friendly congestion control, in the event of poor network conditions, data are protected at the expense of video [56]. If necessary, WebRTC applications can use more sophisticated streaming media congestion control methods, such as RTP translators or simulcast, and both RTP streams and SCTP flows can use differentiated services (DiffServ) to ensure quality of service (QoS).

3. Materials and Methods

This section introduces the flying monitoring system used in the field experiments, highlighting the time-sensitive aspects; defines the end-to-end delays measured at both the logical channel level and transport protocol level; shows the method for creating of a series of end-to-end delays; and finally describes the extreme values and measures of location calculated from these series.

3.1. System

In all experiments, a flying monitoring system built on the basis of the framework in [28] was used. Structurally, the system consists of an air station and a ground station, and functionally of an IoT system and an IoT carrier. The air station was an unmanned quadcopter, operating as the IoT carrier, with an IoT system on board, i.e., environmental sensors connected to a single-board computer (SBC) Raspberry Pi 4 Model B running the authors monitoring application. The environmental sensors included four weather sensors previously used to build a mobile weather station [9] and a gas sensor used in a pollution monitoring system [7], which allowed for the reuse of existing sensor-dependent code. The monitoring application was written in the JavaScript language as part of a web page, and its runtime environment was the Chromium browser in headless mode and run on the Raspberry Pi OS operating system. The authors' analysis of the Chromium browser implementation showed that the browser has a limited send buffer, which allows transmissions to reduce buffering times, and the stream socket was set to disable Nagle's algorithm, so there was no need to additionally set these parameters.

The monitoring application included the WebRTC video service, positioning service, and sensor service. The WebRTC video service was built as a classic WebRTC video application. The positioning service and the sensor service were built in a browser-driven manner [27], typical for IoT systems. In the experiments presented in this paper, the video service was turned off, the positioning service sent its data to the sensor service, and only the sensor service sent its data to the ground. Data from sensor service were transmitted in message queuing telemetry transport (MQ telemetry transport, or MQTT) messages bearing the MQTT topic, which identified each datum. Example topics used in the experiments and the method of creating them were described in the authors previous paper [7]. MQTT messages were transmitted over a web logical channel, using both the WebRTC data channel and the WebSocket. In the latter case, to improve the time properties of the TCP, the PUSH option was set in each TCP packet carrying the MQTT message, which means immediate pushing of the received data to the application. Because the correct operation of the monitoring application required that the central processing unit (CPU) always had a sufficient reserve of resources, this had to be monitored during the performance of all tests.

Unlike the air station, which was one device, the ground station was divided into two separate devices (Figure 1): the command and control console (CCC), and the WebRTC multimedia and monitoring station (WMMS). The CCC was used to pilot the IoT carrier. It was connected to the UAV via the control network (yellow lightning in Figure 1). The WMMS is designed for IoT purposes. It was connected to the monitoring software via the IEEE 802.11ac production network (red lightning in Figure 1), which was built as a heterogeneous extended service set (ESS). As an effect, the transmission between the air station and the WMMS was carried out both in a wireless environment and in a mixed wired/wireless one, in which the access points were connected to each other via a gigabit Ethernet (IEEE 802.3ab) network. The intermediate devices used in the ESS were NETGEAR Nighthawk X4 R7500 AC2350 access points (AP1, AP2, and AP3) and an HP 3500-24G-PoE+ yl Switch (SW1). The AP1 and the SW1 were placed close to the corners of a rectangular 70 m × 70 m parking lot, which was the test area. The AP2 and the AP3 were located at 50 m from the AP1 and the SW1, respectively.

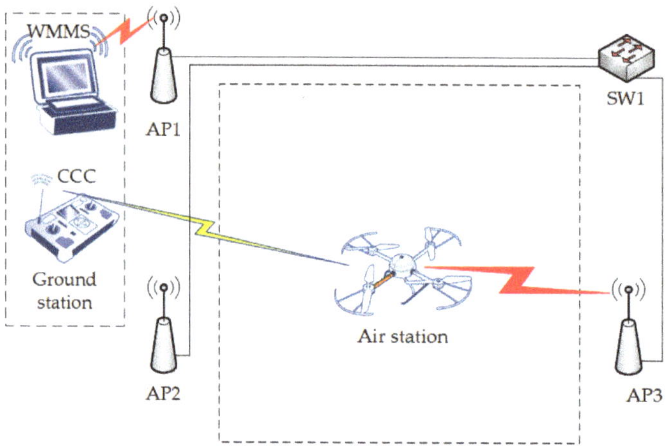

Figure 1. The testbed.

3.2. Series of End-to-End Delays

During the flights, time parameters were collected, both at the air station and at the ground station. These parameters were used to determine a pair of delays: one at the transport level, the other at the level of the web logical channel.

Definition 1. *The end-to-end delay d_i^t of the i-th IoT datum transmitted between the air station and the ground station, measured at the transport level, is defined as*

$$d_i^t = t_i^t - t_i^{ilc} \tag{1}$$

where t_i^t is the reception time from the transport protocol and t_i^{ilc} is the entry time into the logical channel.

Definition 2. *The end-to-end delay d_i^t of the i-th IoT datum transmitted between the air station and the ground station, measured at the logical channel level, is defined as*

$$d_i^{lc} = t_i^{olc} - t_i^{ilc} \tag{2}$$

where t_i^{olc} is the reception time from the logical channel.

In the case of transmissions carried out over the WebRTC Data Channel, the times t_i^{ilc}, t_i^t and t_i^{olc} were measured for each transmitted IoT datum, where *i* was the sequence number of this datum. From these times, the delays d_i^t and d_i^{lc} were then calculated according to Formulas (1) and (2), respectively. The difference between corresponding delays resulted from the processing of the payload of the SCTP packet (MQTT message) placed in the receive buffer of the logical channel. In the case of transmissions conducted over the WebSocket logical channel, performed for comparison purposes, for each transmitted

IoT datum, the times t_i^{ilc} and t_i^{olc} were measured, from which the delays d_i^{lc} were then determined, according to Formula (2).

Let t_1^{ilc} be the starting time, defined as the instant of time when the air station was directly above the ground station, D^{lc} be the series of $N = 40,000$ end-to-end delays measured at the logical channel level, starting at time t_1^{ilc}, and D^t be the corresponding series of end-to-end delays of the same amount measured at the transport level, starting at time t_1^{ilc}. The N value of 40,000 provided a high delivery rate of 0.999975 for a single error. After each pair of flights, one series of delays measured at the transport level, D^t_{WRTC}, and two series of delays measured at the logical channel level, D^{lc}_{WRTC} and D^{lc}_{WS}, were generated. The times t_1^{ilc}, at which the series D^{lc}_{WRTC} and D^{lc}_{WS} started, were shifted relative to each other by the time of the first flight of the pair and the service time of the second flight. The WebRTC and WS indexes indicate which web logical channel was used in a given measurement series (WebRTC data channel and WebSockets, respectively). The series D^t_{WRTC}, D^{lc}_{WRTC} and D^{lc}_{WS} were subjected to statistical processing.

For the analysis of the impact of single outliers, in particular the impact of the delay of IoT datum conveyed in the retransmitted packet, truncated measures were used. For this purpose, the series D^t_{WRTC}, D^{lc}_{WRTC}, and D^{lc}_{WS} of end-to-end delays d_i, $i = 1, 2, \ldots, 40,000$ were sorted in non-decreasing order. Then, the two extreme delays (minimum and maximum delay) were discarded from the series of end-to-end delays. This resulted in new, shorter series DT^t_{WRTC}, DT^{lc}_{WRTC}, and DT^{lc}_{WS} of end-to-end delays d_j, $j = 1, 2, \ldots, 39,998$. These series were subjected to the same statistical processing as the series from which they were derived.

3.3. Statistics

For time-sensitive applications, the main key performance indicator (KPI) is latency, defined as the end-to-end transmission delay. As a result, of all the related works, only Ref. [41] took latency and jitter into account, while the rest only focused on latency. Following this lead, extremes and measures of location were calculated in statistical processing. In particular, after each flight achieved

- the minimum value in each series: $min(D^t_{WRTC})$, $min(D^{lc}_{WRTC})$, $min(D^{lc}_{WS})$,
- the maximum value in each series: $max(D^t_{WRTC})$, $max(D^{lc}_{WRTC})$, $max(D^{lc}_{WS})$.

From the measures of location, both the classic measure of location, namely arithmetic mean, and the measures of position were calculated:

- arithmetic mean: $\mu(D^t_{WRTC})$, $\mu(D^{lc}_{WRTC})$, $\mu(D^{lc}_{WS})$:

$$\mu(D) = \frac{1}{card(D)} \sum_{i=1}^{card(D)} d_i, d_i \in D \tag{3}$$

- median: $med(D^t_{WRTC})$, $med(D^{lc}_{WRTC})$, $med(D^{lc}_{WS})$,
- mode: $mod(D^t_{WRTC})$, $mod(D^{lc}_{WRTC})$, $mod(D^{lc}_{WS})$,
- lower quartile: $Q1(D^t_{WRTC})$, $Q1(D^{lc}_{WRTC})$, $Q1(D^{lc}_{WS})$,
- upper quartile: $Q3(D^t_{WRTC})$, $Q3(D^{lc}_{WRTC})$, $Q3(D^{lc}_{WS})$.

The same statistics, calculated from the truncated series of end-to-end delays, namely DT^t_{WRTC}, DT^{lc}_{WRTC} and DT^{lc}_{WS}, produced truncated statistics, such as the truncated minimum $min(DT^t_{WRTC})$, truncated maximum $max(DT^t_{WRTC})$, truncated mean $\mu(DT^t_{WRTC})$, etc. These statistics were used to assess whether and to what extent a single retransmission affected the statistical properties of the analyzed end-to-end delays.

4. Experiments

Document [26] introduces a number of requirements that a WebRTC-based IoT must meet. The aim of the experiments described in this section was to check whether and to what extent the UAV-borne IoT, based on the current WebRTC standard, was able to meet the N15 requirement of [26], i.e., was able to provide low and consistent latencies under varying network conditions. As mentioned in Section 1, the challenge is in time-sensitive applications that require end-to-end delays measured in single-digit milliseconds at the application level. To meet this challenge, a WebRTC-based UAV-borne IoT should communicate with the ground station through a highly reliable, low-latency network. Since a delivery rate of 99.99% to 99.999% is considered high reliability, at most 1 packet error detected in the transport layer per 40,000 IoT data sent was assumed, i.e., a minimum packet delivery rate of 99.9975%.

The second assumption was that variable network conditions should result from both deterministic and random factors. Classic deterministic factors include the network heterogeneity (wired and wireless links), handovers, signal strength decrease with distance from access points, and UAV behavior (moving, hovering). Random factors include the different weather conditions and the different times of conducting experiments, which results in different user activities in co-existing networks in the same area, which in turn results in different loads on co-existing networks. The source of any transmission errors should be random factors.

The rest of this section presents the location of the field experiments and the course of the experiments; and discusses the flight days and sessions, network operating conditions, and the number of errors detected in the medium access control (MAC) sublayer. Finally, the measurement series selected for statistical analysis and the reasons why these series were selected and not others are described.

4.1. Location of the Experiments

The field experiments were carried out in a square parking lot 70 m long and 70 m wide, located on the campus of the AGH University of Krakow, Poland. The location of the experiment site between the university's teaching buildings and the dormitory made it possible to conduct experiments at times of the day when the students' Internet activity was low and high, generating low and high loads on the wireless networks coexisting with the air-to-ground production network in the test area. The high load on co-existing networks was a factor contributing to the occurrence of single transmission errors in the transport layer.

4.2. Course of Experiments

During the experiments, the air station performed automatic flights, sweeping the same 70 m × 70 m test area, zigzagging over the parking lot along the same flight path, at the same speed (1.67 m/s), and at the same altitude of 15 m. Air-to-ground transmissions were conducted both on the fly and hovering, and flight phases were intertwined with hovering phases. The summary flight time was about 460 s, and the summary hover time was about 280 s. This gave a total of just over 740 s (about 12.5 min) mission duration. The hover point locations and hover times were always the same. As the air station swept the entire test area, it switched between access points transmitting data through the 802.11ac production network described in the previous section. To ensure a seamless handover, the production network used the fast handover technique, which is part of the IEEE 802.11ac standard.

The source of the IoT data was the five environmental sensors that the air station was equipped with. During each flight, the sensors cyclically performed 9 measurements of the environmental parameters in a given time interval (0.5 s). Since each measurement datum was accompanied by two metadata (time and position), the air station sent a burst of 27 packets to the ground every half a second. This was over 1480 bursts, i.e., over 40,000 data packets, per flight. During each flight, the entry times into the logical channel, the reception times from the transport protocol (only IoT transmissions over WebRTC), and the reception times from the logical channel were collected.

During the experiments, data were sent over a web logical channel. In order to compare the IoT transmission over the WebRTC data channel with the classic solution, each evaluation flight in which IoT data were transmitted over the WebRTC data channel was followed without undue delay by a comparison flight in which IoT data were transmitted via WebSockets. This required developing a procedure for quickly replacing the web pages that included the monitoring applications, which were downloaded from a web server and run in the Chromium browser environment, which is the web server run at the WMSS. The use of a buffer power supply for both the SBC and the flight controller made it possible to change the software and replace the battery in parallel. As a result, the total elapsed time for maintenance between the evaluation flight and the next comparison flight was about 1 min.

4.3. Flight Days, Flight Sessions, and Pairs of Flights

The experiments were conducted from the end of January to the end of May, on separate days, on average every two weeks with an interval of at least one week, and on the same day of the week. The separation of experiments into individual days allowed the authors to run tests under different environmental conditions, such as the temperature, relative humidity, and time of day. Because the experiments started in midwinter and ended at the turn of spring to summer, transmissions were carried out from mild winter days, when the temperature rose above 1 degree Celsius, to warm late spring days, when the temperature rose to 25 degrees Celsius, and from dry weather, with a relative humidity above 40%, to rainy weather, with a relative humidity below 90%.

Experiments were organized into flight sessions. Before each session, the clocks at the air station and the ground station were synchronized. After each flight session, check-ups were performed to check for time drift, detected as a mismatch between the air and ground station clocks after the end of the session. Due to the detection of a time drift, the results collected during one flight session were rejected. Each flight session lasted up to two hours. Morning flight sessions began after 6:00 a.m. and ended before the start of classes at the University, no later than approximately 7:50 a.m. Midday flight sessions started around noon, and the evening ones started around 5 p.m. Since the experiments were conducted during the semester on campus, the time of day was related to the degree of load on the IEEE 802.11 networks coexisting with the production network on the AGH University campus and using the 5 GHz band.

The flight sessions were organized into pairs of test flights, with the evaluation flight (IoT transmission over WebRTC) immediately followed by a comparison flight (IoT transmission over WebSocket). Breaks between test flights belonging to the same pair could not be longer than would result from normal operation of the monitoring system. On a flight day, one flight session was conducted, and at least three pairs of test flights were performed during each flight session.

4.4. Network Conditions

Network conditions can be roughly expressed by the number of errors in the MAC sublayer: the fewer errors, the better the network conditions. The environmental conditions, especially the time of day, affected the network conditions, which were manifested in the different numbers of lost IEEE 802.11 frames per 40,000 transmitted IoT data. The number of errors in the MAC sublayer was reported by the network interface during the experiments.

Based on the number of frames lost, the network conditions were divided into good, medium, and poor. Less than two-fifths of the transmissions took place under good conditions, with just over 30 MAC frames lost per 40,000 IoT data sent. More than two-fifths were carried out under medium conditions, with more than 40 and no more than about 95 frames lost. More than one fifth of the transmissions took place under poor network conditions, when approximately 100 frames or more were lost per 40,000 IoT data transmitted.

The IEEE 802.11ac error control mechanism successfully retransmitted almost all lost frames detected by the MAC sublayer. Under both good and medium network conditions, the MAC sublayer was always able to correct the transmission errors. As an effect, no errors were detected at the transport layer. In poor network conditions, the underlying network was always unable to successfully retransmit one lost frame. As a result, a single transmission error (one lost packet per 40,000 IoT data sent) was detected at the transport layer. Errors in the transport layer usually appeared during both flights from a given pair. There was only one registered exception to this rule, when the transmission of IoT data over the WebRTC data channel was error-free at the transport layer, while during the transmission of IoT data over the WebSocket, the TCP detected a single transmission error.

4.5. Selection of Measurement Series

Out of 35 pairs of flights, we selected five, conducted on five different flight days, when transmissions were carried out under the three different network conditions (good, medium, poor):

- On day 1, transmissions were carried out in good network conditions. No errors were detected in the transport layer for both IoT data transmission over the WebRTC Data Channel and over the WebSocket. Thus, the packet error rate (PER) in the transport layer was $PER_{WRTC} = PER_{WS} = 0$.
- On day 2, network conditions were on the border between medium and poor. During the first flight, the exception described in previous section occurred: no errors were detected in the transport layer when transmitting IoT data over the WebRTC data channel, and one error was detected during transmission over WebSocket. PER_{WRTC} was 0 and PER_{WS} was 0.0025%.
- On day 3 transmissions were again carried out under good network conditions. No errors were detected in the transport layer during both transmissions ($PER_{WRTC} = PER_{WS} = 0$).
- On the fourth day, transmissions took place under poor network conditions. Each transport protocol detected one transmission error ($PER_{WRTC} = PER_{WS} = 0.0025\%$).
- On day 5, transmissions were conducted under medium network conditions. No errors were detected in the transport layer during both transmissions ($PER_{WRTC} = PER_{WS} = 0$).

The five selected pairs of flights were conducted during different flight sessions (a morning session, a midday session, and an evening session) and under different weather conditions: from cold days to warm days (1.5 to 25 degrees Celsius), during dry, wet, and just after rainy weather (relative humidities from 46% up to 85%).

5. Results

This section presents and discusses the results of the field experiments, to verify whether a UAV and WebRTC-based IoT is suitable for time-sensitive applications operating in a highly mobile outdoor environment when the underlying network is capable of providing a reliable, low-latency communication service.

5.1. WebRTC Data Channel: Minimum and Maximum of the End-to-End Delays Measured at the Transport Level

Table 3 includes the minimum $min(D^t_{WRTC})$ and maximum $max(D^t_{WRTC})$ of the end-to-end delays measured at the transport level when the IoT data were transmitted over the WebRTC data channel, and the truncated maximum $max(DT^t_{WRTC})$, calculated after discarding the extreme values from the series of end-to-end delays D^t_{WRTC}. Day 4 was the only day on which the PER was not equal to zero, and the large maximum delay recorded on that day was the retransmitted packet delay.

Table 3. Minimum and maximum of the end-to-end delays measured at the transport level when the IoT data were transmitted over the WebRTC data channel.

Days	$min(D^t_{WRTC})$	$max(D^t_{WRTC})$	$max(DT^t_{WRTC})$ [1]
day 1	3394 µs	3469 µs	3469 µs
day 2	3384 µs	3469 µs	3469 µs
day 3	3404 µs	3469 µs	3469 µs
day 4	3181 µs	10,120 µs	3497 µs
day 5	3377 µs	3469 µs	3469 µs

[1] Maximum of the series of end-to-end delays truncated by the extremes.

During the IoT transmissions over the WebRTC data channel carried out on day 1 to day 3 and day 5, where no transmission errors were detected at the transport layer, the maximum end-to-end delay measured at the transport level always achieved a single-digit millisecond value (Table 3). Each of the four transmissions achieved a maximum end-to-end delay of 3.5 ms (3469 µs). The maxima of the truncated series were also 3469 µs. This extremely high repeatability of the values of maxima and truncated maxima obtained on different days, when the values were repeated with an accuracy of one microsecond, may indicate an exceptionally high stability of the transmissions conducted under good and average network conditions.

The end-to-end delay minima did not show such outstanding stability, in the sense of the repeatability of results, over the different experiments carried out in the lossless environment. But even here, when the PER was zero, the differences between the results obtained on the different days did not exceeded 30 µs (values from 3377 µs to 3404 µs), which is less than 1% of the minimum values. As an effect, compared to the extremes calculated for the transmission error experiment conducted on day 4, both the maximum and the minimum can be considered stable across the error-free experiments carried out on the same network, but under different network conditions.

The single transmission error that occurred in the experiment conducted on day 4 affected both the maximum value of end-to-end delay measured at the transport level and the minimum value. Because the transmission error led to the retransmission of the lost packet, the maximum end-to-end delay was 10,120 µs, which is more than three times higher than the maxima obtained during the error-free transmissions (Figure 2a). When the delay of the retransmitted packet was discarded from the series of end-to-end delays, the maximum value dropped to 3497 µs. This is less than 1% above the maximum obtained during the error-free transmissions (Figure 2b). This shows that, at the transport level, this large increase in delay was local and its impact was limited to a single error correction via selective retransmission. The SCTP packets, except the retransmitted packet, were transmitted with delays suitable for time-sensitive applications.

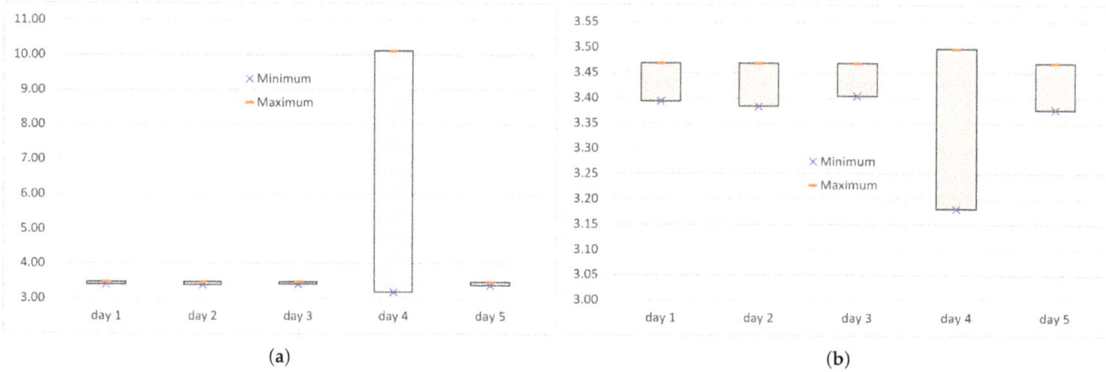

Figure 2. The range of end-to-end delays (in ms) measured at the transport level when the IoT data were transmitted over the WebRTC data channel: (**a**) full series D^{t}_{WRTC}; (**b**) truncated series DT^{t}_{WRTC}.

The occurrence of an error not only increased the maximum, but also lowered the minimum (Figure 2). For day 4, the minimum was 3181 µs, which is about 10% (300 µs) less than on the other days. Analysis of the instantaneous values (in some publications, e.g., Ref. [37], also called real-time values) of the end-to-end delay shows that the delay of the next burst after the packet loss decreased, then started to increase, and after a few seconds returned to the level observed before the transmission error. The truncated minimum, calculated after discarding the minimum and the maximum delay from the series of delays, was the same as the minimum (i.e., 3181 µs).

5.2. WebRTC Data Channel: Minimum and Maximum of the End-to-End Delays Measured at the Logical Channel Level

While delays measured at the transport level refer to the moment at which the MQTT message decapsulated from the SCTP packet is placed in the receive buffer of the logical channel, delays measured at the logical channel level refer to the moment at which the MQTT is informed that the MQTT message is ready for reception. Table 4 includes the minimum $min(D^{lc}_{WRTC})$, maximum $max(D^{lc}_{WRTC})$, and truncated maximum $max(DT^{lc}_{WRTC})$ of the end-to-end delays measured at the logical channel level when the IoT data embedded in MQTT messages were transmitted over the WebRTC data channel.

Table 4. Minimum, maximum, and truncated maximum of end-to-end delays measured at the logical channel level when the IoT data were transmitted through the WebRTC data channel.

Days	$min(D^{lc}_{WRTC})$	$max(D^{lc}_{WRTC})$	$max(DT^{lc}_{WRTC})$ [1]
day 1	3396 µs	3471 µs	3471 µs
day 2	3386 µs	3471 µs	3471 µs
day 3	3406 µs	3471 µs	3471 µs
day 4	3183 µs	10,122 µs	9187 µs
day 5	3379 µs	3471 µs	3471 µs

[1] Maximum of the series of end-to-end delays truncated by the extremes.

The difference of two microseconds between almost all of the statistics listed in Table 3, except the truncated maximum, and the corresponding statistics listed in Table 4 is the processing time of WebRTC data channel that processed the payload of the SCTP packet buffered in the receive buffer.

A large difference is visible on day 4 between the maxima of the truncated end-to-end delay series measured at the transport level and at the logical channel level. While, at the transport level, the truncated maximum was about 3.5 ms (precisely: 3497 µs), at the logical channel level, it was about 9 ms (9187 µs). The truncated maximum calculated at the logical channel level (Figure 3b) was more than two and a half times larger than the truncated maximum obtained at the transport level (Figure 2b), and only about 10% less than the end-to-end delay of the retransmitted packet (10,122 µs), as presented in Figure 3a. Such a large difference resulted from the fact that the packets already received, but sent after the lost packet, were waiting for the retransmission of the lost packet. Only when the retransmitted packet was transferred to the receive buffer of the logical channel did the MQTT protocol receive information that these packets were ready.

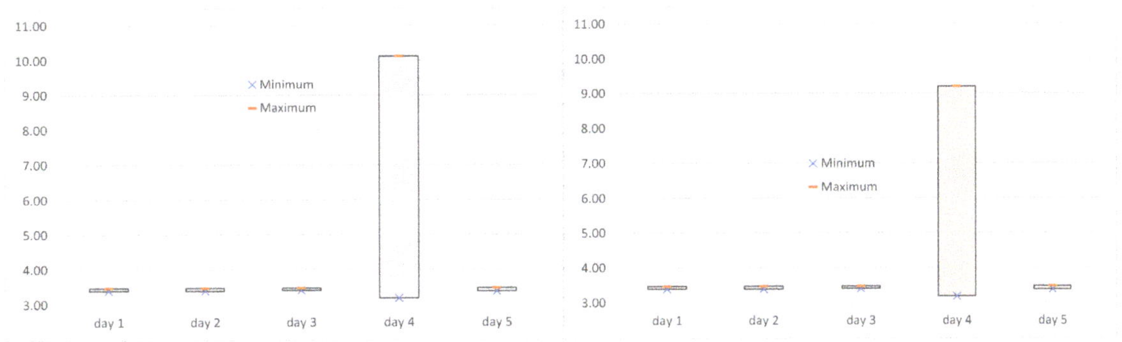

(**a**) (**b**)

Figure 3. The range of end-to-end delays (in ms) measured at the logical channel level when the IoT data were transmitted over the WebRTC data channel: (**a**) full series D^{lc}_{WRTC}; (**b**) truncated series DT^{lc}_{WRTC}.

5.3. WebSocket: Minimum and Maximum of End-to-End Delays Measured at the Logical Channel Level

Table 5 lists the minimum $min(D^{lc}_{WS})$, maximum $max(D^{lc}_{WS})$, and the truncated maximum $max(DT^{lc}_{WS})$ values of the end-to-end delays measured at the logical channel level when the MQTT messages were transmitted over the WebSocket. The end-to-end delays of the transmission of IoT data over the WebSocket, measured during experiments conducted on days 1 to 5, had relatively small minima (2515 µs to 2550 µs), smaller than minima of the delays of transmissions over WebRTC (Table 4). However, the maxima were very large, at 65,723 µs to 87,145 µs.

Table 5. Minimum and maximum of the end-to-end delays measured at the logical channel level when the IoT data were transmitted through the WebSocket logical channel.

Days	$min(D^{lc}_{WS})$	$max(D^{lc}_{WS})$	$max(DT^{lc}_{WS})$ [1]
day 1	2550 µs	69,865 µs	69,854 µs
day 2	2520 µs	83,652 µs	83,649 µs
day 3	2545 µs	65,723 µs	65,489 µs
day 4	2515 µs	87,145 µs	87,143 µs
day 5	2538 µs	74,722 µs	74,549 µs

[1] Maximum of the series of end-to-end delays truncated by the extremes.

The occurrence of packet loss on days 2 and 4 slightly (by 20–30 µs) lowered the minimum and significantly (by 10–20 ms) increased the maximum compared to the experiments in which PER was equal to 0. Unlike the end-to-end delays measured during transmissions over WebRTC (Figure 3), discarding the extreme delays from the series of delays did not lower the maximum value enough to be practical (Figure 4). When the PER was non-zero, the truncated maximum was 2 µs (day 4) to 13 µs (day 2) less than the maximum, and when the PER was zero, the truncated maximum was 11 µs (day 1) to 234 µs (day 3).

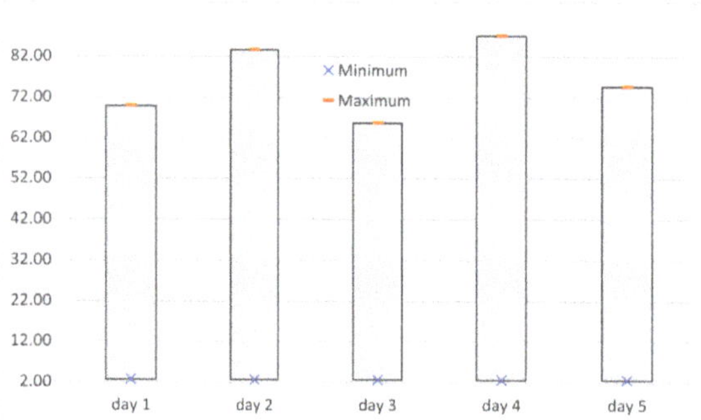

(**a**)

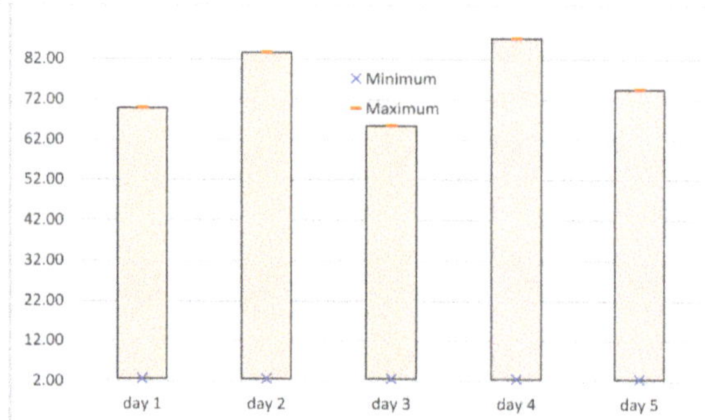

(**b**)

Figure 4. The range of end-to-end delays (in ms) measured at the logical channel level when the IoT data were transmitted over the WebSocket: (**a**) full series D^{lc}_{WS}; (**b**) truncated series DT^{lc}_{WS}.

5.4. WebRTC: Measures of Location

Tables 6 and 7 present the mean $\mu(D^{t}_{WRTC})$ and $\mu(D^{lc}_{WRTC})$, median $med(D^{t}_{WRTC})$ and $med(D^{lc}_{WRTC})$, mode $mod(D^{t}_{WRTC})$ and $mod(D^{lc}_{WRTC})$, upper quartile $Q3(D^{t}_{WRTC})$ and $Q3(D^{lc}_{WRTC})$, and lower quartile $Q1(D^{t}_{WRTC})$ and $Q1(D^{lc}_{WRTC})$ of the end-to-end delays measured at the transport level and at the logical channel level, respectively, when the MQTT messages were transmitted over the WebRTC data channel. The results in Tables 6 and 7 differ by 2 microseconds. This was a delay resulting from the processing in the receive buffer of the logical channel.

Table 6. Measures of location (mean, median, mode, upper quartile, and lower quartile) of end-to-end delays measured at the transport level when the IoT data were transmitted through the WebRTC data channel.

Days	$\mu(D^{t}_{WRTC})$	$med(D^{t}_{WRTC})$	$mod(D^{t}_{WRTC})$	$Q3(D^{t}_{WRTC})$	$Q1(D^{t}_{WRTC})$
day 1	3460 µs	3467 µs	3468 µs	3468 µs	3458 µs
day 2	3459 µs	3466 µs	3468 µs	3468 µs	3454 µs
day 3	3460 µs	3466 µs	3468 µs	3468 µs	3455 µs
day 4	3492 µs	3495 µs	3496 µs	3496 µs	3486 µs
day 5	3460 µs	3467 µs	3468 µs	3468 µs	3457 µs

Because for days 1–3 and day 5, when $PER = 0$, the maxima of end-to-end delays were less than 10 ms, the measures of location for these days were also less than 10 ms (Tables 6 and 7). Because the non-zero PER that occurred on day 4 was small (one packet lost per 40,000 packets sent), a single outlier (transport level) or a small group of outliers (logical channel level) were unable to influence either the arithmetic mean or measures of position. As an effect, all measures of location presented in Tables 6 and 7 are one-digit milliseconds.

Table 7. Measures of location of end-to-end delays measured at the logical channel level when the IoT data were transmitted over the WebRTC data channel.

Days	$\mu(D_{WRTC}^{lc})$	$med(D_{WRTC}^{lc})$	$mod(D_{WRTC}^{lc})$	$Q3(D_{WRTC}^{lc})$	$Q1(D_{WRTC}^{lc})$
day 1	3462 µs	3469 µs	3470 µs	3470 µs	3460 µs
day 2	3461 µs	3468 µs	3470 µs	3470 µs	3456 µs
day 3	3462 µs	3468 µs	3470 µs	3470 µs	3457 µs
day 4	3494 µs	3497 µs	3498 µs	3498 µs	3488 µs
day 5	3462 µs	3469 µs	3470 µs	3470 µs	3459 µs

In the case of error-free transmission in the transport layer (PER equal to 0), which was carried out on day 1 to 3 and day 5, the measures of location calculated both at the transport level and at the logical channel level showed similar extremely high repeatability as the maximums of error-free transmissions shown in the previous section. At the logical channel level, the arithmetic mean of the end-to-end delays was 3461 to 3462 µs (i.e., 3461.5 $\pm$ 0.5 µs), the median was 3468 to 3469 µs (i.e., 3468.5 $\pm$ 0.5 µs), the mode was 3470 µs and equaled the upper quartile, and the lower quartile was 3456 to 3460 µs (i.e., 3458 $\pm$ 2 µs). At the transport level, the measures of location were reduced by 2 µs, and the abovementioned numerical relationships between the statistical measures were the same.

The stability of the measures of location over the different experiments, observed for $PER = 0$, was accompanied by very small differences between the measures (Figure 5b). The upper quartile was 1–2 microseconds greater than the median, and the lower quartile was about 10 ms lower than the median. The differences between the maximums and medians were also of a few microseconds. At the logical channel level, the maximum end-to-end delay was 3471 µs, while the median of these delays was 2 to 3 µs smaller. Because the above numerical relationships were preserved at the transport level, at least 50% of the end-to-end delays were maximum at both considered levels (within 3 microseconds). This indicated a strong stability for the transmissions performed on days 1–3 and day 5, with a very small jitter.

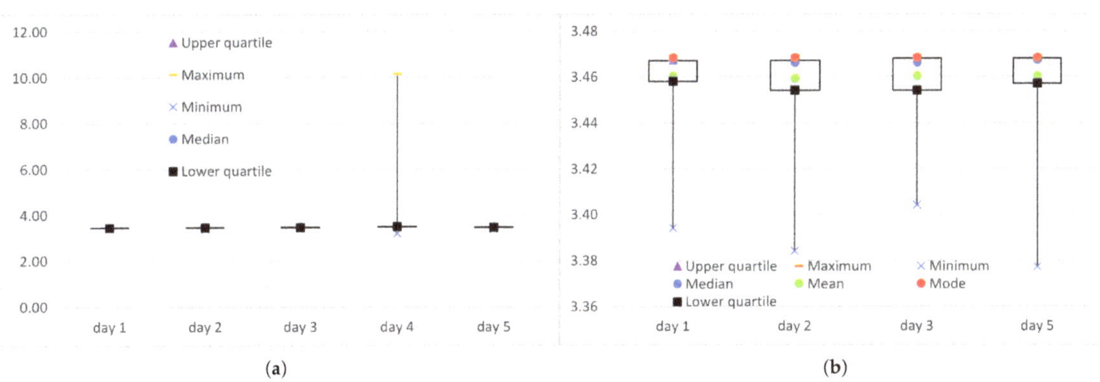

Figure 5. The five-number summary of the end-to-end delays (in ms) measured at the logical channel level when the IoT data were transmitted over WebRTC: (**a**) all experiments ($PER >= 0$); (**b**) the transport layer considered the transmission to be error-free ($PER = 0$). For $PER = 0$, the arithmetic mean and the mode are also shown.

The poor network condition on day 4 caused a single transmission error that was detected by the transport protocol, and then the lost packet was retransmitted. The end-to-end delays measured on this day had higher values for all measures of location, both the mean and quartiles, by about 30 µs in relation to the measures calculated for error-free transmissions. Due to this uniform shift, the numerical relationships between measures of position were the same as observed in the case of error-free transmissions: the mode was equal to the upper quartile, the upper quartile was 1–2 microseconds (here, 1 µs) greater than the median, and the lower quartile was about 10 microseconds (here, 9 µs) below the median (Table 6 and 7). Since the maximum end-to-end delay was the delay of the IoT datum sent in the retransmitted packet, the difference between the median and the maximum value was more than 6.5 ms (Figure 5a). The values of the maximum and the median end-to-end delay cannot therefore be considered close. However, in the case of delays measured at the transport level, the truncated median (3495 µs) and the truncated maximum (3497 µs) were close to each other, differing by only 2 µs. In the case of delays measured at the logical channel level, due to the long waiting time for packets to be sorted out in the receive buffer after retransmitting a lost packet, the truncated median (3.497 µs) and the truncated maximum (9.187 µs) differed by 5.690 µs.

5.5. WebSockets: Measures of Location

Table 8 summarizes the measures of location: mean $\mu(D^{lc}_{WS})$, median $med(D^{lc}_{WS})$, mode $mod(D^{lc}_{WS})$, upper quartile $Q3(D^{lc}_{WS})$, and lower quartile $Q1(D^{lc}_{WS})$ of the end-to-end delays measured at the logical channel level when MQTT transmissions were carried out over the WebSocket. While all measures of the location of end-to-end delays of IoT data transmitted over WebRTC satisfied the single-digit millisecond requirements of time-sensitive applications (Table 7, Figure 5a), the lower quartile and the mode were the only measures of location that always met this requirement when transmissions were carried out over the classic web logical channel (Table 8, Figure 6a). The median only met this requirement on days 1, 3, and 5, when the underlying IEEE 802.11 network was able to ensure reliable transmission in the transport layer. The arithmetic mean and upper quartile only met it when the error rate in the MAC sublayer was not greater than 0.1 percent (day 1 and day 2).

The comparison of the results presented in Tables 7 and 8 shows that the mode was the only measure of location in terms of which the IoT transmissions over WebSocket were superior to IoT transmissions over WebRTC Data Channel. Two such measures were therefore found: the mode and minimum (see previous sections). In IoT transmissions using WebSocket, the mode was not equal to the upper quartile, as in the IoT transmissions using WebRTC (Figure 5b), but to the minimum (Figure 6b). However, the relatively small number of delays whose value was a modal value (WebRTC: 12–15 thousand on days 1–3 and 5, over 10 thousand on day 4; WebSocket: about 1 thousand on days 1, 2, and 5, about 500 on day 2, no modal value on day 4) shows that this modal superiority of transmissions over WebSocket does not matter much in practice.

Table 8. Measures of location (mean, median, mode, upper quartile, and lower quartile) of end-to-end delays measured at the logical channel level when the IoT data were transmitted over the WebSocket.

Days	$\mu(D^{lc}_{WS})$	$med(D^{lc}_{WS})$	$mod(D^{lc}_{WS})$	$Q3(D^{lc}_{WS})$	$Q1(D^{lc}_{WS})$
day 1	7947 µs	5235 µs	2550 µs	8247 µs	4431 µs
day 2	17,605 µs	12,233 µs	2520 µs	26,445 µs	4721 µs
day 3	7260 µs	5123 µs	2545 µs	6919 µs	4403 µs
day 4	22,584 µs	16335 µs	-	32,448 µs	8833 µs
day 5	13,030 µs	9563 µs	2538 µs	19,389 µs	4537 µs

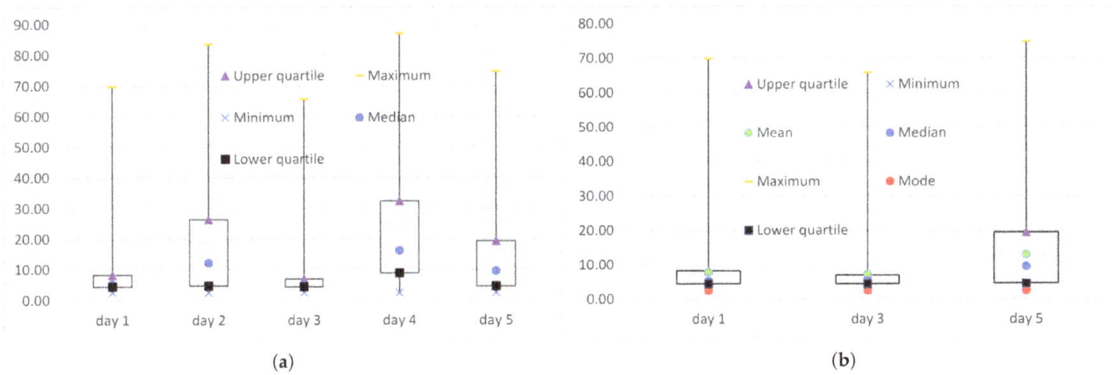

Figure 6. The five-number summary of end-to-end delays (in ms) measured at the logical channel level when IoT data were transmitted over WebSocket: (**a**) all experiments ($PER >= 0$); (**b**) the transport layer considered the transmission to be error-free ($PER = 0$). For $PER = 0$ the arithmetic mean and the mode are also shown.

6. Discussion

The previous section presented and analyzed the minimum, maximum, arithmetic mean, mode, and quartiles (upper, median, and lower quartile) of the end-to-end delays of air-to-ground IoT transmissions carried out using the WebRTC data channel or WebSocket. Table 9 contains a comparison of the statistics collected at the logical channel level, in the form of the ratio of the value of a given statistical measure calculated for IoT transmission via WebSocket (Tables 5 and 8) to the value of the same measure calculated for IoT transmission via WebRTC (Tables 4 and 7). If the values of the same statistical measure calculated for transmissions using WebRTC Data Channel and WebSocket are equal, the ratio will be 1. In such a situation, it will not matter, from the point of view of a given statistical measure, through which of the web logical channels the IoT transmission is carried out between the UAV and the ground. This is a purely hypothetical case and does not appear in Table 9.

Table 9. Comparison of extremes (minimum and maximum) and measures of location (mean, median, mode, upper quartile, and lower quartile) of the end-to-end delays measured at the logical channel level when the IoT data were transmitted using WebRTC data channel and WebSocket.

Days	$\frac{min(D^{lc}_{WS})}{min(D^{lc}_{WRTC})}$	$\frac{max(D^{lc}_{WS})}{max(D^{lc}_{WRTC})}$	$\frac{\mu(D^{lc}_{WS})}{\mu(D^{lc}_{WRTC})}$	$\frac{med(D^{lc}_{WS})}{med(D^{lc}_{WRTC})}$	$\frac{mod(D^{lc}_{WS})}{mod(D^{lc}_{WRTC})}$	$\frac{Q3(D^{lc}_{WS})}{Q3(D^{lc}_{WRTC})}$	$\frac{Q1(D^{lc}_{WS})}{Q1(D^{lc}_{WRTC})}$
day 1	0.75	20.14	2.3	1.51	0.74	2.38	1.28
day 2	0.74	24.11	5.09	3.53	0.73	7.62	1.37
day 3	0.75	18.95	2.1	1.48	0.73	1.99	1.27
day 4	0.79	8.61	6.47	4.67	-	9.28	2.53
day 5	0.75	21.54	3.77	2.76	0.73	5.59	1.31

A ratio of statistical measures less than 1 indicates the superiority of the WebSocket web logical channel over the WebRTC data channel. This ratio appears in Table 9 twice: for minimum values and for modal values. The ratio of minima, $min(D^{lc}_{WS})$ to $min(D^{lc}_{WRTC})$, ranged from 0.74 for day 2 to 0.79 for day 4. This means that, when using WebRTC, the minimum end-to-end delays were approximately one third greater under good and medium network conditions and approximately a quarter greater under poor network conditions than the minimum delays achieved when transmitting using WebSockets. The ratio of modal values, $mod(D^{lc}_{WS})$ to $mod(D^{lc}_{WRTC})$, amounted to 0.73–0.74 for IoT transmissions under good and medium network conditions. Under the poor network conditions, no modal value was observed during transmission using WebSockets. However, the relatively small number of minimally delayed packets made this advantage of WebSockets over WebRTC relatively minor.

Figures 7 and 8 show scatter plots drawn for the end-to-end delay statistics calculated at the logical channel level and presented in the previous section. The values obtained for transmissions using WebSockets (Tables 5 and 8) are plotted as a function of the corresponding values obtained for the transmissions using WebRTC (Tables 4 and 7). The markers denote statistics calculated for transmissions under good (x), average (+), and poor (o) network conditions. The diagonal of each plot (dashed line) illustrates the hypothetical case of a ratio of a given statistical measure equal to 1. Below the diagonal, there were only minimum and mode markers (Figure 7a). The values of these statistics for transmissions using WebRTC were higher than for transmissions using WebSockets, so the ratio given in Table 9 is less than 1. The highest minimum latencies and highest latency modes were observed under good network conditions. As the network conditions deteriorated, the minimum and mode began to decrease, although the observed differences were small, and when using WebRTC, there was no difference between the modes calculated under good and medium network conditions. The lowest minimum delay occurred under poor network conditions, but only in the case of WebRTC was the reduction in the minimum significant.

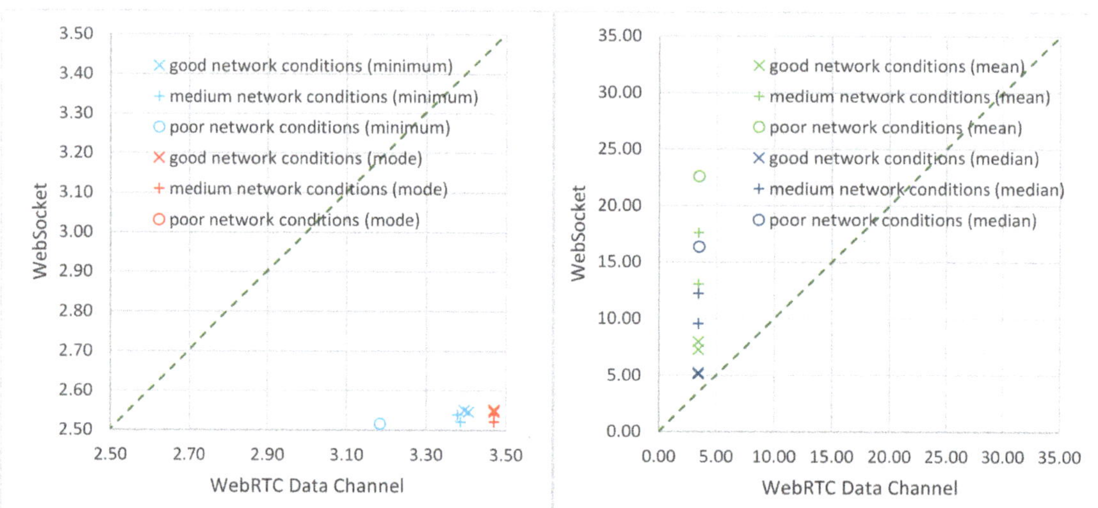

(a) (b)

Figure 7. Scatter plots for the statistics of the end-to-end delays (in ms) measured at the logical channel level during air-to-ground transmissions using the WebRTC data channel and using the WebSocket logical channel: (**a**) minimum and mode; (**b**) mean and median.

A ratio of statistical measures greater than 1 indicates the superiority of the WebRTC data channel over WebSocket. This ratio appears in Table 9 for the arithmetic mean, quartiles, and maximum value. Except for the latter, unlike the case of ratios smaller than 1, the least ratios greater than 1 were obtained under good network conditions, and as the network conditions deteriorated, the ratio value increased. As an example, the ratio of arithmetic means, $\mu(D^{lc}_{WS})$ to $\mu(D^{lc}_{WRTC})$, ranged from 2.1 for day 3 to 6.47 for day 4. Under good network conditions, the mean delay of air-to-ground IoT transmissions using WebSockets was more than twice the mean delay of transmissions using WebRTC.

Under medium network conditions, it was almost four times greater. When the network conditions were on the verge of medium to poor, the mean delay of transmissions using WebSocket was just over five times greater, and when the network conditions were poor, it was well over six times greater than the mean delay of transmissions using WebRTC (Table 9, Figure 7b). In addition, in the case of the ratio of quartiles, the greater the distance (in the number of samples, or position) of a given measure from the minimum, the greater the ratio values. The ratio of lower quartiles, $Q1(D^{lc}_{WS})$ to $Q1(D^{lc}_{WRTC})$, ranged from 1.27 for day 3 to 2.53 for day 4 (Table 9, Figure 8a). The ratio of medians, $med(D^{lc}_{WS})$ to $med(D^{lc}_{WRTC})$ ranged from 1.48 for day 3 to 4.67 for day 4 (Table 9, Figure 7b). The ratio of upper quartiles, $Q3(D^{lc}_{WS})$ to $Q3(D^{lc}_{WRTC})$, ranged from 1.27 for day 3 to 9.28 for day 4 (Table 9, Figure 8a).

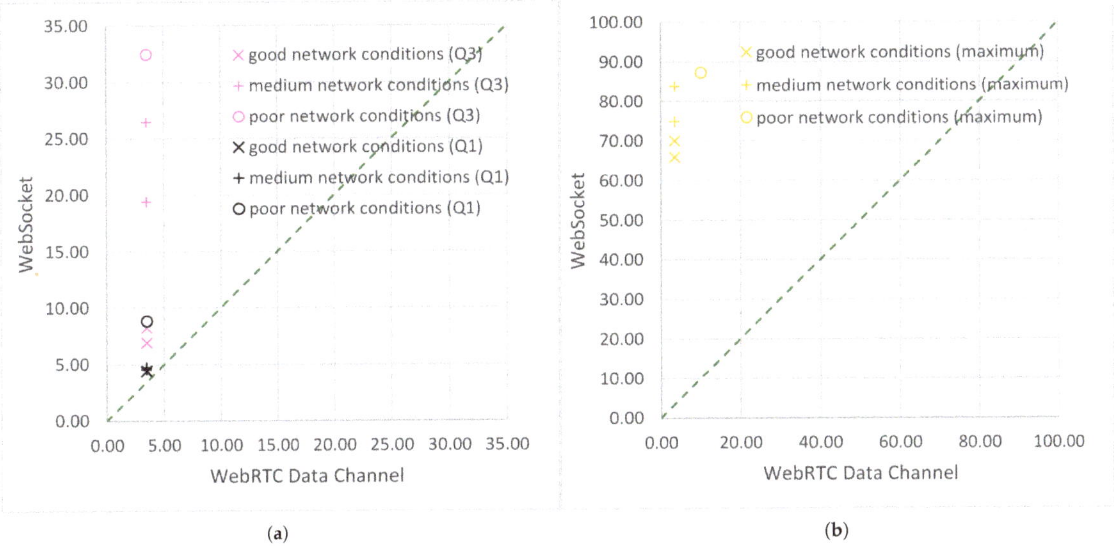

Figure 8. Scatter plots for statistics of end-to-end delays (in ms) measured at the logical channel level during air-to-ground transmissions using WebRTC data channel and using the WebSocket logical channel: (**a**) lower quartile and upper quartile; (**b**) maximum.

In the case of end-to-end delay maxima ratios, $max(D^{lc}_{WS})$ to $max(D^{lc}_{WRTC})$, the above observations were true for days 1 to 3 and day 5, when transmission at the transport layer was error-free. Under good network conditions, the maximum delay of transmission using WebSocket was approximately 20 times greater than the maximum delay of transmission using WebRTC. When the network conditions were medium, it was about 22 times greater, and when the network conditions were between medium and poor, it was more than 24 times greater than the maximum delay of transmission using WebRTC. In the case of a single transmission error (day 4), the ratio of maxima dropped to over 8 (Table 9), due to the large increase in the maximum delay in transmission using WebRTC caused by packet retransmission. As with all other measures for which the ratio was greater than 1 (Figures 7b and 8a), the maximum transmission delay using the WebSocket was the highest under poor network conditions (Figure 8b).

When comparing the obtained results with those reported in related works, the delay introduced by the implementation of the MQTT protocol should be taken into account. Additional laboratory experiments showed that the classic Eclipse Paho JavaScript Client implementation of the MQTT protocol introduced delays averaging approximately 1.5 ms. For WebSocket-based IoT, this resulted in average application-level end-to-end delays of approximately 9 ms to 24 ms. Taking into account that similar delays during transmission from the UAV to the ground station in systems using WebSocket have been reported in the literature (e.g., 23 ms [30], 20 ms to 25 ms in the IEEE 802.11 network [34]), it can be concluded that that the end-to-end delay values for UAV-borne IoT using WebSocket were comparable to those reported in related works.

In the case of solutions other than WebSocket, but still based on the TCP protocol, the situation is similar. In [32], replacing single-path transmission using TCP with a multi-path transmission using the improved MPTCP reduced the latency from 920.4 ms to 568.1 ms (i.e., 1.62 times). This was achieved at the expense of the parallel transmission of cloned packets. However, in this paper, the use of WebRTC provided a greater relative improvement than [32] and without as much computational and energy cost. Significantly, modifying the MPTCP so that the UDP was used as the underlying protocol instead of the TCP [31] allowed for a relative improvement similar to that shown in this paper, and at lower computational costs than in [32] due to the simplicity of the UDP mechanisms. However, the energy cost of the solution proposed in [31] remained significant. Moreover, since the STCP implements multihoming, multipath transmissions of cloned sensor data can also be used in WebRTC-based IoT.

It can be expected that the advantages of the WebRTC data channel used in UAV communication may be comparable to those of using any other UDP-based solution. Comparing the results of experiments on MQTT over WebRTC data channel presented in this work with the results of the experiments on MQTT over QUIC presented in [38], it can be seen that, in the case of error-free transmissions, the results were approximately similar. If the latencies introduced by the underlying IEEE 802.11 networks (2 ms in this article and 25 ms in [38]), as well as the estimated delays introduced by the Paho implementation of the MQTT protocol (1.5 ms), are subtracted from the average results, the approximate average delays obtained in this paper and in [38] are the same and equal 1.5 ms. However, the spread of end-to-end delay values was much larger in [38] than in this work. Due to the significant differences in the test environment (in this paper, a mobile and highly variable real-world environment using a low-latency network was employed; a static environment using an emulated high-latency network was employed in [38]), it is impossible to say with certainty how beneficial it would be to use a WebRTC data channel in UAV-IoT communication instead of QUIC.

The second example of a UDP-based solution is presented in [37], which compared IoT transmissions over SCTP with IoT transmissions over TCP. The results of simulation experiments showed the better performance of SCTP and better stability of TCP in heterogeneous networks (wired and wireless). During error-free transmissions, the performance difference between SCTP and TCP shown in [37] was not as large as the performance difference estimated from the results presented in this paper. The issue of stability was also different: in this paper, the SCTP was extremely stable, and much more stable than the TCP. It is worth emphasizing here that, in [37], an older version of the SCTP was discussed, and the research conducted in this paper used a new, WebRTC-oriented version of the SCTP that is currently implemented in web browsers.

7. Conclusions

In recent years, wireless communication for time-sensitive IoT applications has become a hot research topic, including applications that require end-to-end delays measured in single-digit milliseconds. One of the problems encountered in these applications is the processing in higher network layers: even if the underlying network is capable of providing highly reliable, low-latency communications, the delays introduced at the transport layer and above may prove too great to meet stringent time requirements. The aim of this paper was to show that, in the case of IoT carried by UAV, the use of WebRTC can help solve this problem.

The paper used high-resolution time measurement procedures and timer synchronization to perform delay measurements at the level of the transport protocol and at the level of the network logical channel of the WebRTC IoT application, run on board the UAV. During the field experiments, air-to-ground IoT transmissions were carried out under various network conditions, followed by statistical analysis of these delays, focusing on extreme values and location measures. The obtained results were compared with those obtained for IoT transmission via the WebSocket logical channel, under the same circumstances.

The statistical characteristics of the end-to-end delays showed that, during air-to-ground transmission, the WebRTC-based IoT was able to achieve single-digit-millisecond end-to-end delays on both the transport protocol level and the logical channel level. When the WebRTC transmission was error-free, stable end-to-end delays well below 10 ms were achieved. When a single transmission error occurred, higher end-to-end delays were observed in the immediate vicinity of the retransmitted packet, although they were still below 10 ms. Only the delay of the retransmitted packet slightly exceeded 10 ms.

The results of the same IoT transmissions performed via WebSocket under the same circumstances showed that the WebRTC-based UAV-borne IoT had 8.5 to 24 times lower maximum delays and 2 to 6.5 times lower mean delays than the same IoT using WebSocket. The smallest differences between the maximum values and the largest differences between the arithmetic means were associated with the occurrence of a transmission error. The results therefore indicated the superiority of the WebRTC logical channel over the classic web logical channel.

Future research will focus on analyzing WebRTC-based UAV-borne IoT transmissions in Wi-Fi 6e and Wi-Fi 7 networks, as well as UAV swarm tests in a 5G test network.

Author Contributions: Conceptualization, A.C. and R.R.C.; formal analysis, A.C.; investigation, R.R.C.; software, A.C. and R.R.C.; visualization, A.C.; writing—original draft, A.C. and R.R.C.; writing—review and editing, A.C.; funding acquisition, A.C. All authors have read and agreed to the published version of the manuscript.

Funding: This research received no external funding.

Institutional Review Board Statement: Not applicable.

Informed Consent Statement: Not applicable.

Data Availability Statement: The data supporting the conclusions of this article will be made available by the authors on request.

Conflicts of Interest: The authors declare no conflicts of interest.

Abbreviations

The following abbreviations are used in this manuscript:

Abbreviations	
4G	Fourth-generation (technology for broadband cellular networks)
5G	Fifth-generation (technology for broadband cellular networks)
AMEC	Aerial mobile edge computing
AMQP	Advanced Message Queuing Protocol
API	Application programming interface
CPU	Central processing unit
CCC	Command and control console
DiffServ	Differentiated Services
ESS	Extended Service Set

HSPA+	Evolved High Speed Packet Access (Evolved HSPA, HSPA Evolution)
HTTP	Hypertext Transfer Protocol
IEEE	Institute of Electrical and Electronics Engineers
IoT	Internet of Things
ISP	Internet service provider
KPI	Key performance indicator
LRT	Low-latency Reliable Transmission
LTE	Long Term Evolution
MAC	Medium access control
MPTCP	Multipath Transmission Control Protocol (Multipath TCP)
MQTT	Message Queuing Telemetry Transport (MQ Telemetry Transport)
PAV	Personal air vehicle
PER	Packet error rate
QoS	Quality of Service
QUIC	Quick UDP Internet Connections
RTP	Real-time Transport Protocol
SBC	Single-board computers
SCTP	Stream Control Transmission Protocol
SDN	Software-defined networking
SDR	Software-defined radio
SIP	Session Initiation Protocol
TCP	Transmission Control Protocol
TNS	Time-Sensitive Networking
UAV	Unmanned aerial vehicle
UDP	User Datagram Protocol
W3C	World Wide Web Consortium
WebRTC	Web real-time communication
WLAN	Wireless local network
WMMS	WebRTC multimedia and monitoring station
WoT	Web of Things
XMPP	Extensible Messaging and Presence Protocol
Indexes	
i	i-th IoT datum
j	j-th IoT datum
ilc	input of logical channel
lc	logical channel level
olc	output of logical channel
t	transport level (output of transport protocol)
$WRTC$	WebRTC's web logical channel (Data Channel)
WS	WebSocket web logical channel
Symbols	
d	end-to-end delay
d^{lc}	end-to-end delay measured at the logical channel level
d^t	end-to-end delay measured at the transport level
D	series of end-to-end delays
D^{lc}	series of end-to-end delays at the logical channel level
D^t	series of end-to-end delays at the transport level
DT	truncated series of end-to-end delays
DT^{lc}	truncated series of end-to-end delays at the logical channel level
DT^t	truncated series of end-to-end delays at the transport level
t	time
t_1	starting time
t^{ilc}	entry time into the logical channel
t^{olc}	reception time from the logical channel

t^t	reception time from the transport protocol
(D)	arithmetic mean of the time series D
(DT)	truncated arithmetic mean of the time series D
$max(D)$	maximum value in series D
$max(DT)$	truncated maximum value in series D
$med(D)$	median of the time series D
$med(DT)$	truncated median of the time series D
$min(D)$	minimum value in series D
$min(DT)$	truncated minimum value in series D
$mod(D)$	mode of the time series D
$mod(DT)$	truncated mode of the time series D
$Q1(D)$	lower quartile of the time series D
$Q1(DT)$	truncated lower quartile of the time series D
$Q3(D)$	upper quartile of the time series D
$Q3(DT)$	truncated upper quartile of the time series D

References

1. Mohsan, S.A.H.; Othman, N.Q.H.; Li, Y.; Alsharif, M.H.; Khan, M.A. Unmanned aerial vehicles (UAVs): Practical aspects, applications, open challenges, security issues, and future trends. *Intell. Serv. Robot.* **2023**, *16*, 109–137. [CrossRef] [PubMed]
2. Unmanned Aircraft System. FAA Aerospace Forecast Fiscal Years 2022–2042. FAA. 2022. Available online: https://www.faa.gov/sites/faa.gov/files/2022-06/Unmanned_Aircraft_Systems.pdf (accessed on 23 December 2024).
3. Wikarek, J.; Sitek, P.; Zawarczyński, Ł. An Integer Programming Model for the Capacitated Vehicle Routing Problem with Drones. In *Computational Collective Intelligence. ICCCI 2019*; Nguyen, N., Chbeir, R., Exposito, E., Aniorté, P., Trawiński, B., Eds.; Lecture Notes in Computer Science; Springer: Cham, Switzerland, 2019; Volume 11683, pp. 511–520. [CrossRef]
4. Sitek, P.; Wikarek, J.; Rutczyńska-Wdowiak, K. Capacitated Vehicle Routing Problem with Pick-Up, Alternative Delivery and Time Windows (CVRPPADTW): A Hybrid Approach. In Proceedings of the Distributed Computing and Artificial Intelligence, 16th International Conference, Special Sessions, Avila, Spain, 26–28 June 2019; Springer International Publishing: Cham, Switzerland, 2019; pp. 33–40. [CrossRef]
5. Moradi, N.; Wang, C.; Mafakheri, F. Urban Air Mobility for Last-Mile Transportation: A Review. *Vehicles* **2024**, *6*, 1383–1414. [CrossRef]
6. Motlagh, N.H.; Kortoçi, P.; Su, X.; Lovén, L.; Hoel, H.K.; Haugsvær, S.B.; Srivastava, V.; Gulbrandsen, C.F.; Nurmi, P.; Tarkoma, S. Unmanned aerial vehicles for air pollution monitoring: A survey. *IEEE Internet Things J.* **2023**, *10*, 21687–21704. [CrossRef]
7. Chodorek, A.; Chodorek, R.R.; Yastrebov, A. The Prototype Monitoring System for Pollution Sensing and Online Visualization with the Use of a UAV and a WebRTC-Based Platform. *Sensors* **2022**, *22*, 1578. [CrossRef] [PubMed]
8. Sziroczak, D.; Rohacs, D.; Rohacs, J. Review of using small UAV based meteorological measurements for road weather management. *Prog. Aerosp. Sci.* **2022**, *134*, 100859. [CrossRef]
9. Chodorek, A.; Chodorek, R.R.; Sitek, P. Response Time and Intrinsic Information Quality as Criteria for the Selection of Low-Cost Sensors for Use in Mobile Weather Stations. *Electronics* **2022**, *11*, 2448. [CrossRef]
10. Elloumi, M.; Dhaou, R.; Escrig, B.; Idoudi, H.; Saidane, L.A. Monitoring road traffic with a UAV-based system. In Proceedings of the 2018 IEEE Wireless Communications and Networking Conference (WCNC), Barcelona, Spain, 15–18 April 2018; pp. 1–6. [CrossRef]
11. Phade, G.; Kishore, A.T.; Omkar, S.; Suresh Kumar, M. IoT-Enabled Unmanned Aerial Vehicle: An Emerging Trend in Precision Farming. In *Drone Technology: Future Trends and Practical Applications*; Mohanty, S.N., Ravindra, J.V.R., Surya Narayana, G., Pattnaik. C.R., Mohamed Sirajudeen, Y., Eds.; Scrivener Publishing LLC.: Beverly, MA, USA, 2023; pp. 301–324. [CrossRef]
12. Ju, C.; Son, H.I. Multiple UAV Systems for Agricultural Applications: Control, Implementation, and Evaluation. *Electronics* **2018**, *7*, 162. [CrossRef]
13. Ganesh, S.; Gopalasamy, V.; Shibu, N.S. Architecture for drone assisted emergency ad-hoc network for disaster rescue operations. In Proceedings of the 2021 International Conference on COMmunication Systems & NETworkS (COMSNETS), Bangalore, India, 5–9 January 2021; pp. 44–49. [CrossRef]
14. Chand, G.S.L.K.; Lee, M.; Shin, S.Y. Drone based wireless mesh network for disaster/military environment. *J. Comput. Commun.* **2018**, *6*, 44–52. [CrossRef]
15. Mohammed, A.; Erbad, A.; Nahom, H.; Albaseer, A.; Abdallah, M.; Guizani, M. FDRL Approach for Association and Resource Allocation in Multi-UAV Air-To-Ground IoMT Network. In Proceedings of the GLOBECOM 2022–2022 IEEE Global Communications Conference, Rio de Janeiro, Brazil, 4–8 December 2022; pp. 1417–1422. [CrossRef]

16. Liu, X.; Liu, H.; Zheng, K.; Liu, J.; Taleb, T.; Shiratori, N. AoI-minimal clustering, transmission and trajectory co-design for UAV-assisted WPCNs. *IEEE Trans. Veh. Technol.* **2024**, 1–16. [CrossRef]
17. Fu, X.; Huang, X.; Pan, Q. Collaborative relay for achieving long-term and low-AoI data collection in UAV-aided IoT systems. *Veh. Commun.* **2024**, *45*, 100719. [CrossRef]
18. Rahimi, O.; Shafieinejad, A. Minimizing age of information in multi-UAV-assisted IoT networks: A graph theoretical approach. *Wirel. Netw.* **2024**, *30*, 533–555. [CrossRef]
19. Deng, C.; Fu, X. Low-AoI Data Collection in UAV-Aided Wireless-Powered IoT Based on Aerial Collaborative Relay. *IEEE Sens. J.* **2024**, *24*, 33506–33521. [CrossRef]
20. Chodorek, A.; Chodorek, R.R.; Krempa, A. An analysis of elastic and inelastic traffic in shared link. In Proceedings of the 2008 Conference on Human System Interactions, Krakow, Poland, 25–27 May 2008; pp. 873–878. [CrossRef]
21. Giordani, M.; Polese, M.; Mezzavilla, M.; Rangan, S.; Zorzi, M. Toward 6G Networks: Use Cases and Technologies. *IEEE Commun. Mag.* **2020**, *58*, 55–61. [CrossRef]
22. Han, M.; Lee, J.; Rim, M.; Kang, C.G. Dynamic Bandwidth Part Allocation in 5G Ultra Reliable Low Latency Communication for Unmanned Aerial Vehicles with High Data Rate Traffic. *Sensors* **2021**, *21*, 1308. [CrossRef] [PubMed]
23. Rico, D.; Merino, P. A Survey of End-to-End Solutions for Reliable Low-Latency Communications in 5G Networks. *IEEE Access* **2020**, *8*, 192808–192834. [CrossRef]
24. Purucker, P.; Schmid, J.; Höß, A.; Schuller, B.W. System Requirements Specification for Unmanned Aerial Vehicle (UAV) to Server Communication. In Proceedings of the 2021 International Conference on Unmanned Aircraft Systems (ICUAS), Athens, Greece, 15–18 June 2021; pp. 1499–1508. [CrossRef]
25. Mourtzis, D.; Angelopoulos, J.; Panopoulos, N. Smart Manufacturing and Tactile Internet Based on 5G in Industry 4.0: Challenges, Applications and New Trends. *Electronics* **2021**, *10*, 3175. [CrossRef]
26. WebRTC Extended Use Cases W3C Group Draft Note 14 December 2023. Available online: https://www.w3.org/TR/webrtc-nv-use-cases/ (accessed on 23 December 2024).
27. Chodorek, A.; Chodorek, R.R. Work-in-Progress: A Browser-Driven Sensor Service for Embedded IoT. In Proceedings of the 2022 International Conference on Embedded Software (EMSOFT), Shanghai, China, 7–14 October 2022; pp. 15–16. [CrossRef]
28. Chodorek, A.; Chodorek, R.R.; Sitek, P. UAV-Based and WebRTC-Based Open Universal Framework to Monitor Urban and Industrial Areas. *Sensors* **2021**, *21*, 4061. [CrossRef]
29. Herrero, R. MQTT-SN, CoAP, and RTP in wireless IoT real-time communications. *Multimed. Syst.* **2020**, *26*, 643–654. [CrossRef]
30. Yigit, Y.; Nguyen, L.D.; Ozdem, M.; Kinaci, O.K.; Hoang, T.; Canberk, B.; Duong, T.Q. TwinPort: 5G drone-assisted data collection with digital twin for smart seaports. *Sci. Rep.* **2023**, *13*, 12310. [CrossRef]
31. Lee, W. Enabling Reliable UAV Control by Utilizing Multiple Protocols and Paths for Transmitting Duplicated Control Packets. *Sensors* **2021**, *21*, 3295. [CrossRef]
32. Lee, W.; Lee, J.Y.; Joo, H.; Kim, H. An MPTCP-Based Transmission Scheme for Improving the Control Stability of Unmanned Aerial Vehicles. *Sensors* **2021**, *21*, 2791. [CrossRef] [PubMed]
33. Jung, W.S.; Yim, J.; Ko, Y.B.; Singh, S. ACODS: Adaptive computation offloading for drone surveillance system. In Proceedings of the 16th Annual Mediterranean Ad Hoc Networking Workshop (Med-Hoc-Net), Budva, Montenegro, 28–30 June 2017; pp. 1–6. [CrossRef]
34. Ramos, J.; Ribeiro, R.; Safadinho, D.; Barroso, J.; Pereira, A. Communication Protocol for Unmanned Vehicles: An Architectural Approach. In Proceedings of the 2020 Global Internet of Things Summit (GIoTS), Dublin, Ireland, 3 June 2020; pp. 1–7. [CrossRef]
35. Ramos, J.; Ribeiro, R.; Safadinho, D.; Barroso, J.; Rabadão, C.; Pereira, A. Distributed Architecture for Unmanned Vehicle Services. *Sensors* **2021**, *21*, 1477. [CrossRef] [PubMed]
36. Lee, H.; Yoon, J.; Jang, M.-S.; Park, K.-J. A Robot Operating System Framework for Secure UAV Communications. *Sensors* **2021**, *21*, 1369. [CrossRef]
37. Sun, W.; Yu, S.; Xing, Y.; Qin, Z. Parallel Transmission of Distributed Sensor Based on SCTP and TCP for Heterogeneous Wireless Networks in IoT. *Sensors* **2019**, *19*, 2005. [CrossRef]
38. Jeddou, S.; Fernández, F.; Diez, L.; Baina, A.; Abdallah, N.; Agüero, R. Delay and Energy Consumption of MQTT over QUIC: An Empirical Characterization Using Commercial-Off-The-Shelf Devices. *Sensors* **2022**, *22*, 3694. [CrossRef]
39. Fernández, F.; Zverev, M.; Garrido, P.; Juárez, J.R.; Bilbao, J.; Agüero, R. Even Lower Latency in IIoT: Evaluation of QUIC in Industrial IoT Scenarios. *Sensors* **2021**, *21*, 5737. [CrossRef]
40. Kilic, F.; Hassan, M.; Hardt, W. Prototype for Multi-UAV Monitoring–Control System Using WebRTC. *Drones* **2024**, *8*, 551. [CrossRef]
41. Jiang, S.; Zhang, Q.; Wu, A.; Liu, Q.; Wu, J.; Xia, P. A low-latency reliable transport solution for network-connected UAV. In Proceedings of the 2018 10th International Conference on Communication Software and Networks, Ponta Delgada, Portugal, 6–9 July 2018; pp. 511–515. [CrossRef]

42. *IEEE Std 802.11-2020*; IEEE Standard for Information Technology—Telecommunications and Information Exchange between Systems—Local and Metropolitan Area Networks—Specific Requirements—Part 11: Wireless LAN Medium Access Control (MAC) and Physical Layer (PHY) Specifications. IEEE: New York, NY, USA, 2021; 4379p. [CrossRef]
43. Wireless Technologies and Use Cases in Industrial IOT. 2020. Available online: https://web.archive.org/web/20240502073912/https://www.ciscolive.com/c/dam/r/ciscolive/emea/docs/2020/pdf/BRKIOT-1775.pdf (accessed on 23 December 2024).
44. Anand, B.; Kambhampaty, H.R.; Rajalakshmi, P. A Novel Real-Time LiDAR Data Streaming Framework. *IEEE Sens. J.* **2022**, *22*, 23476–23485. [CrossRef]
45. Moorthy, S.K.; Lu, C.; Guan, Z.; Mastronarde, N.; Sklivanitis, G.; Pados, D.; Bentley, E.S.; Medley, M. CloudRAFT: A Cloud-based Framework for Remote Experimentation for Mobile Networks. In Proceedings of the 2022 IEEE 19th Annual Consumer Communications & Networking Conference (CCNC), Las Vegas, NV, USA, 8–11 January 2022; pp. 1–6. [CrossRef]
46. Tiberkak, A.; Hentout, A.; Belkhir, A. WebRTC-based MOSR remote control of mobile manipulators. *Int. J. Intell. Robot. Appl.* **2023**, *7*, 304–320. [CrossRef]
47. Samdanis, K.; Taleb, T. The road beyond 5G: A vision and insight of the key technologies. *IEEE Netw.* **2020**, *34*, 135–141. [CrossRef]
48. Vu, V.A.; Walker, B. Redundant multipath-tcp scheduling with desired packet latency. In Proceedings of the 14th Workshop on Challenged Networks, Los Cabos, Mexico, 25 October 2019; pp. 7–12. [CrossRef]
49. Chodorek, R.R.; Chodorek, A. MPTCP protocol misbehaviour in high-speed, uncongested network. *J. Mar. Eng. Technol.* **2017**, *16*, 248–256. [CrossRef]
50. Stewart, R.; Tüxen, M.; Nielsen, K. Stream Control Transmission Protocol. RFC 9260. 2022. Available online: https://www.rfc-editor.org/info/rfc9260 (accessed on 23 December 2024).
51. Kumar, P.; Dezfouli, B. Implementation and analysis of QUIC for MQTT. *Comput. Netw.* **2019**, *150*, 28–45. [CrossRef]
52. Martalò, M.; Pettorru, G.; Atzori, L. A Cross-Layer Survey on Secure and Low-Latency Communications in Next-Generation IoT. *IEEE Trans. Netw. Serv. Manag.* **2024**, *21*, 4669–4685. [CrossRef]
53. Web of Things (WoT) Architecture 1.1 W3C Recommendation 05 December 2023. Available online: https://www.w3.org/TR/wot-architecture11/ (accessed on 23 December 2024).
54. Chodorek, A.; Chodorek, R.R. Model warstwowy ustanawiania sesji WebRTC. *Stud. Inform.* **2016**, *37*, 117–126. Available online: https://scholar.archive.org/work/lucrhn2ykrcppbcknqshbg2mum/access/wayback/http://studiainformatica.polsl.pl:80/index.php/SI/article/download/767/729 (accessed on 23 December 2024).
55. Grigorik, I. *High Performance Browser Networking: What Every Web Developer Should Know About Networking and Web Performance*; O'Reilly Media, Inc.: Newton, MA, USA, 2013.
56. Chodorek, A.; Chodorek, R.R.; Wajda, K. Benefits of Using WebRTC Technology for Building of Flying IoT Systems. In Proceedings of the 35th International Conference on Advanced Information Networking and Applications (AINA-2021), Toronto, ON, Canada, 12–14 May 2021; pp. 310–322. [CrossRef]

MDPI AG
Grosspeteranlage 5
4052 Basel
Switzerland
Tel.: +41 61 683 77 34

Sensors Editorial Office
E-mail: sensors@mdpi.com
www.mdpi.com/journal/sensors

www.ingramcontent.com/pod-product-compliance
Lightning Source LLC
Chambersburg PA
CBHW041948260926
51850CB00018B/1957
* 9 7 8 3 7 2 5 8 5 0 2 3 5 *